Lecture Notes in Computer Science 1190

Edited by G. Goos, J. Hartmanis and J. van Leeuwen

Springer

Berlin
Heidelberg
New York
Barcelona
Budapest
Hong Kong
London
Milan
Paris
Santa Clara
Singapore
Tokyo

Stephen North   (Ed.)

Graph Drawing

Symposium on Graph Drawing, GD '96
Berkeley, California, USA
September 18-20, 1996
Proceedings

 Springer

Series Editors

Gerhard Goos, Karlsruhe University, Germany

Juris Hartmanis, Cornell University, NY, USA

Jan van Leeuwen, Utrecht University, The Netherlands

Volume Editor

Stephen North
AT&T Research
600 Mountain Avenue, Murray Hill, NJ 07974, USA
E-mail: north@research.att.com

Cataloging-in-Publication data applied for

Die Deutsche Bibliothek - CIP-Einheitsaufnahme

Graph drawing : proceedings / Symposium on Graph Drawing,
GD '96, Berkeley, California, USA, September 18 - 20, 1996.
Stephen North (ed.). - Berlin ; Heidelberg ; New York ;
Barcelona ; Budapest ; Hong Kong ; London ; Milan ; Paris ;
Santa Clara ; Singapore ; Tokyo : Springer, 1997
  (Lecture notes in computer science ; Vol. 1190)
  ISBN 3-540-62495-3
NE: North, Stephen [Hrsg.]; Symposium on Graph Drawing <1996,
  Berkeley, Calif.>; GT

CR Subject Classification (1991): G.2.2, F.2.2, D.2.2, I.3.5, J.6

1991 Mathematical Subject Classification: 05CXX, 68R10, 90C35, 94C15

ISSN 0302-9743
ISBN 3-540-62495-3 Springer-Verlag Berlin Heidelberg New York

© Springer-Verlag Berlin Heidelberg 1997
Printed in Germany

Typesetting: Camera-ready by author
SPIN 10549894   06/3142 – 5 4 3 2 1 0   Printed on acid-free paper

Preface

Graph Drawing '96 was held at the Mathematical Sciences Research Institute at the University of California in Berkeley, USA, September 18-20.

Graph drawing is the focus of considerable interest in the mathematics, computer science, and graphics communities. The problems investigated include properties of graph embeddings and drawings, algorithms, complexity, order theory, systems, and applications in diverse fields including information visualization, software engineering, circuit layout, network management, and user interfaces. The annual symposium is an important forum for presenting recent results and fostering new research collaborations.

This year a strong technical program of 24 papers and 8 demos was selected from the 50 papers and 21 demos submitted. The quality of these presentations made the symposium a fine success. The invited talk, "Graph Drawing from a User's Perspective" given by Donald Knuth, was attended by an overflow audience. The main technical program that followed in the next two and a half days incorporated 30-minute presentations of papers and 15-minute system demonstrations. More than 90 registered participants attended from 13 countries. Session topics included planarity, upward and orthogonal drawing, heuristics, experimental results, and graph drawing systems. A report on the judging of the annual graph drawing competition was again a highlight of the meeting, and a written summary can be found in these proceedings.

Thanks are in order for the generous financial support received from the NSF and corporate sponsors. Their donations made it possible for a number of students and other academic colleagues to attend and give presentations. We are also grateful to the director of MSRI, Bill Thurston, for hosting the symposium and to MSRI's staff for their diligent efforts.

The real credit, of course, is due to the presenters for their technical contributions. The members of the program committee, listed below, also played a key role.

On a personal note, I warmly thank Emden Gansner for his collegial advice and friendship, and Kimberly Garrett for organizing the meeting arrangements at MSRI so capably. AT&T staff members Carolyn Heaps and Lillian Linnell came to the rescue in distributing the submissions on a tight schedule. The leadership of AT&T Laboratories also deserves recognition for its long-term support of this research area.

Graph Drawing symposia are now being planned through 1999. Because a number of of the submissions that could not be accepted this year appeared to be preliminary descriptions of very promising work, we look forward to reports on the outcome of these efforts at future meetings.

Califon, New Jersey   *Stephen North*
November 1996

Organizer
Stephen C. North, AT&T Laboratories

Program Committee
Franz J. Brandenburg, University of Passau, Germany
Giuseppe Di Battista, University of Rome III, Italy
Emden Gansner, AT&T Laboratories, USA
Tomihisa Kamada, ACCESS Co., Japan
David Kirkpatrick, University of British Columbia, Canada
Stephen North, AT&T Laboratories, USA
János Pach, CUNY and Courant Institutes, USA
Pierre Rosenstiehl, EHESS, Paris, France

Steering Committee
Franz J. Brandenburg, University of Passau, Germany
Giuseppe Di Battista, University of Rome III, Italy
Peter Eades, University of Newcastle, Australia
Takao Nishizeki, Tohoku University, Japan
Stephen North, AT&T Laboratories, USA
Pierre Rosenstiehl, EHESS, Paris, France
Roberto Tamassia, Brown University, USA
Ioannis G. Tollis, University of Texas at Dallas, USA

Sponsors
National Science Foundation (through MSRI)
AT&T Laboratories
Sun Microsystems
Tom Sawyer Software

Meeting Arrangements
Kimberly Garrett, MSRI

Demo Arrangements at MSRI
Joe Christy
Neal Cassidy
Jim Hoffman
Rachelle Summers
Dave Wright
Computers provided by Silicon Graphics and Tom Sawyer Software

Graph Drawing Competition Organizers
Joe Marks, Mitsubishi Electric Research Laboratories
Peter Eades, University of Newcastle

External Referees
James Abello, AT&T Laboratories
Edith Cohen, AT&T Laboratories

Walter Didimo, University of Rome III
Peter Eades, University of Newcastle
Qingwen Feng, University of Newcastle
Sandra Follaro, University of Rome III
Patrick Garvan, University of Newcastle
Ervin Győri, Mathematical Institute, Hungarian Academy of Sciences
Keisuke Hara, Access Co.
Yusuke Higuchi, Showa University
Michael Himsolt, University of Passau
Michael Kaufmann, University of Tübingen
Antonio Leonforte, Integra Spa
Joe Marks, Mitsubishi Electric Research Laboratories
Katsuhiro Ota, Keio University
Brian Regan, University of Newcastle
Mauricio Resende, AT&T Laboratories
Miklós Simonovits, Mathematical Institute, Hungarian Academy of Sciences
Steve Skienna, SUNY Stony Brook
Andreas Stübinger, University of Passau
Géza Tóth, Courant Institute, NYU
Francesco Vargiu, Aipa
Richard Webber, University of Newcastle

Contents

Bipartite Embeddings of Trees in the Plane

M. Abellanas[1] J. García[2] G. Hernández[1] M. Noy[3] and P. Ramos[4]

[1] Facultad de Informática, Univ. Politécnica de Madrid, Boadilla del Monte, 28660 Madrid, SPAIN. E-mail: mavellanas@fi.upm.es, gregorio@fi.upm.es.
[2] Escuela Universitaria de Informática, Univ. Politécnica de Madrid, Crta. de Valencia, km 7, 28031 Madrid, SPAIN. E-mail: jglopez@eui.upm.es.
[3] Dep. de Matemàtica Aplicada II, Univ. Politècnica de Catalunya, Pau Gargallo 5, 08028 Barcelona, SPAIN. E-mail: noy@ma2.upc.es.
[4] Escuela Universitaria de I.T. Aeronáutica, Univ. Politécnica de Madrid, Pza. Cardenal Cisneros s/n, 28040 Madrid, SPAIN. E-mail: pramos@fi.upm.es.

Abstract. Given a tree T on n vertices and a set P of n points in the plane in general position, it is known that T can be straight line embedded in P without crossings. Now imagine the set P is partitioned into two disjoint subsets R and B, and we ask for an embedding of T in P without crossings and with the property that all edges join a point in R (red) and a point in B (blue). In this case we say that T admits a *bipartite embedding* with respect to the bipartition (R, B). Examples show that the problem in its full generality is not solvable. In view of this fact we consider several embedding problems and study for which bipartitions they can be solved. We present several results that are valid for any bipartition (R, B) in general position, and some other results that hold for particular configurations of points.

1 Introduction

Given a tree T on n vertices and a set P of n points in the plane in general position, it is known that T can be straight line embedded in P without crossings. The problem becomes more difficult if T is rooted and we want to root it at any particular point of P. The problem in this form was posed by Perles and partially solved by Pach and Torosick [6]. A complete solution was found by Ikebe et al. [5]. A related result by A. Tamura and Y. Tamura [7] is that, given a point set $P = \{p_1, \ldots, p_n\}$ and a sequence $d = (d_1, \ldots d_n)$ of positive integers with $\sum d_i = 2n - 2$, there exists an embedding of some tree in P such that the degree of p_i is equal to d_i. Optimal algorithms for solving the above problems have been found by Bose, McAllister and Snoeyink [2].

In this paper we consider the following embedding problem. A point set P in the plane in general position (no three points collinear) is partitioned into two disjoint sets R and B (the *red* and the *blue* points), and we are asked to embed a tree T in P without crossings and with the additional property that all the edges are red/blue, i.e. all edges connect a point in R to another point in B. We call such an embedding a *bipartite embedding* of T with respect to the bipartition (R, B).

Any tree T is a bipartite graph and the bipartition $V(T) = (V_1, V_2)$ induced in the vertex set is in fact unique. An obvious necessary condition for the existence of a bipartite embedding of T in P is that the cardinalities of both bipartitions, those of T and of P, match correctly. However, simple examples show that this is not always sufficient. It is then natural to relax the requirements of the problem and to ask: given a bipartition (R, B), is it always possible to find a bipartite embedding of *some* tree with respect to (R, B)? It is straightforward to prove that the answer is affirmative. Take any red point and join it to all the blue points. It is then clear that the remaining red points can be connected to suitable blue points wihtout creating crossings. This simple solution has the shortcoming of producing trees with very large maximum degree.

The approach taken in this paper is to consider several natural embedding problems and to investigate for which bipartitions they can be solved. The problems we study (considering (R, B) as the input) are:

1) find a bipartite embedding of a tree with bounded degree;

2) find a bipartite embedding of a tree where the degrees of the vertices in one of the parts, say R, are prescribed in advance;

3) find a bipartite embedding of a spanning path.

The paper is organized as follows. In Section 2 we discuss in detail a collection of problems about bipartite embeddings. Section 3 contains the results for points in general position, while in Section 4 we restrict our attention to particular configurations of points. We conclude with a brief discussion of some open problems.

2   Discussion of the problems

In this section we fix the terminology we use in the remainder of the paper, and we state and discuss in detail three embedding problems.

P denotes a set of n points in the plane in general position (no three of them collinear), partitioned into two subsets $R = \{p_1, \ldots, p_r\}$ and $B = \{q_1, \ldots, q_b\}$ with cardinalities $r = |R|$ and $b = |B|$. The bipartition is denoted (R, B). The set of vertices in the convex hull of P is denoted $CH(P)$.

Let T be a spanning tree of the complete bipartite graph $K_{r,b}$, and let $(d_1, \ldots, d_r)$ and $(d'_1, \ldots, d'_b)$ be the degree sequences corresponding to the two parts of the bipartition. Since the number of edges in T is $r + b - 1$ we have that

$$\sum_{i=1}^{r} d_i = \sum_{j=1}^{b} d'_j = r + b - 1.$$

Given a tree T on n vertices, we say that T admits a *bipartite embedding* with respect to (R, B) if T can be straight-line embedded into P without crossings, i.e. edges can only intersect at the vertices, and with the additional property that an edge always joins a point in R and a point in B. An obvious necessary condition for the existence of such an embedding is that T is a spanning tree of $K_{r,b}$. In other words, if $V(T) = V_1 \cup V_2$ is the unique bipartition of the vertex set of T, then we must have either $|V_1| = r, |V_2| = b$ or $|V_1| = b, |V_2| = r$.

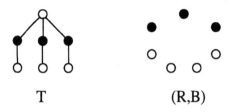

T (R,B)

Fig. 1. T does not admit a bipartite embedding with respect to (R, B).

This condition is not always sufficient, as the example in Figure 1 shows. Because of this fact we adopt the following point of view. The bipartition (R, B) is given as input, and the problem is to find a bipartite embedding of *some* tree with respect to (R, B) satisfying a given condition. We introduce three conditions that give rise to the three problems discussed below.

Bounded degree embeddings. Let T be a spanning tree of $K_{r,b}$, and let $d_1, \ldots, d_r$ and $d'_1, \ldots, d'_b$ be the degree sequences of the vertices of the two parts. The equation $\sum d_i = r + b - 1$ implies that T has a vertex whose degree is at least $1 + \lceil (b - 1)/r \rceil$ (this is the red mean degree). Our first problem asks for the existence of a bipartite embedding attaining a bound of this order of magnitude (we show later that the exact bound cannot always be achieved).

Problem 1. Given a bipartition (R, B) with $r \le b$, find a bipartite embedding of a tree with respect to (R, B) having maximum degree $\Delta = O(b/r)$.

Fixed degree embeddings. Let $\mathbf{d} = (d_1, \ldots, d_r)$ and $\mathbf{d}' = (d'_1, \ldots, d'_b)$ be sequences of positive integers satisfying $\sum d_i = \sum d'_j = r + b - 1$. It is easy to see that $K_{r,b}$ admits a spanning tree whose degree sequences are equal to $\mathbf{d}$ and $\mathbf{d}'$, respectively. However, given (R, B) it is not always possible to find a bipartite embedding realizing both degree sequences (take $\mathbf{d} = (2, 2, 2)$ and $\mathbf{d}' = (3, 1, 1, 1)$ in the example of Figure 1). Our second problem asks for a bipartite embedding in which the degree of every vertex in R is fixed in advance.

Problem 2. Given a bipartition (R, B), $R = \{p_1, \ldots, p_r\}$, and a sequence of positive integers $(d_1, \ldots, d_r)$ with $\sum d_i = r + b - 1$, find a bipartite embedding of a tree with respect to (R, B) such that the degree of p_i is d_i.

Embedding a spanning path. If we take $\Delta = 2$ in Problem 1 then we are asking for a bipartite embedding of a spanning path. An obvious necessary condition is $|r - b| \leq 1$. The example in Figure 2 shows that it is not sufficient. To see why note that the first edge of the path has to be an edge of the convex hull. The fact that the cardinalities of consecutive red and blue chains differ always in more than two units, prevents the path from spanning all the vertices. Our last problem asks for sufficient conditions that guarantee the existence of such a path.

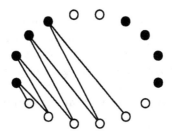

Fig. 2. The bipartite path cannot be completed to a spanning path.

Problem 3. Characterize those bipartitions (R, B) with $|r - b| \leq 1$ admitting a bipartite embedding of a spanning path.

We find a complete solution to Problem 2, and prove several results concerning Problems 1 and 3. These are presented in the next two sections.

3   Points in general position

In this section we present two results that hold for any bipartition in general position. The first one provides an answer to Problem 1 and the second one solves completely Problem 2.

Theorem 4. *Given a bipartition (R, B) in the plane with $r \leq b$, one can find a bipartite embedding of a tree with respect to (R, B) such that the maximum degree Δ is $O(b/r + \log r)$.*

Proof. We use the ham-sandwich theorem [3], that asserts that the sets R and B can be simultaneously bisected by a straight line. Assume for simplicity that $r = 2^k$ and that $b = \alpha r = \alpha 2^k$. Applying repeatedly the ham-sandwich theorem we arrive, after k steps, to a partition of the plane into convex polygonal regions, each one of them containing exactly one red point and α blue points. Join every red point to the corresponding α blue points to obtain a collection of r disjoint copies of a bipartite embedding of a star $K_{1,r}$.

Next we merge these r partial trees into a single tree, in the opposite order as they have been produced, preserving the bipartite character of the embedding. To do that we need the following lemma, whose easy proof is omitted.

Lemma. Given two disjoint bipartite embeddings T_1 and T_2 separated by a straight line, one can add an additional edge between T_1 and T_2, resulting in a global bipartite embedding of a tree.

We have to perform k merging steps, every time reducing by half the number of trees. It is clear that at every step the maximum degree increases at most by one. Since initially $\Delta = \alpha$, at the end $\Delta \leq \alpha + k = b/r + \log_2 r$. The cases where r is not a power of two or b/r is not an integer are treated similarly.

In the next section it will be shown that the optimal bound $O(b/r)$ can be achieved for several particular configurations.

Theorem 5. *Given a bipartition (R, B) in the plane, with $R = \{p_1, \ldots, p_r\}$ and a sequence $(d_1, \ldots, d_r)$ of positive integers with $\sum d_i = r + b - 1$, there exists a bipartite embedding of a tree with respect to (R, B) such that the degree of p_i is equal to d_i.*

Proof. The proof is by induction on $r + b$. Assume without loss of generality that $d_1 = \max d_i$, and let l be an oriented line through p_1 not containing any other point from R or B. Let H^+ and H^- be the right and left open halfspaces in which the line l divides the plane. Let $r^+ = |R \cap H^+|$ and $b^+ = |B \cap H^+|$, and let also $r^- = |R \cap H^-|$ and $b^- = |B \cap H^-|$. Finally, define two functions (that depend on l) f^+ and f^- as follows

$$f^+ = \sum_{p_i \in H^+} d_i - r^+ - b^+;$$

$$f^- = \sum_{p_i \in H^-} d_i - r^- - b^-,$$

and observe that $f^+ + f^- = -d_1$.

We claim that there exists some position of l in which $-d_1 < f^+ < 0$. To prove the claim assume that initially $f^+ \geq 0$ and consider the changes in f^+ as l turns around p_1.

If a red point p_i enters H^+, then f^+ increases by $d_i - 1$;
If a red point p_i exits H^+, then f^+ decreases by $d_i - 1$;
If a blue point enters H^+, then f^+ decreases by 1;
If a blue point exits H^+, then f^+ increases by 1.

In any case the change in absolute value is at most $d_i - 1 \leq d_1 - 1$. Since after a turn of 180 degrees the values of f^+ and f^- are interchanged, and $f^+ = -d_1 - f^-$, it follows that $f^+ \leq -d_1$. All this implies that for some intermediate value we have $-d_1 < f^+ < 0$. If we assume instead that initially $f^+ \leq -d_1$ we proceed in the same way, and the claim is proved.

Now by induction we can find a bipartite embedding of a tree with respect to the bipartition $((R \cap H^+) \cup \{p_1\}, B \cap H^+)$ in which p_1 has degree $-f^+$ and p_i has degree d_i for $p_i \in R \cap H^+$. Similarly we get a tree on H^- in which p_1 has degree $-f^-$, and the union of the two trees does the job.

4 Points in restricted positions

We have already mentioned that a bipartition does not always admit a spanning path, and that Theorem 4 does not give the best possible bound for the maximum degree. It is then natural to restrict the geometry of the problem in order to obtain positive results. We consider three such restrictions, or particular positions, that arise naturally. Firstly, when R and B are separated by a straight line. Secondly, when $R \cup B$ is a set in convex position. And finally when the vertices of R define a convex polygon containing all the vertices in B.

4.1 Linearly separable partitions

We say that a bipartition (R, B) is *linearly separable* if there exists a straight line separating R and B. Equivalently, if the convex hulls of R and B are disjoint.

Theorem 6. *Every linearly separable bipartition $P = R \cup B$ with $|r - b| \leq 1$ admits a bipartite embedding of a spanning path.*

Proof. Assume without loss of generality that R and B are separated by a horizontal line, and let $p_1 \in R$ and $q_1 \in B$ be such that $p_1 q_1$ is the left red/blue edge of the convex hull of P. We say that $p_1 q_1$ is the left *bridge* of P. The initial point of the spanning path will be p_1 if $b = r - 1$, q_1 if $b = r + 1$, and either p_1 or q_1 if $b = r$. Assume we start at p_1 and set $C = \{p_1\}$ (C is an ordered list that corresponds to the spanning path as it is constructed). At every step compute the left bridge pq of $P \backslash C$, and add p to C if the last point in C is in B, or add q if the last point in C is in R. In this way we get a bipartite embedding of a path that has no crossings because C is disjoint from the convex hull of $P \setminus C$ and hence from all edges added to it during the algorithm.

The technique in the above proof can also be used to prove the following result.

Theorem 7. *Every linearly separable bipartition $P = R \cup B$ with $r \leq b$ admits a bipartite embedding of a tree T with $\Delta(T) \leq 1 + \lceil (b-1)/r \rceil$.*

Proof. Let $\bar{d} = 1 + \lceil (b-1)/r \rceil$. The idea is to find a bipartite embedding of a tree with respect to (R, B) in which the degrees in R are equal to $\bar{d}$ or to $\bar{d} - 1$ and, at the same time, be able to bound the degrees in B.

First we can suppose that $\bar{d} > 2$, that is $b - 1 > r$. Otherwise, since we are assuming $b \geq r$, Theorem 6 implies the existence of a bipartite embedding with $\Delta = 2$ and the theorem holds. Now find a sequence $(d_1, \ldots, d_r)$ of positive integers such that $\sum d_i = r + b - 1$ and $d_i = \bar{d}$ or $d_i = \bar{d} - 1$. We could use Theorem 5 in order to find a bipartite embedding realizing this degree sequence, but then we would not be able to control the degrees in B.

Instead we proceed as follows. Find a point p_1 in R such that there is a line separating p_1 and d_1 points $q_1, \ldots, q_{d_1}$ in B from the remaining points, where the q_j are sorted in polar order with respecto to p_1. This is always possible taking left bridges as in the proof of the previous theorem. Join p_1 to $q_1, \ldots, q_{d_1}$, remove all these points except q_{d_1} from the bipartition and find a new point p_2 in R that can be separated together with to d_2 points in B. If we repeat this process, at the end we get a single point p_r in R and d_r points in B (see Figure 3 for an illustration). The fact that $d_i > 1$ for every i, implies that the degrees of the points in B are equal to one or two.

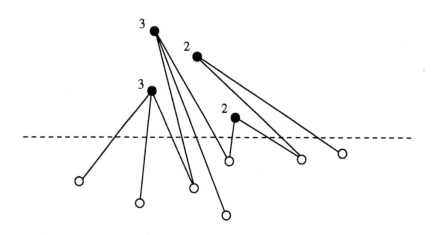

Fig. 3. Realizing the degree sequence $(3, 3, 2, 2)$.

4.2 Sets in convex position

We say that a set of points is in convex position if it is the vertex set of a convex polygon. We have already seen an example of a bipartition in convex position with $|r - b| \leq 1$ which does not admit a spanning path, and the example can be generalized to any number of points (larger than 15). On the positive side we prove that Problem 1 can be solved in this case. If $R \cup B$ is in convex position and $r = b$, Problem 3 was solved by Akiyama and Urrutia [1]. They gave an algorithm that determines if $R \cup B$ admits a spanning path in $O(n^2)$ time finding such path if it exists.

Theorem 8. *Let $R \cup B$ be a set in convex position with $r \leq b$. Then the bipartition (R, B) admits a bipartite embedding of a tree T with $\Delta(T) \leq \lceil b/r \rceil + 2$.*

Proof. Assume for simplicity that $r = 2^k$ and that $b = \alpha 2^k$. Using ham-sandwich cuts as in the proof of Theorem 4 we can obtain r disjoint red/blue copies of a star $K_{1,\alpha}$. The key point in this case is that we can control the degrees in the merging step.

Set initially T equal to any of the r stars. For every edge e of the convex hull of T that is not an edge of the convex hull of $R \cup B$, consider the trees $T_1, \ldots, T_j$ that are visible from e and lie on the halfspace determined by e not containing T, ordered clockwise (this makes sense because the set is in convex position).

Next select one of the vertices of e and construct a bipartite polygonal chain connecting T and the trees $T_1, \ldots, T_j$ (see Figure 4). Because the set is in convex position, we can construct this chain in such a way that the degree of any vertex increases by at most 2. Set T equal to the tree obtained with the above costruction and iterate the process until T is a spanning tree.

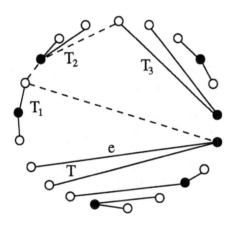

Fig. 4. Proof of Theorem 4.3.

To see that the condition on the degree is satisfied, observe that the new edges of $CH(T)$ that are not in $CH(R \cup B)$ are determined by two vertices that belong to one of the trees $T_1, \ldots, T_j$. Since only one vertex of each tree is used when constructing the polygonal chain, we can guarantee that we always have a free vertex to iterate the process. Therefore, points in R have degree at most $\alpha + 2 = b/r + 2$, and points in B have degree at most 3.

4.3 Bipartitions in which $R = CH(R \cup B)$

The situation can be described in this way: the points of R are the vertices of a convex polygon containing the points of B. We need the following result by García and Tejel [4].

Lemma 9. *Let P be a set of points in general position, and assume that $CH(P) = \{p_1, \ldots, p_k\}$ and that there are n interior points. Let $n = n_1 + \ldots + n_k$, where the n_i are positive integers. Then the convex hull of P can be partitioned into k convex polygons $Q_1, \ldots, Q_k$ such that Q_i contains n_i points and $p_i p_{i+1}$ is an edge of Q_i.*

Proof. (Sketch of the proof by García and Tejel) The proof is by induction on k. If $k = 3$, by continuity arguments and due to the generic position of points, it is easy to prove that it exists a point $q \notin P$ inside the triangle $p_1 p_2 p_3$ such that triangles $p_1 p_2 q$, $p_2 p_3 q$ and $p_3 p_1 q$ contains n_1, n_2, n_3 points respectively.

The induction step begin by considering an arbitrary diagonal, for example $p_1 p_j$, $1 < j < k$. Without loss of generality one can asume that the polygon $p_1, \ldots, p_j$ contains $n_1 + \ldots + n_{j-1}$ or more points. In this polygon we can apply the induction hypothesis. Let $p_1, q_1, \ldots, q_i, p_j$ the polygon obtained in the previous decomposition corresponding to the edge $p_1 p_j$. If this polygon contains n_j or more points, connecting p_{j+1} with a point q' in the polygonal chain $p_1, q_1, \ldots, q_i, p_j$ the polygon splits into two parts satisfying the induction hypothesis. In other case it is the polygon $p_1, p_{j+1}, \ldots, p_k$ which verify such condition.

Using the above lemma it is not difficult to prove the following results.

Theorem 10. *Let (R, B) be a bipartition in which $R = CH(R \cup B)$ and with $r \leq b$. Then it admits a bipartite embedding of a tree T with $\Delta(T) \leq 1 + \lceil (b-1)/r \rceil$.*

Proof. If $b = r\alpha + 1$ we can use lemma 4.5 to decompose the convex hull of R into r convex polygons $Q_1, \ldots, Q_r$ such that Q_j , $j \neq 1$, contains α points of B , Q_1 contains $\alpha + 1$ points and $p_i p_{i+1}$ is an edge of Q_i.

Let $s_j \in Q_j \cap B$ such that $dist(s_j, p_j) = \min\{dist(s, p_j) : \quad s \in Q_j \cap B\}$. We join every red point p_j to the corresponding α blue points in Q_{j-1} except s_1. Next we merge these r partial trees, by connecting s_j to p_j. In this way we obtain a tree T with $\Delta(T) = 1 + \alpha$.

If $\frac{b-1}{r}$ is not an integer then we obtain a tree with $\Delta(T) \leq 1 + \lceil \frac{b-1}{r} \rceil$.

Theorem 11. *Let (R, B) be a bipartition with $R = CH(R \cup B)$ and $|r - b| \leq 1$. Then it admits a bipartite embedding of a spanning path.*

Proof. If $r = b$, in the partition of the lemma 4.5 every Q_j contains only one blue point. Hence it is easy to obtain a bipartite embedding of a spanning path. If $|r - b| = 1$ we proceed in a similar way.

5 Concluding remarks

In this paper we have introduced the problem of finding bipartite embeddings of trees in the plane and have obtained a number of results. Some of our results are valid for any set of points in general position, and some of them only apply to particular configurations of points. There are several interesting questions left open. One is whether one can always achieve the optimal bound in Problem 1, that is, whether any bipartition (R, B) with $r \leq b$ admits a bipartite embedding of a tree with degree bounded by $O(b/r)$. Finally, we remark that most of our proofs are constructive and actually provide algorithms for solving the various problems. A direct analysis of the proofs shows, for example, that the construction in Theorem 3.2 can be done in $O(n^2 \log n)$ time, where $n = r + b$, and those in Theorems 4.1 and 4.2 can be done in time $O(n \log^2 n)$. We leave as an open question to find optimal algorithms for these constructions.

Acknowledgements

Thanks are due to Sue Withesides, Jorge Urrutia and to the referees for their suggestions which contributed to the final version.

References

1. J. Akiyama and J. Urrutia, Simple Alternating Path Problems, *Discrete Mathematics* 84 (1990), pp. 101-103.
2. P. Bose, M. McAllister and J. Snoeyink, Optimal Algorithms to Embed Trees in the Plane, in *Proc. Graph Drawing 95*, Springer Verlag LNCS Vol. 1027, pp.64-75.
3. H. Edelsbrunner, *Algorithms in Combinatorial Geometry*, Springer Verlag (Berlin, 1987).
4. A. García and J. Tejel, Dividiendo una nube de puntos en regiones convexas, *Actas VI Encuentros de Geometría Computacional*, pp. 169-174, 1995.
5. Y. Ikebe, M. Perles, A. Tamura and S. Tokunaga, The Rooted Tree Embedding Problem into Points in the Plane, *Discrete and Computational Geometry* 11 (1994), pp. 51-63.
6. J. Pach and J. Töröcsik, Layout of Rooted trees, in *Planar Graphs* (W.T. Trotter, ed.), DIMACS Series, Vol. 9, Amer. Math. Soc., pp. 131-137.
7. A. Tamura and Y. Tamura, Degree Constrained Tree Embedding Into Points in the Plane, *Information Proc. Letters* 44 (1992), pp. 211-214.

Series-Parallel Planar Ordered Sets Have Pagenumber Two

(Extended Abstract)

Mohammad Alzohairi[1]   Ivan Rival[2]

[1] Department of Mathematics and Statistics
University of Ottawa
Ottawa K1N 6N5, Canada
alzohair@csi.uottawa.ca

[2] Department of Computer Science
University of Ottawa
Ottawa K1N 6N5 Canada
rival@csi.uottawa.ca

Abstract. *The pagenumber of a series-parallel planar P is at most two.* We present an $O(n^3)$ algorithm to construct a two-page embedding in the case that it is a lattice. One consequence of independent interest, is a characterization of series-parallel planar ordered sets.

1 Introduction

A *book embedding* of a graph G consists of an embedding of its nodes along the spine of a book (i.e., a linear ordering of the nodes), and an embedding of its edges on pages so that edges embedded on the same page do not intersect. In a *book embedding* for an ordered set P the vertices of P on the spine form a linear extension (a total order $L = \{x_1 < x_2 < \cdots < x_n\}$ of the elements of P is a *linear extension* if $x < y$ in L whenever $x < y$ in P).

We say *a covers b* (or *b covered by a*) in the ordered set P, and write $a \succ b$ (or $b \prec a$), if whenever $a > c \geq b$ then $c = b$. Also, we say a is an *upper cover* of b, or b is a *lower cover* of a, or (a, b) is an edge in P. We say a is a *minimal* (respectively, *maximal*) element of P if a has no lower covers (respectively, a has no upper covers). We denote the set of all minimals (respectively, maximals) of P, $min(P)$ (respectively, $max(P)$). The *covering graph* of P, $cov(P)$, is the graph whose vertices are the elements of P, and the pair $\{a, b\}$ forms an edge in $cov(P)$ if $a \succ b$ or $a \prec b$. It is possible to orient $cov(P)$ in such a way the y-coordinate of a is less than the y-coordinate of b if $a \prec b$ and the edge (a, b)

does not pass through any other element of P . We call such drawing an *upward drawing* of P.

The *pagenumber* in both cases ($page(G)$, respectively $page(P)$) is the minimum number of pages needed taken over all linear layouts for graphs and all linear extensions for an ordered set. For instance, $page(P) = 2$ for the ordered set illustrated in Figure 1, while $page(cov(P)) = 1$. On the other hand the planar lattice in Figure 2 required three pages (this example is due to J. Czyzowicz [7]).

The pagenumber was first defined for graphs by Bernhart and Kainen [1], who conjectured that planar graphs may require an arbitrary large number of pages. In a series of attempts, it was finally established by Yannakakis [11], that $page(G) \leq 4$ for every planar graph G, and this upper bound is achieved. Fraysseix, Mendez and Pach [4] have shown that the pagenumber of any planar graph with quadrilateral faces is at most two.

The page number for ordered sets has been introduced by Nowakowski and Parker [7], who show that $page(P) = 1$ if and only if $cov(P)$ is a forest. Also, they derive a general lower bound on the page number of ordered sets and upper bounds for special classes of ordered sets. Hung [3] shows that there exists a 48-element planar ordered set which needs four pages (see Figure 3). Moreover, no planar ordered set with pagenumber five is known. Sysło [9] provides a lower bound on the page number in terms of its bump number. He also shows that, $page(P) \leq 2$ if the jump number of P is one. Ordered sets with jump number two can have an arbitrarily large page number. Later, Heath and Pemmaraju [8] gave a sequence of ordered sets each with planar covering graph and with unbounded page number. Computationally, we recently proved that finding the minimum number of pages required for a fixed linear extension of an ordered set is NP-complete.

In section 2 we study the structure of series-parallel planar lattices. In section 3 we will construct, for a series-parallel planar lattice P, an $O(n^3)$ two-page algorithm where n is the number of the elements of P. In section 4 we continue the study of the structure of series-parallel planar ordered sets. In section 5 we exploit the fact that the completion $\overline{P}$ of a series-parallel planar ordered set P is itself a series-parallel planar lattice. We use the result in section 3 to obtain a two-page linear extension $\overline{L}$ of $\overline{P}$, which we transfer to a two-page linear extension of P. In section 6 we give three open problems related to the pagenumber problem.

2 Structure of series-parallel planar lattices

The *linear sum* $P \oplus Q$ of the two disjoint ordered set P, Q is an ordered set on $P \cup Q$, that is, $a \leq b$ if

    1. $a \leq b$ in P, or   2. $a \leq b$ in Q, or   3. $a \in P$ and $b \in Q$.

If we eliminate the third condition of the definition of linear sum, we will have the *disjoint sum* $P + Q$ of P, Q.

An ordered set P is *series-parallel* if P can be constructed from singletons using only the constructions of disjoint sum $+$ and linear sum $\oplus$. In other words, P can be decomposed into singletons using only disjoint sum and linear sum. For instance, the the series-parallel lattice illustrated in Figure 4 can be decomposed into

$$1\oplus(((2+6)\oplus3\oplus(4+(7\oplus(8+(10\oplus11)+12+13)\oplus9)))+(14\oplus(15+17)\oplus16))\oplus5$$

For $a \neq b$ in the ordered set P we say a is *comparable* to b if either $a < b$ or $a > b$. Otherwise, a is *noncomparable* to b, write $a \parallel b$. An *antichain* is a subset A of an ordered set P such that any two distinct elements of A are noncomparable. Dually, a *chain* of P is a subset C of P where, each pair of C are comparable.

A four-element subset $\{a, b, c, d\}$ of an ordered set P forms an **N** if the only comparabilities among them in P are $a < c$, $b < c$ and $b < d$. It is known that an ordered set is a series-parallel if and only if it contains no such **N** [10].

Fix a planar embedding of P, and let $C = \{x_1 < x_2 < ... < x_n\}$ be the left boundary chain. For each $x \in P - C$ define the interval $I(x) = (x_i, x_j)$, where

$$x_i = \max_{1 \leq k < n} \{x > x_k\} \qquad\qquad x_j = \min_{1 < k \leq n} \{x < x_k\}$$

Of course, $j \geq i + 1$. Notice that, $j > i + 1$ because if $j = i + 1$ then the edge (x_{i+1}, x_i) will not be an essential edge. (An edge (a, b) is not essential if there is c such that $a < c < b$.)

Notice that, every pair of these intervals is either disjoint or one contains the other. Hence the set of intervals ordered by inclusion is a forest (an ordered set P is a forest if the graph $cov(P)$ is a forest). For $y, z \in P - C$, say $y \sim z$ if $I(y) = I(z)$. It is clear that this relation is an equivalence relation. Call the equivalence classes *components*.

For example, the components of the series-parallel order in Figure 4 are:
$C_1 = \{7, 8, 9, 10, 11, 12, 13\}$ which corresponds to the interval (3,5);
$C_2 = \{6\}$ which corresponds to the interval (1,3);
$C_3 = \{14, 15, 16, 17\}$ which corresponds to the interval (1,5).
The forest obtained by ordering the intervals by inclusion is shown in Figure 6. We can show that there are no edges between the components.

Here are a few elementary terms. Fix a lattice P and fix a planar upward drawing of it. For noncomparable element $a, b \in P$ such that $a \succ c$ and $b \succ c$, we say a is *left* of b if any horizontal segment (moving from left to right) which cuts both edges, always cuts the edge (c, a) before the edge (c, b). For arbitrary noncomparable elements a and b ($a \parallel b$) in P say that a is *left* of b, denoted $a\lambda b$, if a' is left of b', where $a \geq a' \succ inf(\{a, b\})$ and $b \geq b' \succ inf(\{a, b\})$. An element a, which does not belong to the maximal chain C is left of C if there is $b \in C$ such that $a\lambda b$. In fact, a is left of b if a is left to any maximal chain containing b. (Of course, all of these ideas are ambidextrous. If a is left of b then b is right of a, etc.)(For details see [5].)

Once equipped with the equality relation, λ becomes an order relation on P, denoted P_λ. (This result is due to J. Zilber see [2] page 32, ex. 7(c).) For example, the ordered set in Figure 5 is P_λ where P is the planar lattice in Figure 4.

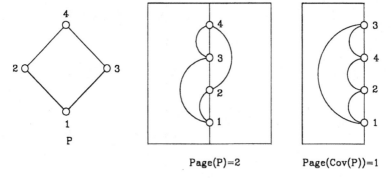

Page(P)=2   Page(Cov(P))=1

Figure 1

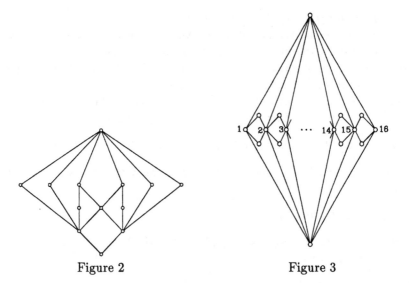

Figure 2   Figure 3

For a series-parallel planar lattice P, fix a planar upward drawing of P, and define the sequence of *peels* of P as follows:

$L_0 = \{x \in P : x \text{ belongs to the left boundary}\}$.

$L_1 = \{x \in P - L_0 : \text{ if } y \text{ lies to the left of } x, \text{ then } y \in L_0\}$.

$L_2 = \{x \in P - (L_0 \cup L_1) : \text{ if } y \text{ lies to the left of } x, \text{ then } y \in L_0 \cup L_1\}$.

$\vdots$

$L_t = \{x \in P - (L_0 \cup L_1 \cup \cdots \cup L_{t-1}) : \text{ if } y \text{ lies to the left of } x, \text{ then } y \in L_0 \cup L_1 \cup \cdots \cup L_{t-1}\}$.

We call any L_i a *peel* of P. Actually, the peels of a planar lattice P are the levels of P_λ, where P_λ the underlying set P ordered by λ.

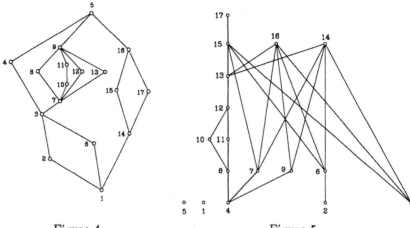

Figure 4   Figure 5

Thus, $L_i = \min(P_\lambda - (\bigcup_{j=0}^{i-1} L_j, 0 \le j \le t)$

Of course, t is equal the height of P_λ, where the height of an ordered set is less one than the maximum number of elements of a chain.

For example, in the series-parallel ordered set P with respect to the upward drawing shown in Figure 4

$L_0 = \{1,2,3,4,5\}$ $L_1 = \{6,7,8,9\}$ $L_2 = \{10,11\}$ $L_3 = \{12\}$
$L_4 = \{13\}$ $L_5 = \{14,15,16\}$ $L_6 = \{17\}$

Lemma 1 *Let P be a series-parallel planar lattice. If $0 \le i \le t$, then*

1. *for any $x \in L_i, i > 0$, there exist $y \in L_{i-1}$ such that x lies to the right of y,*

2. *the peel L_i forms a chain,*

3. *the number of peels equals width(P)–1 (width(P) is the maximum size of antichain in P).*

Call a chain C in P is *saturated* if all of its covering relations, are covering relations in P. Each chain decomposes into its (maximal) saturated chains.

In a series-parallel planar lattice P each peel L_i can be decomposed into maximal saturated subchains $C_{i1}, C_{i2}, \ldots, C_{in_i}$ called the *clamped* chains for P. For a clamped chain $C_{ij} \in L_i, i \ge 1$ define:

$$l(C_{ij}) = \{y \in L_0 \cup L_1 \cup \cdots \cup L_{i-1} : inf(C_{ij}) \succ y\}$$
$$u(C_{ij}) = \{y \in L_0 \cup L_1 \cup \cdots \cup L_{i-1} : sup(C_{ij}) \prec y\}$$

For example the table below shows the clamped chains in the series parallel planar lattice in Figure 4.

Clamped chain C_{ij}	$l(C_{ij})$	$u(C_{ij})$
$C_0 = \{1,2,3,4,5\}$	$-$	$-$
$C_{11} = \{6\}$	1	3
$C_{12} = \{7,8,9\}$	3	5
$C_{21} = \{10,11\}$	7	9
$C_{31} = \{12\}$	7	11
$C_{41} = \{13\}$	7	9
$C_{51} = \{14,15,16\}$	1	5
$C_{61} = \{17\}$	14	16

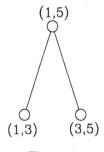

Figure 6

Lemma 2 *Let C_{ij} be a clamped chain in a series-parallel planar lattice P.*

1. *Each $l(C_{ij})$ and $u(C_{ij})$ is unique.*

2. *Each $x \in C_{ij} - \{inf C_{ij}, sup C_{ij}\}$ has neither lower covers nor upper covers in $L_0 \cup L_1 \cup \cdots \cup L_{i-1}$. Also, if $inf C_{ij} \neq sup C_{ij}$ then $inf C_{ij}$ (respectively, $sup(C_{ij})$) has no lower (respectively, upper) covers in $L_0 \cup L_1 \cup \cdots \cup L_{i-1}$,*

3. *If $u(C_{ij}) \in C_{km}$, then $l(C_{ij}) \in C_{km} \cup \{l(C_{km})\}$.*

4. *$inf C_{ij}$ (respectively, $sup C_{ij}$) has a unique lower (respectively, upper) cover in P.*

3   Two pages are enough

In this section we will give an $O(n^3)$ two-page algorithm for a series-parallel planar lattice P, where n is the number of elements of P.

To obtain a two-page linear extension of a series-parallel planar lattice P

(i) Fix a planar upward drawing for P.

(ii) List the clamped chains of P in the following order
$C_0, C_{11}, C_{12}, \ldots, C_{1n_1}, C_{21}, C_{22}, \ldots, C_{2n_2}, \ldots, C_{w1}, C_{22}, \ldots, C_{wn_w}$.
We will process chain by chain according to the above order.

(iii) Put C_0 on the spine of the book. Draw the bottom edge on the right page and draw all other edges on the left page.

(iv) Suppose two pages are enough up to C_{ij-1}. For C_{ij} put all the elements of C_{ij} right below $u(C_{ij})$. Draw the edge $(inf(C_{ij}), l(C_{ij}))$ on the right page and draw all C_{ij} edges and the edge $(u(C_{ij}), sup(C_{ij}))$ on the left page.

Call this algorithm the *two-page* algorithm.

Figure 7, illustrates the steps of the two-page algorithm applied on the series-parallel planar lattice P in Figure 4.

A *greedy linear extension* of an ordered set P is a linear extension $x_1 < x_2 < \cdots < x_n$ of P such that $x_1 \in min(P)$ and, for $i \geq 1, x_{i+1} \in min(P - \{x_1, x_2, \cdots, x_i\})$ and, if possible, $x_{j+1} > x_j$. Thus, a greedy linear extension obtained by following "the rule climb as heigh as you can".

A *left greedy linear extension* of P is that greedy linear extension whose the ith element x_{i+1} is the (unique) left-most element belonging to $min(P - \{x_1, x_2, \cdots, x_i\})$

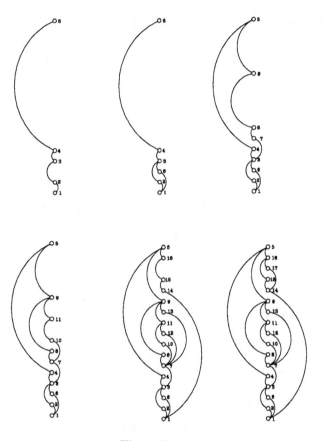

Figure 7

Lemma 3 *Let P be a series-parallel planar lattice. If L is the permutation obtained by the two-page algorithm, then*
1. L is a linear extension of P,
2. if $x \parallel y$ in P, and y lies to the left of x, then $y < x$ in L. (i.e., L is a left greedy linear extension).

Theorem 4 *The two-page algorithm for an n-element series-parallel planar lattice produces a two-page linear extension L in $O(n^3)$ time.*

For the complexity, we can find the peel C_0 by checking for each $x \in P$ if there is $y \in P - \{x\}$ such that $y \parallel x$ and y lies to the left of x. Thus, we need at most n^3 comparison operations to obtain the peels of P. To obtain the clamped chains of a certain peel we need first to sort it in $O(n \log n)$ comparisons, then determine the covering relations in this peel and that can be done in $O(n-1)$ comparisons.

Therefore, we can find all clamped chains in $O(n^2 \log n)$ comparisons. For each clamped chain C_{ij} we can find $u(C_{ij})$ by find the element in L_{i-1} which covers $inf(C_{ij})$ and this can be done in $O(n)$ comparisons. Thus, we can find $u(C_{ij})$ and $l(C_{ij})$ for all clamped chains C_{ij} in $O(n^2)$ comparisons. For the distribution of the edges among the two pages we process each edge just one time; thus, we can decide the page for each edge in $O(n^2)$ comparisons. Thus, the whole algorithm can be done in $O(n^3)$ comparisons.

4   Structure of series-parallel planar ordered sets

The completion of an ordered set P is the smallest lattice $\overline{P}$ contains P as suborder. Notice that $\overline{P}$ exists and called MacNeille completion (cf. [6].)

First, we will show that the completion $\overline{P}$ of series-parallel ordered set is series-parallel planar lattice.

The question may arise now whether we can transfer the two-page linear extension $\overline{L}$ of $\overline{P}$ (obtained by Theorem 4) to a two-page linear extension L of P ?

For example, we consider the series-parallel planar ordered set P and its completion $\overline{P}$ in Figure 8. In Figure 9, $\overline{L}$ is the two-page linear extension of $\overline{P}$ obtained by the two-page algorithm for series-parallel planar lattice. Let L be the linear extension obtained from $\overline{L}$ by removing the elements in $\overline{P}$–P. Notice that, the linear extension L needs at least three pages.

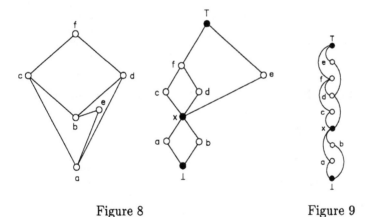

Figure 8   Figure 9

But if we redraw $\overline{P}$ in a different planar embedding as it is in Figure 12, then using the two-page algorithm for series-parallel lattices we will obtain the two-page embedding $\overline{L}$ as it illstrates in Figure 10. In Figure 10, we also, see that the linear extension L of P induced by $\overline{L}$ is a two-page linear extension.

This leads us to this question, whether we can always find a planar embedding of the completion $\overline{P}$ of the series-parallel planar ordered which can lead finally to a two-page linear extension of the ordered set ? The answer is yes.

Lemma 5 *If P is a series-parallel planar ordered set and $\overline{P}$ its completion then,*

    (i) $\overline{P}$ is series-parallel,   *(ii) $\overline{P}$ is a planar lattice.*

We say the ordered set P contains $K_{m,n}$, $m,n \geq 2$ if it contains a subset $\{a_1, a_2, \cdots, a_m, b_1,, b_2, \cdots, b_n\}$ satisfying $a_i \prec b_j$ for $i = 1, 2, \ldots, m$ and $j = 1, 2, \ldots, n$. (See Figure 11). Notice that, if P contains $K_{m,n}, m, n \geq 2$ then, the sets $\{a_1, a_2, \cdots, a_m\}$ and $\{b_1, b_2, \cdots, b_n\}$ are antichins.

We say $K_{m,n} = \{a_1, a_2, \cdots, a_m, b_1,, b_2, \cdots, b_n\}$ is *maximal* in P if there is neither $a_{m+1} \neq a_i, 1 \leq i \leq m$ satisfying $a_{m+1} \prec b_j$ for every $1 \leq j \leq n$, nor $b_{n+1} \neq b_j, 1 \leq j \leq n$ satisfying $a_i \prec b_{n+1}$ for every $1 \leq i \leq m$. If a planar ordered set contains $K_{m,n}, m, n \geq 2$, then either $m = 2$ or $n = 2$.

After series of Lemma's, we conclude that if the series-parallel planar ordered set P is not a lattice, then the only obstacle to be a lattice (except the top and the bottom) is existing the maximals of $K_{2,n}$ and/or $K_{m,2}, m, n \geq 2$.

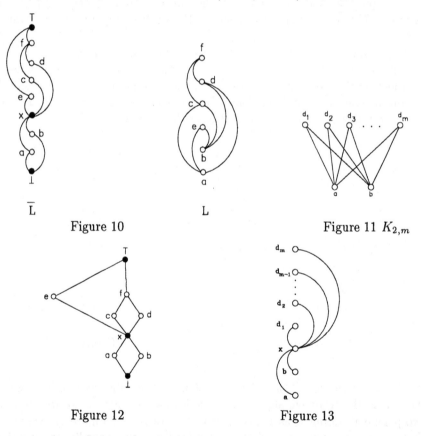

Figure 10   Figure 11 $K_{2,m}$

Figure 12   Figure 13

Lemma 6 *If the ordered set P contains $K_{2,m} = \{a, b, d_1, \ldots, d_m\}, m \geq 2$, and if P satisfies one of the following conditions, then P is not planar.*

    i) There is an upper bound of some three-element subset of $\{d_1, \ldots, d_m\}$.

ii) a and b have a common lower bound and some two-element subset of $\{d_1, \ldots, d_m\}$ has a common upper bound.

iii) There are two different two-element subsets of $\{d_1, \ldots, d_m\}$ each of which has an upper bound.

As we indicate in the beginning of this section, obtaining the two-page linear extension for P depends on the planar embedding of the completion $\overline{P}$ of P. The next lemma describe such planar embedding.

Lemma 7 *Let P be a series-parallel planar ordered set. For each maximal $K_{2,m} = \{a, b, d_1, \cdots, d_m\}, m \geq 3$ and each maximal $K_{n,2} = \{d'_1, \cdot, d'_n\}, n \geq 3$ such that $\{d_1, d_2\}$ has a minimal upper bound d and $\{d'_1, d'_2\}$ has a maximal lower bound d', there is a planar upward drawing of the completion lattice $\overline{P}$ of P in which d lies to the right of $d_3, d_4, \cdots, d_m$ and d' lies to the left of $d'_3, d'_4, \cdots, d'_n$.*

5 The Main result

In this section we will prove our main result . We will first prove that two pages are enough for a series-parallel planar ordered set. As a consequence, we will give a characterization of series-parallel planar ordered sets.

Theorem 8 *If P is series-parallel planar ordered set then, page(P) ≤ 2.*

The transformation algorithm Let $\overline{P}$ be the completion of P. By Lemma 5, $\overline{P}$ is series-parallel planar lattice. Fix a planar embedding of $\overline{P}$ satisfying

1. Whenever P contains a maximal $K_{2,m} = \{a, b, d_1, \ldots, d_m\}, m \geq 3$, such that d is an upper bound of $\{d_{m-1}, d_m\}$ then, d lies to the right of $\{d_3, \ldots, d_m\}$.

2. Whenever P contains a maximal $K_{m,2} = \{d_1, \ldots, d_m, a, b\}, m \geq 3$, such that d is a lower bound of $\{d_1, d_2\}$ then d lies to the left of $\{d_3, \ldots, d_m\}$.

This is possible according to Lemma 7. If P contains either a maximal $K_{2,m}$ or a maximal $K_{m,2}, m \geq 2$, we may assume that a lies to the left of b and d_i lies to the left of d_{i+1} for $1 \leq i \leq m-1$, in $\overline{P}$.

Notice that, if P contains a maximal $K_{2,m}, m \geq 2$, then the set of the upper covers of a is $\{d_1, \ldots, d_m\}$ which also is the set of the upper covers of b. Also, the set of the lower covers of d_i is $\{a, b\}$ for each $i = 1, \ldots, m$. Dually for $K_{m,2}$.

Since $\overline{P}$ is a series-parallel parallel planar lattice, by Theorem 4 there exists a two-page linear extension $\overline{L}$ of $\overline{P}$. We will transfer it to a two-page linear extension L for P.

For a four-cycle $C = \{a < c > b < d > a\}$ in an ordered set, a *splitting element* x satisfying $a, b \leq x \leq c, d$.

If P contains a maximal $K_{2,m} = \{a, b, d_1, \ldots, d_m\}, m \geq 2$, such that x is the splitting element of $K_{2,m}$ in $\overline{P}$, then we have $a < b < x < d_1 < d_2 < \ldots < d_m$ in $\overline{L}$ and the edges distributed as in Figure 13.

Also, if P has a maximal $K_{m,2} = \{d_1, \ldots, d_m, a, b\}, m \geq 2$ such that x is the splitting element of $K_{m,2}$ in $\overline{P}$, then we have $d_1 < d_2 < \ldots < d_m < x < a < b$ in $\overline{L}$ and the edges distributed as in Figure 14.

Since P is planar, by Lemma 6, if $\{a, b\}$ has a lower (respectively, an upper) bound of $K_{2,m}$ (respectively, $K_{m,2}$) in P, then there is no subset of two elements or more of the set $\{d_1, \ldots, d_m\}$ which has an upper (respectively, a lower) bound.

To obtain a two-page linear extension L of P from $\overline{L}$

1. Remove the set $\overline{P} - P$ from $\overline{L}$ and all edges connected to its vertcies.

2. For each maximal $K_{2,m} = \{a, b, d_1, \ldots, d_m\}, m \geq 2$, in P

    i) If $\{a, b\}$ *has a lower bound* in P draw the edges (a, d_i) on the left page and the edges (b, d_i) on the right page (see Figure 15).

    ii) If $\{d_{m-1}, d_{m-1}\}$ *has an upper bound* in P draw the edges $\{(b, d_1), (a, d_i) : 1 \leq i \leq m - 1\}$ on the left page and draw the edges $\{(a, d_m), (b, d_i) : 2 \leq i \leq m\}$ on the right page for each $1 \leq i \leq m$ (see Figure 16).

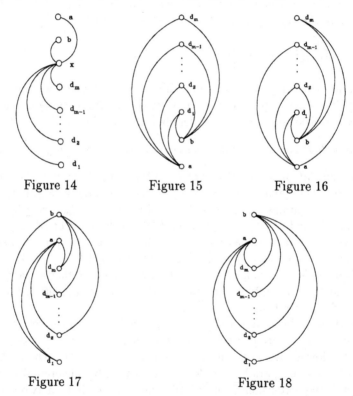

Figure 14   Figure 15   Figure 16

Figure 17   Figure 18

3. For each maximal $K_{m,2} = \{d_1, \ldots, d_m, a, b\}, m \geq 2$, in P

    i) If $\{d_1, d_2\}$ *has a lower bound* in P draw the edges $\{(d_1, b), (d_i, a) : 1 \leq i \leq m - 1\}$ on the left page and draw the edges $\{(d_m, a), (d_i, b) : 2 \leq i \leq m\}$ on the right page (see Figure 17).

    ii) If $\{a, b\}$ *has an upper bound* in P draw the edges $\{(d_i, a) : 1 \leq i \leq m\}$ on the left page and the edges $\{(d_i, b) : 1 \leq i \leq m\}$ on the right page (see Figure 18).

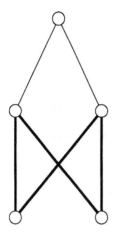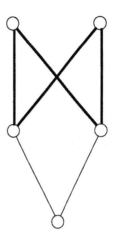

Figure 19 Simple castles.

By Lemma 3, L is greedy linear extension of P. We will show that adding the edges of the maximals $K_{2,m}$ and $K_{m,2}, m \geq 2$, do not create crossing in the same page first for $K_{2,m}$ then for $K_{m,2}$.

A *simple castle* is a covering four-cycle with the top or bottom. (The top, or bottom, need not be in a cover relation with the covering four-cycle.)(See Figure 19) A *castle* is any union of simple castles, which preserves the covering relations of each simple castle. An ordered set P *contains* a castle C if C is a subset of P and P preserves the covering relations of its simple castles.(See Figure 20)

Corollary 9 *Let P be a series-parallel planar ordered set. Then P is planar if and only P contains no $K_{3,3}$ and P contains no nonplanar castle.*

Figure 21 illustrates nonplanar ordered sets each of which contains neither $K_{3,3}$ nor a nonplanar castle. In fact, non is series-parallel.

6 Open problems

1. *Is the pagenumber for planar ordered sets bounded?*

   This question was first asked by Nowakowski and Parker [7]. Hung [3] gave a 48-element planar ordered set which requires four pages (see Figure 44). No planar ordered set required five pages is known.

2. We proved that two pages are enough to embed a series-parallel planar ordered set. Series-parallel ordered sets have dimension two. Is there a positive integer k, such that $page(P) \leq k$, for each planar ordered set of

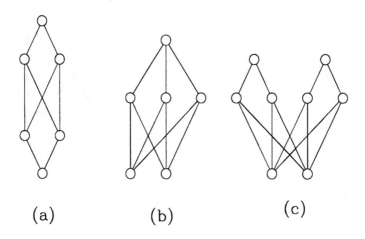

(a)   (b)   (c)

Figure 20 Minimal nonplanar castles.

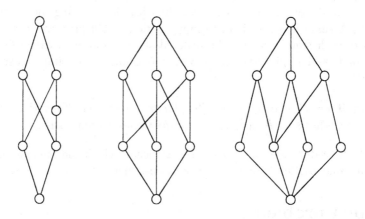

Figure 21

dimension two? ($k \geq 3$ because, $page(P) = 3$ for the planar lattice P in Figure 2). What about planar lattices?

3. Can we extend our result to (nonplanar) series-parallel ordered set?

   What is an upper bound for the (nonplanar) series-parallel ordered set P, depending on the maximal $K_{m,n}$'s in P.

   For positive integers m, n is there a function $f(m, n)$ such that for any series-parallel ordered set P

   $page(P) \leq max \{f(m, n) : K_{m,n}$ is a maximal in $P, m, n \geq 2\}$.

   In particular, is there a positive integer k such that
   $f(m, n) \leq min\{m, n\} + k$ for every maximal $K_{m,n}$ in P?

References

[1] F. Bernhart and P. C. Kainen (1979) *The book thickness of a graph*, Journal of Combinatorial Theory, Series B 27, 320–331.

[2] G. Birkhoff (1967) *Lattice Theory*, American Mathematical Society, Providence, Rhode Island.

[3] L. T. Q. Hung (1993) *A Planar Poset which Requires four Pages*, Ars Combin 35, 291–302

[4] H. de Fraysseix, P. O. de Mendez and J. Pach (1995) *A Left Search Algorithm for Planar Graphs*, Discrete Comput. Geom. 13, 459–468.

[5] D. Kelly and I. Rival (1975) *Planar lattices*, Canad. Journal of Mathematics 27, 636–665.

[6] H. M. MacNeille (1937) *Partially ordered sets*, Trans. Amer. Math. Soc. 42, 416–460.

[7] R. Nowakowski and A. Parker (1989), *Ordered sets, pagenumbers and planarity*, Order 6, 209–218.

[8] S. V. Pemmaraju (1992) *Exploring the Powers of Stacks and Queues via Graph Layouts*, Ph.D. thesis, Virginia Polytechnic Institute and State University at Blacksburg, Virginia.

[9] M. M. Sysło (1990) *Bounds to the Page Number of Partially Ordered Sets*, Graph-theoretic concepts in computer science (Kerkrade, 1989),181–195, Lecture Notes in Comput. Sci. 411. Springer,Berlin.

[10] J. Valdes, R. E. Tarjan and E. L. Lawler (1982) *The recognition of series-parallel digraphs*, SIAM J. Computing 11, 298–314.

[11] M. Yannakakis (1989) *Embedding planar graphs in four pages*, J. Comput. System Sci. 38, 36–67.

On Rectangle Visibility Graphs

Prosenjit Bose[1], Alice Dean[2], Joan Hutchinson[3], and Thomas Shermer[4] *

[1] Université du Québec à Trois-Rivières
[2] Skidmore College
[3] Macalester College
[4] Simon Fraser University

Abstract. We study the problem of drawing a graph in the plane so that the vertices of the graph are rectangles that are aligned with the axes, and the edges of the graph are horizontal or vertical lines-of-sight. Such a drawing is useful, for example, when the vertices of the graph contain information that we wish displayed on the drawing; it is natural to write this information inside the rectangle corresponding to the vertex. We call a graph that can be drawn in this fashion a *rectangle-visibility graph*, or *RVG*. Our goal is to find classes of graphs that are RVGs. We obtain several results:

1. For $1 \leq k \leq 4$, k-trees are RVGs.
2. Any graph that can be decomposed into two caterpillar forests is an RVG.
3. Any graph whose vertices of degree four or more form a distance-two independent set is an RVG.
4. Any graph with maximum degree four is an RVG.

Our proofs are constructive and yield linear-time layout algorithms.

1 Introduction

In this paper we consider the problem of drawing a graph in the plane so that the vertices of the graph are drawn as rectangles and the edges are horizontal or vertical line segments. We are in particular interested in drawings where each of the line segments can be thickened to have positive width, and none of these thickened segments (which we call *bands of visibility*) intersects the interior of any of the rectangles, although the bands may themselves cross or intersect. We call a graph a *rectangle-visibility graph* (or *RVG* for short) if it has such a drawing; we call the drawing itself the *layout* of the graph.

We study three variations on this idea. In the first, we require that the graph be drawn so that every pair of rectangles with a possible band of visibility between them represents a pair of vertices joined by an edge in the graph, and furthermore that no two rectangles have sides contained in the same (horizontal or vertical) line; we call such graphs *noncollinear RVGs*. In the second, we still

* Supported by NSERC Canada Reseach Grant OGP0046218, and DIMACS, Rutgers University, funded by NSF under contract STC–91–19999 and the New Jersey Commission on Science and Technology.

require that every possible visibility band represents an edge, but we allow two rectangles to have collinear edges; we call such graphs *collinear RVGs*. In the third, we do not require that every possible visibility band represents an edge; we call such graphs *weak RVGs*. Collinearities in drawings of weak RVGs can be eliminated by perturbation.

The results that we establish in this paper are that certain classes of graphs are RVGs (of the different types described above). We will establish that:

1. For $k = 1$ or 2, partial k-trees are noncollinear RVGs.
2. For $1 \leq k \leq 4$, k-trees are noncollinear RVGs (and thus partial 3-trees and partial 4-trees are weak RVGs).
3. All graphs that can be decomposed into two forests of caterpillars are noncollinear RVGs.
4. All graphs whose high-degree vertices form a distance-two independent set are weak RVGs.
5. All graphs whose high-degree vertices form a distance-three independent set are collinear RVGs.
6. All graphs whose high-degree vertices form a distance-four independent set (with a slight extra condition) are noncollinear RVGs.
7. All graphs whose maximum vertex degree is three are noncollinear RVGs.
8. All graphs whose maximum vertex degree is four are weak RVGs.

A caterpillar is a tree containing a path with the property that every vertex is at distance at most one from the path. A high-degree vertex is a vertex of degree four or more.

The problem that we are studying has application to a type of VLSI design known as *two-layer routing*. In two-layer routing, one embeds processing components and their connections (sometimes called *wires*) in two layers of silicon (or other VLSI material). The components are embedded in both layers. The wires are also embedded in both layers, but one layer holds only horizontal connections, and the other holds only vertical ones. If a connection must be made between two components that are not cohorizontal or covertical, then new components (called *vias*) are added to connect horizontal and vertical wires together, resulting in bent wires that alternate between the layers. However, vias are large compared to wires and their use should be minimized. In this setting, asking if a graph is a rectangle-visibility graph is the same as asking if a set of components can be embedded so that there is a two-layer routing of their connections that uses no vias. Our requirement that visibility bands have positive width is motivated by the physical constraint that wires must have some minimum width. A similar problem arises in printed-circuit board design, as printed-circuit boards naturally have two sides, and connecting wires from one side to the other (the equivalent of making vias) is relatively expensive [6].

The motivational two-layer problem discussed above abstracts away one feature of most two-layer routing problems: the processing components are often of a specific size. In other words, not only is a graph representing the components and their connections given, but each vertex of the graph also has a specified width and height. We turn our attention to this consideration only briefly in this paper; we show that any graph with linear arboricity two (which includes graphs with maximum degree three), no matter what widths and heights are specified for the vertices, can be laid out as a rectangle-visibility graph where each rectangle is of the specified size.

Rectangle-visibility layouts of graphs are also of use in other areas of graph drawing. One common graph drawing problem is *labelling*: writing information about the vertices and edges of a graph on the drawing, without having labels intersect or overwrite either other labels or graph components. A common solution to the vertex-labelling problem is to draw each vertex as a region (such as a relatively large circle) inside of which the information is written. Rectangle-visibility layouts are particularly suited to this approach. Furthermore, if we have a graph for which we can construct a layout with specified rectangle size, as discussed above, we can dimension the rectangle to fit exactly around the information that we wish to write inside it. Rectangle-visibility layouts are also advantageous when we wish to edge-label a graph; all edges and vertex sides in such a layout are horizontal or vertical (as opposed to layouts where edges can have any slope or curve), so there is no complicated geometry involved in placing an edge label.

One final advantage of rectangle-visibility layouts is in the nature of its edge crossings. Any edge crossing is between a horizontal edge and a vertical edge, and therefore perpendicular; such crossings make it easier for the eye to follow an edge without being distracted by crossing edges. Furthermore, if we have a collinear or noncollinear RVG, then we can eliminate crossings altogether, by *not* drawing the edges. For drawings intended to be read by people, this approach is only feasible for small graphs, or graphs with relatively short edges.

In the remainder of this paper, we establish our results described above. In Section 2, we review what is known about RVGs and related graphs, and give formal definitions necessary for the problems we consider and the techniques that we use. In Section 3, we establish results about which k-trees and partial k-trees are RVGs. In Section 4, we show that graphs that can be decomposed into two caterpillar forests are noncollinear RVGs; this class includes graphs with linear arboricity two, or with maximum degree three. In that section, we also discuss a few extensions of the caterpillar forest result. In Section 5, we present two geometric lemmas that state that graphs that have been decomposed in a certain way are rectangle-visibility graphs. In Section 6, we use the lemmas of Section 5 to show that graphs whose high-degree vertices are "far enough apart" are RVGs. In Section 7, we show that maximum-degree four graphs are weak RVGs; the main argument is an extension of one of the lemmas of Section 5. Finally, in Section 8, we draw our conclusion and discuss future directions for this work.

In this extended abstract, we omit the proofs of several of our results without notice; these proofs appear in the full version of our work [2, 10].

2 Background and Definitions

2.1 Bar-visibility graphs

The study we pursue on rectilinear drawings of graphs in the plane began with Luccio, Mazzone, and Wong [9] and Duchet *et al.* [5] in their studies of horizontal lines in the plane and vertical visibilities between them. Following the terminology of Tamassia and Tollis [12], we call a graph G a *bar-visibility graph* or *BVG* if its vertices can be represented by closed horizontal line segments in the plane, pairwise disjoint except possibly for overlapping endpoints, in such a way that two vertices u and w are adjacent if and only if each of the corresponding segments is *vertically visible* from the other. The vertices u and w are called vertically visible if there is a non-degenerate rectangular region $B_{u,w}$ (the *band of visibility for u and w*) with two opposite sides that are subsets of each of these segments, and $B_{u,w}$ intersects no other segment. A set of segments realizing a BVG is called a *layout* of the BVG.

Note that this definition *does* require non-degenerate visibility bands, and it does *not* require noncollinearity of segments. Bar-visibility graphs are easily seen to be planar; in addition they have been characterized by Wismath [13], and independently by Tamassia and Tollis [12], as those planar graphs that can be drawn in the plane with all cut-vertices on a single face. The question of whether a graph has a bar-visibility layout can be decided in linear time [12]. Figure 1 shows a bar-visibility layout of the 4-cycle and a bar-visibility layout of a forest containing two trees. If each bar in a bar-visibility layout is widened vertically to become a rectangle, we obtain a rectangle-visibility layout, but more can be achieved if both vertical and horizontal visibility is employed, as explained below.

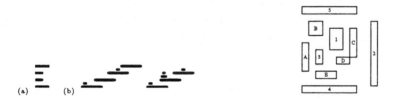

Fig. 1. A collinear bar-visibility layout of C_4, and a bar-visibility layout of a forest.

Fig. 2. A rectangle-visibility layout of $K_{5,5} + e$.

2.2 Rectangle-visibility graphs

Consider a collection $\mathcal{R}$ of rectangles in the plane, where each rectangle has its sides parallel to the axes, and rectangles may share boundary points but not interior points. In this situation, we will call two rectangles u and v *visible* if there is a *band of visibility* $B_{u,v}$ between them: $B_{u,v}$ is a rectangular region with two opposite sides that are subsets of u and v, and such that $B_{u,v}$ intersects no other rectangle of $\mathcal{R}$. The *visibility graph* of $\mathcal{R}$ is the graph of the visibility relation on the elements of $\mathcal{R}$. We call a graph G a *rectangle-visibility graph* or RVG if it is the visibility graph of some collection $\mathcal{R}$ of rectangles; in this situation, $\mathcal{R}$ is called a *layout* of G. The edges of a rectangle-visibility graph G can be partitioned into the two sets representing horizontal and vertical visibility; each of these two edge sets forms a BVG. Thus G, as a union of two planar graphs, is said to have *thickness-two*. Much less is known about thickness-two graphs than about planar ones, although their recognition is known to be NP-complete.

Wismath [14] has shown that every planar graph has a rectangle-visibility layout. Hutchinson, Shermer, and Vince [8] show that a rectangle-visibility graph with n vertices has at most $6n - 20$ edges, in contrast with thickness-two graphs, which may have at most $6n - 12$ edges; in both cases these bounds are attainable. Dean and Hutchinson [4] show that $K_{5,5}$ is *not* a rectangle-visibility graph though $K_{5,5}$ plus any edge is; see Figure 2. Thus rectangle-visibility graphs are not closed under the formation of subgraphs.

Each of the classes of BVGs and RVGs has two important subclasses: graphs with *noncollinear* layouts and those with *strong* layouts, as defined below.

2.3   Collinear and noncollinear layouts

A bar-visibility layout is called *noncollinear* if no two line segments have collinear endpoints; a rectangle-visibility layout is noncollinear if no two rectangles have collinear sides. The 4-cycle (see Figure 1a) does not have a noncollinear bar-visibility layout; Figure 3 shows a (collinear) rectangle-visbility layout of $K_{4,4}$ minus an edge; by a result in [3] it has no noncollinear layout, but a noncollinear layout of $K_{4,4}$ minus two edges is shown in Figure 10.

Fig. 3. A collinear (and strong) layout of $K_{4,4} - e$.

Fig. 4. $K_{4,4}$ minus two edges decomposed into two caterpillars.

2.4   Strong and weak layouts

G is a *strong bar-visibility graph* if its vertices can be represented by *disjoint* closed horizontal line segments in the plane in such a way that two vertices u and w are adjacent if and only if each of the corresponding segments is vertically visible from the other, with visibility permitted along *degenerate* rectangles, i.e., along lines. Similarly, G is a *strong rectangle-visibility graph* if its vertices can be represented by disjoint closed rectangles in the plane, with sides parallel to the axes, in such a way that two vertices u and w are adjacent if and only if each of the corresponding rectangles is vertically or horizontally visible from the other, again with visibility permitted along lines.

It is easy to see that noncollinear bar-visibility graphs are a subclass of strong bar-visibility graphs, which are in turn a subclass of bar-visibility graphs; analogous inclusions hold in the case of rectangle-visibility graphs. Tamassia and Tollis [12] show that this subclass ordering for bar-visibility graphs is strict. In [3] Dean and Hutchinson conjecture that the analogous subclass ordering for rectangle-visibility graphs is also strict, and they show that noncollinear rectangle-visibility graphs form a strict subclass of strong rectangle-visibility graphs. Figure 3 shows a strong layout of a graph that has no noncollinear layout.

Another class of BVGs and RVGs that can be considered is *weak* BVGs and RVGs. A graph is a weak BVG (respectively, weak RVG) if it is a subgraph of some BVG (respectively, RVG). This usage of *weak* and *strong*, which is not antonymous, was introduced by Tamassia and Tollis [12]. In [5] it is shown that every planar graph is a weak BVG, hence the containment from BVG to weak BVG is strict. This containment is also strict for RVGs, for example, from the fact that $K_{5,5}$ is not an RVG but the addition of an edge creates an RVG (see Figure 2).

2.5   Decomposition into trees, caterpillars, and paths

Earlier we noted that the edges of an RVG can be partitioned to form two BVGs. More generally, we will say that graph G can be *decomposed* into graphs $G_1, G_2, \ldots G_k$ if all of $G, G_1, G_2, \ldots G_k$ have the same vertex set and the edge set of G is exactly the union of the edge sets of $G_1, G_2, \ldots G_k$ (which must be disjoint). This can be viewed as coloring the edges of G with k different colors so that one color class forms G_1, one forms G_2, etc. We will use the concepts of decomposition and edge-coloring interchangably; we may, for instance, once we have completed an edge-two-coloring of a graph, say that we have decomposed that graph into two subgraphs. Figure 4 shows a decomposition of $K_{4,4}$ minus two edges into two caterpillars.

When we two-color, we will call the color classes red and blue (and say that red and blue are opposite colors). We will call a cycle monochromatic if all edges of the cycle are the same color. We will call a vertex monochromatic if all incident edges are the same color; vertices of degree zero are considered trivially monochromatic.

Recall that a *caterpillar* is defined as a tree containing a simple path $P(a, b)$ such that every vertex not on $P(a, b)$ is distance one from $P(a, b)$. A vertex on $P(a, b)$ will be called a *path vertex*, and one not on $P(a, b)$ will be called a *foot vertex*. Similarly, we will call an edge that connects a path vertex to a foot vertex a *leg*, and one connecting two path vertices a *path edge*. Figures 5a and 5b show trees that, respectively, are and are not caterpillars. A *linear forest* (respectively, a *caterpillar forest*) is a forest, each of whose components is a simple path (respectively, a caterpillar). The *arboricity* (respectively, *linear arboricity, caterpillar arboricity*) of a graph G is the minimum k such that G can be decomposed into $G_1, G_2, \ldots, G_k$, where each G_i is a forest (respectively, a linear forest, a caterpillar forest). Thus Figure 4 shows that $K_{4,4}$ minus two edges has caterpillar arboricity two. Clearly a linear forest is a caterpillar forest, and we show that a graph with caterpillar arboricity at most two is an RVG. See Figure 10 for a layout of the graph of Figure 4.

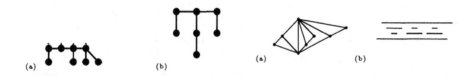

Fig. 5. (a) A graph that is a caterpillar, an interval graph, a 1-tree, and a non-collinear BVG. (b) A graph that is neither a caterpillar nor an interval graph, but is a 1-tree and a noncollinear BVG.

Fig. 6. A 2-tree and its bar-visibility layout.

We use $G \setminus E(H)$ to denote the graph G with the edges of subgraph H removed, and $G \setminus v$ to mean the graph G with the vertex v removed.

It is known [7] that graphs with maximum degree three have linear arboricity two. We present a new proof of this result that leads to a simple, easily-implementable algorithm for finding this decomposition. Then we find a rectangle-visibility layout for these graphs with a property of particular interest to practitioners of graph drawing: a rectangle layout can be constructed where the dimensions of the rectangle corresponding to each vertex is prespecified. For example, this allows each rectangle to be sized to fit exactly around text or information to be inscribed.

3   *k*-Trees and Partial *k*-Trees

A *k-tree* is either a k-vertex complete graph, or a graph formed from another k-tree T by finding a k-vertex clique K of T and adding a new vertex that is adjacent to every vertex in K (and no others). Thus, trees are the same as 1-trees, and every maximal outerplanar graph is a 2-tree; a 2-tree is shown in Figure 6. A *partial k-tree* is a subgraph of a k-tree. For example, partial 1-trees are the same as forests, and every series-parallel graph is a partial 2-tree. (There are at least two nonequivalent definitions of series-parallel graphs in the literature; one of these is equivalent to partial 2-trees and the other is a subclass of partial 2-trees.) A partial 2-tree is shown in Figure 7a; this graph is not a collinear BVG but it is a noncollinear RVG.

The following two theorems can be easily derived from the result of [9].

Theorem 1. *Every 1-tree and 2-tree is a noncollinear bar-visibility graph.*

Theorem 2. *Every partial 1-tree (forest) is a noncollinear bar-visibility graph.*

As every BVG is an RVG, the two previous theorems hold for RVGs as well. In fact, we can extend both of them for RVGs, showing that every k-tree ($1 \leq k \leq 4$) is a noncollinear RVG, as is every partial k-tree ($k = 1$ or 2).

Theorem 3. *For $1 \leq k \leq 4$, every k-tree is a noncollinear rectangle-visibility graph.*

The range of k in this theorem cannot be increased, as there are 5-trees containing $K_{5,13}$, which has thickness three.

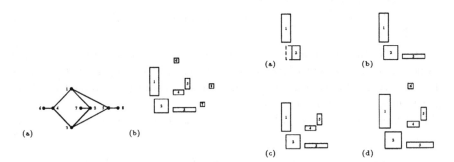

Fig. 7. A partial 2-tree and its rectangle-visibility layout.

Fig. 8. The construction of the layout of Figure 7.

Theorem 4. *Every partial 2-tree is a noncollinear rectangle-visibility graph.*

Proof. Let G be a partial 2-tree, and H be the 2-tree that it is a subgraph of. We proceed by induction on the number of vertices of H.

Let a rectangle R in a layout be called *N-visible* if it can see "to infinity" in a northward direction. We call the unbounded band of visibility from R northward a *band of N-visibility*, and use *E-visible* and *band of E-visibility* for the eastward direction.

In laying out G, we maintain the invariant that there are no collinearities, each rectangle is both N-visible and E-visible, and if two vertices form a clique of the underlying 2-tree, then the band of N-visibility for one intersects the band of E-visibility for the other. The basis, when H has two vertices, is easily handled.

In the general case, let v be a vertex that we can remove from H to get another 2-tree. The graph G' that is formed by removing v from G is a partial 2-tree, and we inductively construct a layout of G' that satisfies the invariant.

Consider reintroducing v to re-form G. If v is not a vertex of G, then we do nothing. Otherwise, the vertex v is connected to two vertices w and x in H. By the invariant, one of the rectangles corresponding to these vertices, wlog $R(x)$, has a band of N-visibility that intersects the band of E-visibility of the other ($R(w)$). We let the intersections of these bands be called WX.

Each of the edges vw and vx may be present or absent in G; we need to show that we can place $R(v)$ so that the correct ones of these edges are present and our invariant is maintained. We examine four cases, omitting the analysis.

Case 1: The edges vw and vx are both present in G. Let $R(v)$ be placed inside the bottom-left quarter of WX so as not to introduce any collinearities; see the placement of rectangle 4 (or rectangle 5) in Figure 8c for an example.

Case 2: The edge vw is present in G, but vx is not. Let T be the line one unit above the top of the rectangle with the highest top in the layout. We place $R(v)$ so that its bottom is contained in T, it is one unit high, and it is contained in the left half of the band of N-visibility from $R(w)$; see the placement of rectangle 6 in Figure 8d for an example.

Case 3: The edge vx is present in G, but vw is not. This case is symmetric to Case 2.

Case 4: Neither vw nor vx is present in G. Let T be a line above the top rectangle as in Case 2, and L be a similar line one unit to the left of the leftmost rectangle. We let $R(v)$ be a unit square whose bottom is contained in T and whose right side is contained in L.

In all cases, we have shown how to embed $R(v)$, and the theorem follows by induction. □

We note here that Biedl [1] has concurrently and independently proven that all series-parallel graphs are RVGs; depending on her definition of series-parallel, her result is either subsumed by or equivalent to our theorem.

Figure 7b shows the construction of the last proof applied to the graph of Figure 7a. This result cannot be extended to partial 3-trees, as $K_{3,5}$ is a partial 3-tree but is not a noncollinear RVG [4]. Theorem 3 implies that partial 3-trees are weak RVGs; the question of whether or not they are collinear RVGs is still open.

4   Caterpillar Forests

Recall that a graph G is called an *interval graph* if each vertex can be represented by an interval on the real line so that two vertices of G are adjacent if and only if the corresponding intervals have nonempty intersection; we call such a representation an *interval layout* of G. For any interval graph, the intervals in the layout can be chosen so that no two share an endpoint. Figure 9 shows an interval layout of the caterpillar shown in Figure 5a.

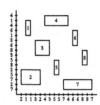

Fig. 9. An interval layout of the graph of Figure 5a.

Fig. 10. A rectangle-visibility layout of the graph of Figure 4.

If each horizontal segment (bar) of a bar-visibility layout of a graph G is vertically projected to an interval on the x-axis, the resulting interval graph is a supergraph of G. Similarly, a rectangle-visibility layout can be projected both vertically and horizontally to obtain two interval graphs, supergraphs of the vertical and horizontal BVGs. The idea of the rectangle-visibility layout constructions in this section is to reverse this process, starting with two interval graphs that are *exactly* the horizontal and vertical BVGs, and combining their interval layouts to obtain a rectangle-visibility layout.

Theorem 5 (The Caterpillar Theorem). *Every graph with caterpillar arboricity two is a noncollinear RVG.*

Proof. We give only the construction.

An arbitrary caterpillar C contains a path $P(x_0, x_k) = x_0, x_1, \ldots, x_k$ of path vertices; to lay out C, we lay $P(x_0, x_k)$ out on an axis with x_i represented by the interval $(2i, 2i + 3)$, $i = 0, 1, \ldots, k$. We then lay out the set of foot vertices adjacent to each path vertex x_i as consecutive disjoint subintervals of $(2i + 1, 2i + 2)$; see Figure 9. A caterpillar forest can then be laid out by putting the interval layouts for the caterpillar forest's components one after another.

Suppose that G can be decomposed into caterpillar forests F_1 and F_2. Lay out F_1 and F_2 as interval graphs on the x-axis and y-axis, respectively, as described above. For each vertex v of G, let $R(v)$ be the rectangle that is the cartesian product of the two intervals (one on the x-axis and one on the y-axis) that correspond to v in the layouts of F_1 and F_2. This gives a layout of G; see Figure 10 for an example. □

In subsequent work, Shermer [11] has shown that determining if a graph has caterpillar arboricity two is NP-complete, and we suspect that the other decomposition problems are also NP-complete. However, it is still possible that some natural, easily-recognizable graph classes are subclasses of caterpillar arboricity two. The graphs with maximum degree three form one such class; any such graph has linear arboricity two [7]. In the full paper, we present a new, algorithmically-useful proof of this fact. This establishes the following theorem:

Theorem 6. *Every maximum-degree 3 graph is a noncollinear RVG.*

Consider the interval layout that we have specified for a path. By making the overlap of consecutive intervals small enough, we can contract and expand the center portion of any interval (moving the rest of the layout) until that interval has a desired length. Applying this to all intervals, we see that we can lay out a path as an interval graph where each vertex of the path has the length of its interval prespecified.

We can obviously extend this idea to linear forests. This implies that we can also prespecify the interval sizes for the two linear forest layouts used in the layout construction algorithm (the proof of Theorem 5) for graphs with linear arboricity two. We can use this, for instance, to create a layout of all unit squares, or to size each rectangle to fit exactly around text or graphics that we wish to inscribe in it.

5 Technical Extensions of the Caterpillar Theorem

In this section, we present two technical lemmas that are extensions of the Caterpillar Theorem (Theorem 5). These lemmas will be used in the proof of the results in the next section, and contain the essential geometry for those results; the proofs in the next section are mainly graph-theoretic.

Lemma 7. *Let G be a graph such that it can be decomposed (edge-colored) into two graphs whose connected components are either caterpillars or cycles, and furthermore, no two cycles of different colors have more than one vertex in common. The graph G is a weak RVG, and a weak layout of G can be constructed in linear time.*

Proof. We proceed somewhat as in the proof of the Caterpillar Theorem, laying out each colored subgraph as intervals on a line, and taking the cartesian product for each vertex. Caterpillars are laid out on the line as before, and cycles are first laid out so that all vertices of the cycle have exactly the same interval (and this interval overlaps no other intervals). Figure 11a shows a five-cycle and its interval layout.

Before actually taking the cartesian products to form rectangles, we adjust the intervals for the vertices of each cycle. Consider a cycle laid out as intervals on the horizontal line. If we were to take the cartesian products for the vertices in this cycle, there would be exactly one with the highest top edge. In Figure 11b, example cartesian products are shown; the vertex e has the highest top edge.

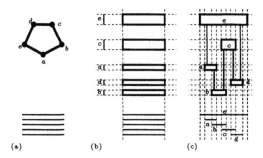

Fig. 11. Laying out a cycle.

We leave the interval for this top vertex t as it is, but lay out the rest of the cycle as a path (legless caterpillar) that starts after t does, and ends before t does; an example is shown on the bottom of Figure 11c. This effectively replaces the cycle in the horizontal layout by a cycle with all of the chords to one of its vertices. Unfortunately, we are not yet finished. These new chords may duplicate edges found in the vertical layout, so we will remove these edges from that layout. The details of this adjustment are omitted here.

Having applied this construction to each cycle, we can now take the cartesian products of the two intervals for each vertex. The proof that we now have a weak layout is straightforward and omitted. □

Let v be a path vertex on a caterpillar, and V' be the adjacent foot vertices. We can enlarge the caterpillar by adding to it a linear forest induced by the vertices of V', and repeating this for each path vertex v in the caterpillar. We call a graph formed in this manner a *caterpillar with footpaths*. A caterpillar with footpaths is shown in Figure 12. We allow the linear forests that we add to have no edges; i.e. a caterpillar is considered to be a caterpillar with footpaths. Each path induced by the vertices adjacent to a path vertex v is called a *footpath*, and is said to have *apex* v. In Figure 12, $bcde$ is a footpath with apex a.

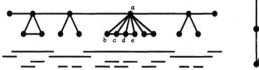

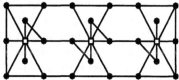

Fig. 12. A caterpillar with footpaths and its interval layout

Fig. 13. A 2-hilly graph.

Suppose that we have a graph G whose edges have been colored so that each color class forms connected components that are caterpillars with footpaths. In this situation, we will call a footpath of one color *safe* if all of its vertices lie in one connected component C of the other color, and furthermore the apex of the footpath is not in C. This condition ensures that the cartesian product construction keeps the rectangles corresponding to the vertices of the footpath together, all on one side of the rectangle corresponding to the apex. A *caterpillar with safe footpaths* is then a caterpillar with footpaths where each footpath is safe.

Lemma 8. *Let G be a graph that can be decomposed into two graphs whose connected components are either caterpillars with safe footpaths or triangles. The graph G is a noncollinear RVG, and a noncollinear layout of G can be constructed in linear time.*

6 k-Hilly graphs

Recall that we call a vertex with degree four or more a *high-degree* vertex; we correspondingly call a vertex with degree three or less a *low-degree* vertex. We call a graph k-*hilly* if the high-degree vertices form a distance-k independent set; i.e. there is no path of length k or less starting at a high-degree vertex and ending at another one. A 4-hilly graph is shown in Figure 13; the high-degree vertices are shown as white squares (a convention that we will follow throughout the section on k-hilly graphs). We will also use a slightly more restrictive version of k-hilly; we call a graph k-*hilly** if it contains no path of length k or less that starts and ends at high-degree vertices. This differs from k-hilly in that it also forbids cycles of length k or less that contain a high-degree vertex.

Theorem 9. *Every 2-hilly graph G is a weak RVG, and a weak layout of G can be constructed in linear time.*

Proof. The following algorithm constructs a layout of G as a weak RVG. The bulk of the algorithm two-colors the edges of G so as to be able to apply Lemma 7. The main idea is in Steps 1, 7, 8, and 13: we remove the high-degree vertices, color the remaining graph so that each color class is a linear forest and each vertex that used to be adjacent to a high-degree vertex is monochromatic, and then reintroduce the high-degree vertices, coloring their edges opposite the color of their monochromatic low-degree end. If this scheme worked perfectly, each color class would then be a collection of paths (through low-degree vertices) and stars (one centered at each high-degree vertex). However, the coloring is not quite possible as stated–but it can be done if cycles are allowed in each color class. The remainder of the steps are for winnowing out and dealing with the cases where cycles are necessary. Some details are omitted in this extended abstract.

Step 1. Remove high-degree vertices. Construct a graph G' from G by removing all high-degree vertices and the edges incident on them. We will use the term *reduced vertex* to refer to a vertex that is present in both G and G' but has smaller degree in G' than it did in G. A reduced vertex has degree at most two in G'.

Step 2. Remove cycles of reduced vertices. Remove all cycles of reduced vertices from G'. Note that such a cycle necessarily forms a connected component of G'.

Step 3. Remove near-cycles of reduced vertices. If there are any cycles in G' that consist of exactly one non-reduced vertex a and several reduced vertices, remove all the elements of these cycles except the vertices a from G'.

Step 4. Contract paths of reduced vertices. Contract all of the edges of any path of reduced vertices in G'. The vertex that such a path contracts to will also be called a reduced vertex. This procedure does not create any doubled edges, as the situations leading to doubled edges were removed in Step 3.

At this point, there are no two adjacent reduced vertices in G'.

Step 5. Contract unbridged reduced vertices. Let v be a reduced vertex with two neighbors c and d. We will call v *bridged* if cd is an edge of G', and *unbridged* otherwise.

Repeat the following procedure until no unbridged reduced vertices remain in G': First, find an unbridged reduced vertex v with neighbors c and d. Then, contract the edge vc (i.e., remove vc, vd, and d, and replace with the edge cd).

Step 6. Set aside Θ-components. We will call a connected component of G' a Θ-component if it is a four-cycle $vcwd$ with chord cd and reduced vertices v and w. We let G_2 be the subgraph of G' that consists of all of the Θ-components, and G_1 be the remaining components of G'.

Now, each pair of degree-2 reduced vertices in G_1 has distance at least three.

Step 7. Color G_1 as two linear forests.

Step 8. Adjust coloring to make reduced vertices monochromatic. We want a coloring of G_1 where each reduced vertex is incident on edges of only one color. If a reduced vertex has degree zero or one, then it is monochromatic in any coloring of the edges; thus we are concerned only with degree-two reduced vertices.

Find a bichromatic reduced vertex v in G'. By Step 5, v, along with its two adjacent vertices c and d, forms a triangle. If c has degree three, then let b be its third neighbor; similarly, if d has degree three, let e be d's third neighbor. The situation is shown in Figure 14a. Neither b nor e can be a degree-2 reduced vertex, as noted in Step 6.

In the coloring of G_1, the triangle cdv has two edges of one color, and one of the other color. Wlog we may assume that the edges cv and cd have the same color, and that color is red, and that dv is blue. If bc is present, it must

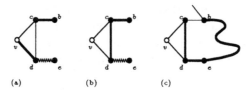

Fig. 14. Making a reduced vertex monochromatic.

be blue, and *de*, if present, may be either color. This coloring is also shown in Figure 14a. It now must be the case that either the recoloring of Figure 14b or Figure 14c leaves the coloring as a decomposition into two linear forests, with *v* monochromatic. Note that the only edges not incident on *v* that change color are *cd* and possibly *cb*. Neither of these edges is incident on a degree-2 reduced vertex, and thus we have not changed any reduced vertex from monochromatic to bichromatic.

We have reduced the number of bichromatic reduced vertices in G_1 by one, and may apply induction to find a coloring with all reduced vertices monochromatic.

Step 9. Decontract unbridged reduced vertices.

Step 10. Decontract paths of reduced vertices. After this step, each component of G_2 is a Θ-graph: a pair of vertices with three disjoint paths between them. Also, all degree-two vertices of these theta-graphs are reduced vertices.

Step 11. Reintroduce cycles and near-cycles of reduced vertices. After this step, each color class of the coloring has components that are either paths or cycles. Furthermore, each reduced vertex that we have reintroduced in this step is monochromatic, so all reduced vertices of G_1 are still monochromatic.

Step 12. Color the Θ-components. Let one color class be a cycle and the other a path; the reduced vertices whose edges are colored in this step are monochromatic.

We now have G' (the full G' constructed in Step 1) edge-colored so that each reduced vertex is monochromatic, and each color class has components that are either paths or cycles.

Step 13. Reintroduce high-degree vertices. Put back all vertices and edges removed in Step 1. Each edge being replaced in this step is given color opposite to the color of the reduced vertex that is its endpoint. This creates a red star and a blue star for each high-degree vertex reintroduced; these stars do not connect to any other components of their color. We now have G colored so that each color class has components that are either paths, cycles, or stars. No two cycles of opposite colors intersect, as such an intersection must take place at a vertex of degree at least four, and no cycle goes through a high-degree vertex.

Step 14. Lay out as rectangles.

□

Theorem 10. *Every 3-hilly graph G is a collinear RVG, and a collinear layout of G can be constructed in linear time.*

The proof of this theorem proceeds as in the proof of Theorem 9, preceding Step 14 with some extra graph manipulation so as to produce a collinear RVG rather than a weak one. This is done by adjusting the two-coloring to satisfy the hypothesis of Lemma 8; we omit the details. The following theorem is proved similarly; the cases where collinearity is used in Theorem 10 cannot occur in a 4-hilly* graph.

Theorem 11. *Every 4-hilly* graph is a noncollinear RVG, and a noncollinear layout of G can be constructed in linear time.*

Note that the technique used in above three theorems can be pushed even farther, to show that RVGs include graphs consisting of a few small clusters of high-degree vertices in a sea of low-degree vertices:

Theorem 12. *Let G be a graph, and H be the graph induced on the high-degree vertices of G. If H is maximum-degree three, and $G \setminus E(H)$ is 2-hilly, 3-hilly, or 4-hilly*, then G is a weak RVG, a collinear RVG, or a noncollinear RVG, respectively. Furthermore, a layout of such a graph can be constructed in linear time.*

We can also show that the results of this section cannot be strengthened to include 1-hilly graphs; there is a 1-hilly graph that is not a weak RVG.

7 Maximum Degree Four

In this section, we show that every maximum-degree 4 graph is a weak RVG. As with the k-hilly proofs, there are two distinct components to the proof: one graph-theoretic, and the other geometric. Here, however, because the graph-theoretic part is less detailed, we present that part before establishing the necessary geometric result.

Theorem 13. *Every maximum-degree 4 graph G is a weak RVG, and a weak layout of G can be computed in linear time.*

Proof. Start by augmenting the graph G with extra edges and vertices so that it becomes a 4-regular graph G'. Next, find an Euler circuit in G'; this is possible, as every vertex of G' has even degree. Alternately color the edges of the circuit red and blue. The Euler circuit will pass through each vertex twice, so each vertex will be incident on two red edges and two blue edges. This means that each color class is a 2-factor, or, more simply, a collection of cycles.

We prove below (as Theorem 14) that any such graph (one that can be decomposed into two collections of cycles) is a weak RVG, with layout taking linear time. Lay out G' as a weak RVG, and then remove any rectangles corresponding to vertices of G' that are not vertices of G. The extra edges of G' and the possible extra visibilities introduced when removing rectangles are inconsequential under the definition of weak RVGs, so the resulting layout is a weak layout of G. □

Theorem 14. *Let G be a graph that can be decomposed (edge-colored) into two graphs whose connected components are either caterpillars or cycles. The graph G is a weak RVG, and a weak layout of G can be computed in linear time.*

Proof. This is Lemma 7 with the restriction that two cycles intersect in at most one vertex removed. That restriction was to make the rightmost or topmost rectangle in a cycle well-defined. Removing it means that we will have to arrange things so that we can safely choose such a rightmost or topmost rectangle.

We assume that the given graph G is connected. Let us label the connected components of the blue subgraph $C_1, C_2, \ldots, C_k$ (C stands for column) and of the red subgraph $R_1, R_2, \ldots, R_l$ (R stands for row). Unlike in previous proofs, we will be careful about which order the rows and columns are placed in.

Whenever we lay out a row, we place it immediately below all rows that have already been laid out, and above all rows yet to be laid out. Similarly, we lay out columns to the left of all columns already laid out. The only real difficulty is getting started; i.e. choosing the rightmost column and topmost row, and showing how to lay out these components. We analyze several cases.

Case 1. There is a component (wlog a column) C_i that is a caterpillar. We let this column be the rightmost, and choose any row R_j that intersects C_i as the topmost row (such a row exists by connectivity of G). Lay out C_i as usual, and if R_j is a caterpillar, lay it out as usual also. If R_j is a cycle, then it has a rightmost vertex v (i.e. one whose horizontal interval's right endpoint is rightmost) in $R_j \cap C_i$, and we can lay it out as we did in Lemma 7, adjusting C_i if necessary.

All further cases assume that all components are cycles.

Case 2. There are two components C_i and R_j whose intersection contains exactly one vertex v. Let this row and column be the topmost and rightmost. Let the rectangle for v be the entire intersection of the horizontal and vertical intervals for C_i and R_j. Both $C_i \setminus v$ and $R_j \setminus v$ are paths and can be laid out as such inside their respective intervals; no visiblities are duplicated and therefore no adjustments need to be made.

Case 3. There are two components C_i and R_j whose intersection contains at least two vertices v and w, and furthermore, v and w are adjacent in one of these components (wlog in C_i). First, let v be placed so that its vertical interval is the vertical interval for R_j, and its horizontal interval is a small interval at the right end of the interval for C_i. We lay out $R_j \setminus v$ as a path as usual.

Let I be the horizontal interval for C_i with the interval for v removed. We lay out $C_i \setminus v$ as a path in I, with w on the left. The visibility between v and w will be horizontal. We must also adjust the layout depending on whether the other vertex u adjacent to v in C_i is in $C_i \cap R_j$: if it is, the interval for u should be contained in I (so that v sees u horizontally), and if not, this interval should extend slightly to the right of I so that u sees v vertically.

Case 4. No other case above holds. Choose a component (wlog a row R_b) whose removal will not disconnect the graph. The row R_b will be the *bottom* row and will be processed last. Let z be a vertex of R_b, and C_i the column containing z. Let y be a vertex adjacent to z in the cycle C_i, and R_j be the row containing z. The rows R_j and R_b are not the same, or else we would be in Case 3.

The components C_i and R_j will be the rightmost and topmost, and we choose $v = y$ and $w = z$ and lay out C_i and R_j as in Case 3, where the the path $C_i \setminus v$ stays strictly within the interval I. An example of this construction is shown in Figure 15. We have now laid out these components, but have not yet established the visibility between v and u, or between v and $w = z$. We will adjust v and lay out R_b at the end of the entire construction so as to establish these visibilities.

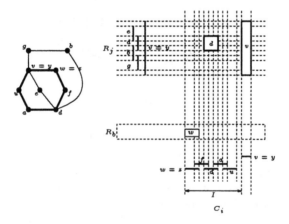

Fig. 15. Case 4 in the proof of Theorem 14.

In any case, we have laid out a topmost and rightmost component. The rest is straightforward:

Until all components have been laid out (except R_b if we were in Case 4), choose a component that is not laid out and that intersects a component that has already been laid out. (Such a component always exists, by connectivity of G or of $G \setminus R_b$.) Lay out this newly chosen component as follows:

Wlog, the component is a row R_j. If R_j is a caterpillar, lay it out as usual. Otherwise, let C_i be the rightmost column that has nonempty intersection with

R_j, and v be the rightmost vertex of this column in their intersection. Let v have the vertical interval corresponding to the entire interval for R_j, and lay out the path $R_j \setminus v$ inside the interval for v. We may have duplicated visibilities already present in C_i; if this is the case then we remove them as in Lemma 7 or by a special-case adjustment, as shown in Figure 16.

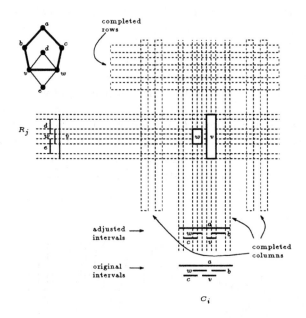

Fig. 16. The general step in Theorem 14.

This ends the general step. If we were in case 4, we have to lay out R_b and adjust the layout to get visibilities between the original v and u and between the original v and w; the details of this are omitted. □

8 Conclusion

We have shown that several classes of graphs are RVGs, including 4-trees, caterpillar arboricity two, maximum degree 4, and 2-hilly. We examined several subclasses of these classes, and established that some of them are collinear or noncollinear RVGs. In each case we have linear layout algorithms. Of particular interest is the class of graphs with maximum degree 3, which can be laid out with all rectangle sizes prespecified.

This abstract summarizes two manuscripts [2, 10]. In subsequent work, Shermer [11] has shown that recognizing RVGs is NP-complete, as is the problem of recognizing graphs with caterpillar arboricity two.

The question of where maximum-degree 4 and maximum-degree 5 graphs truly fall is still unsettled. It may be the case that maximum-degree 4 graphs are collinear or even noncollinear RVGs. On the other hand, Dean and Hutchinson [4] have shown that $K_{5,5}$ is not a collinear RVG, but it is a weak RVG ($K_{5,5}$ plus any edge is a collinear RVG). So maximum-degree 5 graphs either include some graphs that are not RVGs of any sort, or they are a subclass of weak RVGs.

Another question that is open is whether or not every graph with arboricity two is an RVG of some sort. As every forest (graph with arboricity one) is a BVG, one might suspect that this could be true.

In closing, we would like to acknowledge Peter Eades for suggesting some of the motivational ideas, Diane Souvaine and Alan Taylor for helpful discussions, and Stan Wagon for assisting with the figures.

References

1. T. Biedl. personal communication, 1995.
2. P. Bose, A. Dean, J. Hutchinson, and T. Shermer. On rectangle visibility graphs I. k-trees and caterpillar forests. manuscript, 1996.
3. A. M. Dean and J. P. Hutchinson. Combinatorial representations of visibility graphs. preprint, 1996.
4. A. M. Dean and J. P. Hutchinson. Rectangle-visibility representations of bipartite graphs. *Discrete Applied Mathematics*, 1996, to appear.
5. P. Duchet, Y. Hamidoune, M. Las Vergnas, and H. Meyniel. Representing a planar graph by vertical lines joining different levels. *Discrete Mathematics*, 46:319–321, 1983.
6. M. R. Garey, D. S. Johnson, and H. C. So. An application of graph coloring to printed circuit testing. *IEEE Transactions on Circuits and Systems*, CAS-23:591–599, 1976.
7. P. Horak and L. Niepel. A short proof of a linear arboricity theorem for cubic graphs. *Acta Math. Univ. Comenian.*, XL–XLI:255–277, 1982.
8. J. P. Hutchinson, T. Shermer, and A. Vince. On representations of some thickness-two graphs (extended abstract). In F. Brandenburg, editor, *Proc. of Workshop on Graph Drawing*, volume 1027 of *Lecture Notes in Computer Science*. Springer-Verlag, 1995.
9. F. Luccio, S. Mazzone, and C. K. Wong. A note on visibility graphs. *Discrete Mathematics*, 64:209–219, 1987.
10. T. C. Shermer. On rectangle visibility graphs II. k-hilly and maximum-degree 4. manuscript, 1996.
11. T. C. Shermer. On rectangle visibility graphs III. external visibility and complexity. manuscript, 1996.
12. R. Tamassia and I.G. Tollis. A unified approach to visibility representations of planar graphs. *Discrete and Computational Geometry*, 1:321–341, 1986.
13. S. K. Wismath. Characterizing bar line-of-sight graphs. In *Proc. 1st Symp. Comp. Geom.*, pages 147–152. ACM, 1985.
14. S. K. Wismath. *Bar-Representable Visibility Graphs and a Related Network Flow Problem*. PhD thesis, Department of Computer Science, University of British Columbia, 1989.

A Graph Drawing and Translation Service on the WWW*

Stina Bridgeman, Ashim Garg and Roberto Tamassia

Department of Computer Science
Brown University
Providence, RI 02912–1910, USA
{ssb,ag,rt}@cs.brown.edu

Abstract. Both practitioners and researchers can take better advantage of the latest developments in graph drawing if implementations of graph drawing algorithms are made available on the WWW. We envision a graph drawing and translation service for the WWW with dual objectives: drawing user-specified graphs, and translating graph-descriptions and graph drawings from one format to another. As a first step toward realizing this vision, we have developed a prototype service which is available at http://loki.cs.brown.edu:8081/graphserver/home.html.

1 Introduction

Motivated by numerous applications, new graph drawing algorithms are continually being developed. By making implementations of graph drawing algorithms available on the WWW, we can help both practitioners and researchers to use the latest technological innovations. This, however, also requires tackling the problem of the over-abundance of formats for describing graphs and drawings. While there are efforts in this direction [3], there is still no single universally-accepted format. Therefore, researchers typically define their own formats when implementing an algorithm. Because a user cannot be expected to know the format used by each and every implementation, it is advantageous to have translators that can convert the descriptions of graphs and drawings from one format to another. This will allow users to employ a large number of algorithms while knowing only a few formats.

We envision a graph drawing and translation service on the WWW that offers two kinds of services:

- a *drawing service* for constructing a drawing of a graph given by the user, using an algorithm chosen by her, and
- a *translation service* for translating the description of a graph/drawing from one format to another.

Such a service would benefit both practitioners and researchers. From a practitioner's viewpoint, she gets a central facility from where she can "shop around"

---

*Research supported in part by the National Science Foundation under grant CCR–9423847 and a graduate fellowship, by the U.S. Army Research Office under grant DAAH04–96–1–0013, and by a gift from Tom Sawyer Software.

for an algorithm appropriate for her application. From a researcher's viewpoint, she gets a repository of algorithms where she can study and compare their properties and performances. In addition, both practitioners and researchers can use this service just for translation purposes. Potential uses of this service include:

- drawing graphs from user-applications,
- studying and comparing graph drawing algorithms,
- translating between the formats for describing graphs and their drawings,
- creating a database of graphs occurring in user-applications, and
- demonstration purposes in educational settings.

Researchers have recognized the need of making implementations of graph drawing algorithms available on the WWW. A bibliography (and URLs) of many implementations is maintained by Georg Sander at `http://www.cs.uni-sb.de/RW/users/sander/html/gstools.html`. Stephen North has designed a service (visit `http://www.research.att.com/dist/drawdag/mail_server` for details) that accepts a graph sent to it by email and returns, also by email, a drawing constructed using *dot* [4]. This service also maintains a data base consisting of the graphs sent to it by the users [5].

We have developed a prototype graph drawing and translation service on the WWW. To the best of our knowledge, there is no other translation service over the WWW, and the closest approximation to a WWW graph drawing service is the email service provided by Stephen North. The other implementations available on the WWW are source code packages/executables which must be downloaded and installed locally, rather than being interactive servers running on a remote machine.

Our service is located at `http://loki.cs.brown.edu:8081/graphserver/home.html`, and is very easy to use. The user can interact with the service in two ways, by using an HTML form, or a Java-based graph editor embedded in a Web page. In a typical scenario, the user provides the server with the input graph, and selects the format of the input, the type of service desired (drawing/translation), the format of the output, and the drawing algorithm (if the graph is to be drawn). The server receives the request, performs the desired service, and sends back the result — a drawing or a translation — to the user.

In Section 2, we present the software architecture of our prototype service. In Section 3, we give some example interactions between users and our prototype service. Finally, in Section 4, we describe future work.

2 Software Architecture of our Prototype Service

We believe that a graph drawing and translation service should satisfy the following requirements:

- The input/output format used for a graph should be as independent as possible from the algorithm used for drawing it.
- It should be easy to add new formats and algorithms to the service.

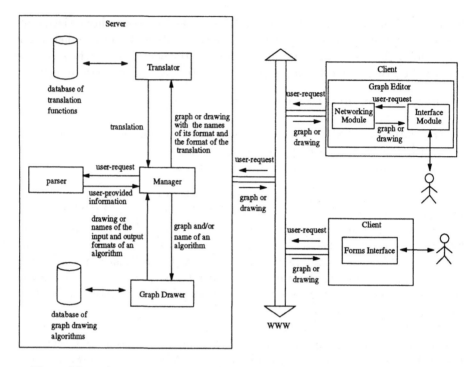

Fig. 1. The software architecture of our graph drawing and translation service.

- The user interface should be simple and intuitive.
- The architecture should be client-server based with the assumption that the client machine and the network may be slow, whereas the server is reasonably fast. This assumption is important for a network such as the WWW.

Figure 1 shows the software architecture of our prototype service. The service has five main logical components: the *client*, the *manager*, the *parser*, the *translator*, and the *graph drawer*. Only the client component runs on the user's machine; the other components all reside on the server.

Client. The client is responsible for maintaining the interface presented to the user. There are currently two versions of the client — the forms interface and the graph editor.

Forms Interface. The forms interface is simply an HTML page with a form that the user can fill to place her request. No code is needed to handle server communications or manage the form, as this is done by the user's Web browser.

Graph Editor. The graph editor is implemented as a Java applet and consists of two modules, the *interface module* and the *networking module*. The interface module is responsible for creating and managing the graph editor window, supporting basic graph editing operations such as insertion, deletion, and movement of vertices, edges, and edge bends, and enabling the user to interactively create a graph to be sent to the server. The networking module is responsible for communicating with the server — it sends the request, provides the interface with updates on the current status of the request, receives the response, and ensures that if the user sends a request, a response will be received even if the server goes down.

Manager. The manager is the central component of the service, and is responsible for coordinating the parser, translator, and graph drawer. It is multithreaded to handle potentially simultaneous requests from multiple clients, and typically spawns many threads for handling the user requests.

Parser. The parser receives from the manager a character string that contains a user request and extracts from it the following information: graph to be drawn, format of the input, type of service desired (drawing/translation), format of the output, and drawing algorithm to use (if the graph is to be drawn).

Translator. Each graph drawing algorithm supported by the service defines its own input/output format which may be different from the format of the graph given by the user. Therefore, there may be a need for translating from one format to another on drawing requests. In addition, the user may explicitly request a translation. The translator takes as input a graph or a drawing and the names of the current format and the format to which translation is to be done, and performs the desired translation. It maintains a database of *basic* translation functions. Given a translation request, it constructs a sequence of basic translation functions which is then applied to the input graph or drawing to carry out the desired translation. Notice that because different formats have different expressive powers in general, some loss of information may occur during a translation. Currently, the sequence of basic translation functions needed is explicitly coded so that the problem of translating from a more powerful format to a less powerful one and then back to a more powerful one is avoided.

Our prototype service currently supports the following input formats: *parenthetic, GDS, Tom Sawyer,* and *MALF,* and the following output formats: *parenthetic, GDS, postscript, GIF, gnuplot, mif, fig* and *MALF.*

The parenthetic format [7] consists of nested lists of keyword-value pairs enclosed by parentheses. It is very flexible in the sense that it allows the users to define their own keywords. GDS, Tom Sawyer, and MALF are minor variations of the formats used by GIOTTO [9], the Tom Sawyer Graph Layout Toolkit, and GD-Workbench [1], respectively. Gnuplot, mif and fig are the formats of **gnuplot**, Frame Maker and **xfig**, respectively.

Graph Drawer. The graph drawer accepts as input the name of a graph drawing algorithm and a graph described in the algorithm's input format, and constructs a drawing of the graph using the algorithm. Currently we support the following three algorithms: GIOTTO, PAIRS, and Planarizer.

GIOTTO [9] is a general-purpose drawing algorithm based on the *planarization* approach and a bend-minimization method [8]. In addition to the original implementation of GIOTTO, we offer a variation of it called GIOTTO-*with-labels,* which expands the vertices to accommodate labels. PAIRS [2] is our implementation of the algorithm of [6], which uses *st*-numberings. Planarizer is not actually a drawing algorithm: it is the first step of GIOTTO, which constructs a planar embedding of the input graph by replacing crossings (if any) with dummy vertices.

Our prototype service does not require any special hardware or software on a client machine other than a commonly available Web browser. The forms interface requires a browser, such as Lynx or Netscape, that supports HTML forms. The graph editor requires a browser that supports the Java 1.0 API, such as Netscape 2.0 or higher for most platforms. Java support is being added to more browsers so that this requirement is becoming less restrictive.

The server should be reasonably fast with some kind of support for multi-threading, and with sufficient memory for storing the users' graphs and drawings.

Our service satisfies the requirements stated earlier in this section as follows:

- The user is free to specify any input/output format independent of the algorithm requested by her. This decoupling is made possible by the presence of the translator.
- It is very easy to add new formats and algorithms. Also notice that an implementation of an algorithm can define its own input/output format provided that it is equipped with a translator to/from a format already supported by service. We believe that in most cases it will be sufficient to give a translator between a new format and the parenthetic format.
- Both the forms interface and the graph editor are simple and intuitive.
- All the computationally intensive work is done by the server. Hence, the service can be used by clients with limited resources as well.

3 Using our Prototype Service

We now give two scenarios to show how a user interacts with our prototype service and how the service satisfies a user request.

Scenario 1. Using the Forms Interface (see Figure 2).

Suppose the user wants to draw a graph described in the MALF input format using the GIOTTO algorithm, and wants the drawing to be in postscript format. She loads the service's home page into her browser, clicks on the *forms interface* link, and gets an HTML form. She can specify a graph in one of two ways — either by giving the URL of the graph or by typing the description of the graph in the form itself. In Figure 2(a) she chooses the latter option and enters the graph (in MALF input format) into the text area provided in the form. She then clicks on the appropriate selection buttons to specify that the input format is MALF input, the drawing algorithm is GIOTTO, and the output format is postscript, and submits the form to the server. Figure 2(a) shows a part of this user-filled form.

The server receives the form and passes it to the manager encoded as a character string. The manager gives this string to the parser which extracts from it the input graph, the type of service requested (DRAWING), the names of the input and output formats (MALF input and postscript, respectively),

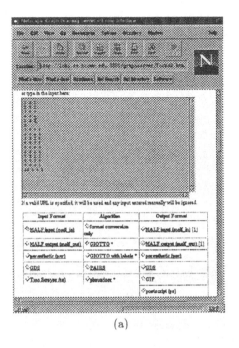

(a)

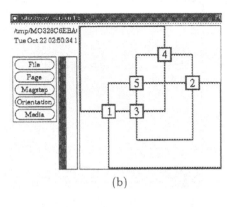

(b)

Fig. 2. *Scenario 1. Using the forms interface.* Drawing a graph described in MALF input format using GIOTTO, with output in postscript format. (a) Part of the user-filled form; (b) Drawing in postscript format returned to the user and displayed by the browser with **ghostview**.

and the drawing algorithm to use (GIOTTO). The manager then consults the graph drawer to determine the graph format required by the drawing algorithm (GIOTTO uses GDS) and calls the translator to convert the graph from MALF input to GDS. The manager sends the graph description in GDS format to the graph drawer, which uses GIOTTO to construct a drawing of the graph. The drawing is returned to the manager in GDS, the output format of GIOTTO, so the manager must once again invoke the translator to convert the drawing to postscript. The final drawing is saved to a file on the server and the URL of this file is returned to the client.

Back on the client, the user's browser receives the URL sent by the server and automatically loads the file, displaying it on the screen as shown in Figure 2(b) (assuming the browser is configured to automatically display postscript files; otherwise the user would be prompted to download the file).

Scenario 2. Using the Graph Editor (see Figure 3).

As before, the user loads the service's homepage into her browser. This time she first clicks on the *interactive applet interface* and then on the *start client* link. This causes a graph editor window to appear. Figure 3(a) shows a graph editor window with a sample graph constructed interactively by the user. To obtain a drawing of the graph the user must specify the algorithm and the output format

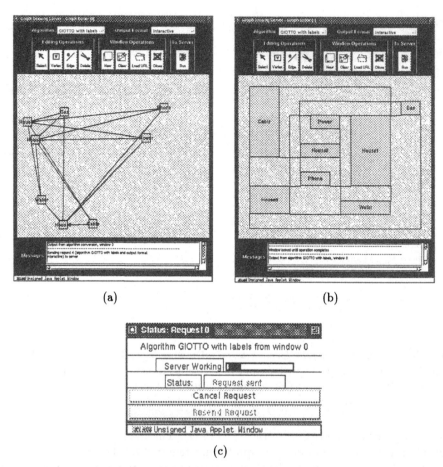

(a) (b)

(c)

Fig. 3. *Scenario 2. Using the graph-editor.* Drawing a graph with GIOTTO-with-labels, where the drawing is to be displayed in an editor window. (a) The interface of the editor and the input graph; (b) Drawing returned by the server; (c) Status window displayed while the request is being handled by the server.

by making the appropriate selections on the pulldown menus at the top of the window. The example shows the algorithm GIOTTO-with-labels and the output format "interactive". (Selecting "interactive" as the output format will cause the drawing to be displayed in another editor window; all of the other formats supported by the forms interface, such as parenthetic notation and GIF, are also supported by this interface and will be displayed in a new browser window.) Once the algorithm and output format have been selected, the user clicks on the "Run" button to send the request.

Sending the request causes a small status window to pop up, providing the user with information about the processing of the request and allowing her to cancel a running request (Figure 3(c)). The processing of the request on the server side follows the same basic pattern as for the forms interface (parsing, translation, drawing, translation); the differences are a slightly different front-end for reading and parsing the request and some extra machinery to ensure that responses destined for the same client are sent back in the proper order.

When the request completes, a new graph editor window appears (since the chosen output format was interactive) with the resulting drawing (Figure 3(b)), scaled to fit in the window. This drawing is fully editable, and can be modified and resubmitted to the server. It can also be saved, by sending a translation request to the server to convert the graph into some other format (presumably a text format), which is displayed in a browser window and can then be saved using the browser's own "Save As" mechanism.

4 Future Work

There are several directions for further work on the service:

- Adding support for new formats such as GML [3].
- Adding new algorithms. We plan to start with implementations already available freely on the Web.
- Adding a programmer's interface, where a small Java or C++ library is provided to handle communications with the server, allowing a user to write a program with a call to the server as a subroutine.
- Adding support for incremental graph drawing.
- Providing a service (accessible through the Web) through which the users can conduct experimental studies on graph drawing algorithms.

References

1. L. Buti, G. Di Battista, G. Liotta, E. Tassinari, F. Vargiu, and L. Vismara. GD-workbench: a system for prototyping and testing graph drawing algorithms. *Graph Drawing (Proc. GD '95). LNCS*, 1027:76–87, 1996. F. J. Brandenburg, Ed.
2. G. Di Battista, A. Garg, G. Liotta, R. Tamassia, E. Tassinari, and F. Vargiu. An experimental comparison of four graph drawing algorithms. *Comput. Geom. Theory Appl.*, 1996. to appear.
3. M. Himsolt. GML: Graph Modelling Language. Manuscript, Universität Passau, Innstraße 33, 94030 Passau, Germany, 1996. Available at http://www.uni-passau.de/~himsolt/Graphlet/GML.
4. E. Koutsofios and S. North. Drawing graphs with *dot*. Technical report, AT&T Bell Laboratories, Murray Hill, NJ., 1995. *dot* user's manual. Available at http://www.research.att.com/dist/drawdag.
5. S. North. 5114 directed graphs, 1995. Manuscript. Available at ftp.research.att.com/dist/drawdag.
6. A. Papakostas and I. G. Tollis. Improved algorithms and bounds for orthogonal drawings. *Graph Drawing (Proc. GD '94). LNCS*, 894:40–51, 1995.
7. S. Singh. Documentation for Paren-to-GDS. Manuscript, Dept. of Comp. Sci., Brown University, 1991.
8. R. Tamassia. On embedding a graph in the grid with the minimum number of bends. *SIAM J. Comput.*, 16(3):421–444, 1987.
9. R. Tamassia, G. Di Battista, and C. Batini. Automatic graph drawing and readability of diagrams. *IEEE Trans. Syst. Man Cybern.*, SMC-18(1):61–79, 1988.

Drawing 2-, 3- and 4-colorable Graphs in $O(n^2)$ Volume

Tiziana Calamoneri and Andrea Sterbini

University of Rome "La Sapienza", Dept. of Computer Science,
Via Salaria 113, 00198 Roma, Italy
{calamo, andrea}@dsi.uniroma1.it

Abstract. A Fary grid drawing of a graph is a drawing on a three-dimensional grid such that vertices are placed at integer coordinates and edges are straight-lines such that no edge crossings are allowed.

In this paper it is proved that each k-colorable graph ($k \geq 2$) needs at least $\Omega(n^{3/2})$ volume to be drawn. Furthermore, it is shown how to draw 2-, 3- and 4-colorable graphs in a Fary grid fashion in $O(n^2)$ volume.

Keywords: k-colorable graphs, Fary grid drawing, 3D drawing.

1 Introduction

In the last years three-dimensional drawing of graphs has increased its interest since the cost of 3D circuits has become reasonably low and recent advances in hardware and software technology for computer graphics have opened the possibility of displaying three-dimensional drawings. At present, the related research tends towards theoretical results understanding of 3D drawings. In particular, the most studied models are the orthogonal grid drawing and the Fary grid drawing, both having large application interest [3, 4, 5, 6]. In this paper we point our attention on the Fary grid drawing and we investigate the volume of such a drawing for some particular classes of graphs.

The *Fary grid drawing* [2] of a graph is a drawing on a three-dimensional grid such that vertices are placed at integer coordinates and edges are straight-lines such that no edge-crossings are allowed.

It is suitable to minimize both the volume and the length of the maximum side of the rectangular prism containing the drawing.

In [1] a method for drawing a general graph in a Fary grid fashion in $O(n^3)$ volume is shown, and it is proved that no general drawing method can achieve smaller size, up to a constant. In the same paper it is shown that for every planar graph and upward binary tree there exists a Fary grid drawing with $O(n^2)$ and $O(n \log n)$ volume, respectively, and it is conjectured that some other classes of graphs need a smaller volume than in the general case.

In this paper we partially solve the previous conjecture by showing how to draw 2-, 3- and 4-colorable graphs in a Fary grid fashion in $O(n^2)$ volume.

Furthermore, it is proved that each k-colorable graph ($k \geq 2$) needs at least $\Omega(n^{3/2})$ volume to be drawn.

2 Fary grid drawing of complete bipartite graphs

In this section we show that $\Omega(n^{3/2})$ volume is required to draw a complete bipartite graph with n vertices even if we conjecture that this lower bound can be improved to $\Omega(n^2)$. Moreover, we show how to draw a complete bipartite graph in $O(n^2)$ volume.

Given any kind of three-dimensional drawing, we call its *rectangular hull* the smallest rectangular prism with sides parallel to the coordinate axes containing the whole drawing; its *size* the length of the longest side and its *volume* the product of its three sides.

Let $K_{r,b} = (R \cup B, E)$ be a complete bipartite graph, that is a graph having $R \cup B$ as vertices set and $E = \{e = (v_1, v_2), \forall v_1 \in R, v_2 \in B\}$ as edges set. From now on, we will call *red* all the vertices belonging to R, and *blue* all the vertices belonging to B; furthermore $|R| = r, |B| = b$ and $r + b = n$.

It is easy to see that the 'worst' case of bipartite graph, in terms of occupied volume, is the complete bipartite graph. Therefore, if we prove that at least $\Omega(n^{3/2})$ volume is needed by a complete bipartite graph in order to be drawn, then this value is a lower bound for the whole class of bipartite graphs.

Lemma 1. *Let D be a three-dimensional Fary grid drawing of the complete bipartite graph $K_{r,b}$, with n vertices, where $r, b \geq 3$, and let D use volume V. Then V is $\Omega(n^{3/2})$.*

Proof. Project along the x axis all red vertices and consider their projections on plane yz. Then, two cases arise:

- The number of different-coordinates projections are exactly r. Then the area of the section of the rectangular hull perpendicular to the x axis is at least r.
- The number of different-coordinates projections are less than r. Then, at least two different red vertices have the same projection along the x axis. On the other hand, no pairs of blue vertices can lie on a straight-line parallel to the x axis, otherwise a crossing would arise. Therefore, the area of the section of the rectangular hull perpendicular to the x axis is at least b.

The same reasoning holds for the sections of the rectangular hull perpendicular to the y, z axes. It is easy to see that the worst case can be obtained when $r = b = \frac{n}{2}$ and when all three sections of the rectangular hull are squares. In such a case each dimension of V is at least $\Omega(n^{1/2})$ and the statement follows. $\square$

Since each graph contains a bipartite graph, then the proof of the previous Lemma can be extended to all k-colorable graphs, for any $k \geq 2$. Therefore, the following result holds:

Theorem 2. *Let D be a three-dimensional Fary grid drawing of a k-colorable graph G, with $n = n_1 + \ldots + n_k$ vertices, where n_i is the number of vertices colored with color k. If there exist $i, j \leq k, i \neq j$, such that $n_i, n_j \geq 3$ and D uses volume V, then V is $\Omega(n^{3/2})$.*

Now we prove that an upper bound of the volume for bipartite graphs is $O(n^2)$ by exhibiting a construction of a Fary grid drawing of a complete bipartite graph.

Theorem 3. *Given a bipartite graph having n vertices, it is possible to draw it as a Fary grid drawing of $O(n^2)$ volume.*

Proof. We prove the theorem by constructing such a drawing.

Consider two skew straight-lines, with parametric equations ($x = p, y = 0, z = 0$) and ($x = 0, y = q, z = 1$). A Fary grid drawing of $K_{r,b}$ is obtained by placing all red vertices on one straight-line, i.e. at coordinates $R_i = (i, 0, 0), i = 1..r$ and all blue vertices on the other line, i.e. at coordinates $B_j = (0, j, 1), j = 1..b$, as shown in fig. 1.

Notice that every edge-crossing requires four co-planar vertices. Since the lines are skew, no four vertices $r_1, r_2 \in R$ and $b_1, b_2 \in B$ can lie on the same plane. This observation garantees that all crossings are avoided. It is easy to see that $O(n^2)$ volume is enough to fit the drawing. □

3 Fary grid drawing of 3- and 4-colorable graphs

The technique used to prove Theorem 3, based on the arrangment of each color on one of some skew straight-lines, can be extended in order to draw 3- and 4-colorable graphs having n vertices in $O(n^2)$ volume.

3.1 3-colorable graphs

Let $K_{r,b,g} = (R \cup B \cup G, E)$ be a complete 3-colorable graph ,where r, b and g are the cardinalities of sets R, B and G, respectively, and $r + b + g = n$.

It is possible to draw $K_{r,b,g}$ in a Fary grid fashion by using one of the following techniques, both consisting in placing three skew lines in such a way that all vertices of one color can be completely lain on each of them:

1. The three lines lie on three parallel planes ($z = 0, 1, 2$), they pass through the origin of the grid and run along the $x = 0, y = -x, y = x$ lines, respectively. Namely, their parametric equations are ($x = p, y = 0, z = 0$), ($x = -q, y = q, z = 1$) and ($x = s, y = s, z = 2$). The vertices will be placed at coordinates $R_i = (i, 0, 0), i = 1..r$, $B_j = (-j, j, 1), j = 1..b$ and $G_k = (-k, -k, 2), k = 1..g$ (see fig. 2). Observe that, in this way, different portions of space are assigned to the three subsets of edges connecting all possible couples of colors (*"separation property"*).

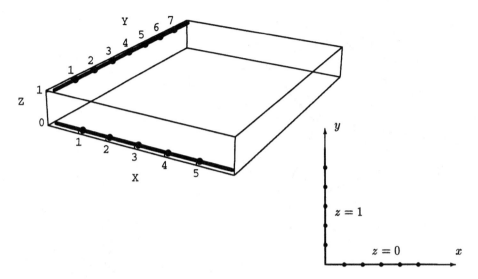

Fig. 1. Drawing of a complete bipartite graph in $O(n^2)$ volume and projection of the same drawing on the xy plane.

2. The three lines lie on three parallel planes ($z = 0, 1, 2$) and their projections on the xy plane form a rectangular triangle. Their parametric equations are $(x = 0, y = p, z = 0)$, $(x = q, y = \max(2g, 2r) - q, z = 1)$ and $(x = s, y = 0, z = 2)$, respectively. (see fig. 3). In this second configuration all the edges lie on the same portion of space.

Theorem 4. *Given a 3-colorable graph having n vertices, both previously described methods give a Fary grid drawing of $O(n^2)$ volume.*

Proof. In both the representations, we will call the three straight-lines by using the capital letters $\mathcal{R}$, $\mathcal{B}$ and $\mathcal{G}$ to mean the lines containing red, blue and green vertices, respectively.

In the first drawing (fig. 2), it is possible to put a vertex on each grid-crossing belonging to one of the three considered straight-lines. No edge-crossings arise because of the *"separation property"* of the space assigned to the edges. Namely, given two edges, if they connect the same couple of colors, they cannot cross because the lines are skew. If the edges connect three colors (let us say, for instance, red-blue and red-green) they cannot cross because they lie in different portions of space.

The volume of such a drawing is $O(n^2)$ because the maximum length of each line is $O(n)$.

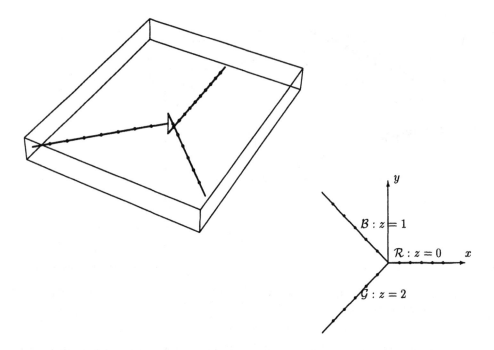

Fig. 2. 3-colorable graphs: first configuration. Drawing in $O(n^2)$ volume and projection of the same drawing on the xy plane.

In the second case it is not possible to put a vertex on each grid-crossing belonging to the straight lines, but we must "forbid" some positions, in order to guarantee the absence of crossings, as it is explained in the following.

Consider fig. 3: $\mathcal{R}$ has null z-coordinate, $\mathcal{B}$ has z-coordinate equal to one and $\mathcal{G}$ has z-coordinate equal to two. Consider two arbitrary green vertices g_1 and g_2, and suppose g_1 adjacent to a red vertex r_1. The plane passing through the three points g_1, g_2 and r_1 intersects $\mathcal{B}$ in one point. If this point coincides with a grid-crossing and a blue vertex b_2 lies on it, then it is possible that the edges (g_1, r_1) and (g_2, b_2) intersect. In order to avoid this problem, consider all the planes containing straight-line $\mathcal{G}$ and passing through a grid-crossing of $\mathcal{B}$. Each of them intersects $\mathcal{R}$ in one point. The set of all such intersections on $\mathcal{R}$ is the set of gridpoints having even y-coordinate on $\mathcal{R}$ (see fig. 4). Then, it is enough to position red vertices only on those grid-crossings of $\mathcal{R}$ having odd y-coordinates. A similar reasoning can be done for green vertices by considering

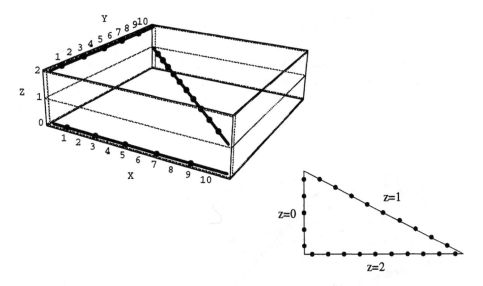

Fig. 3. 3-colorable graphs: second configuration. Drawing in $O(n^2)$ volume and projection of the same drawing on the xy plane.

all planes containing straight-line $\mathcal{R}$. It is not possible that any plane containing straight-line $\mathcal{B}$ intersect both $\mathcal{R}$ and $\mathcal{G}$.

Since the maximum length of each line is $O(n)$ and the height of the rectangular hull is constant, the volume of such a drawing is $O(n^2)$. $\qquad \Box$

Fig. 4. Projection of the second configuration on plane yz.

Notice that in both our constructions, one dimension of the rectangular hull is constant, and it allows to bound the volume in $O(n^2)$. It would be interesting to find a method to draw a 3-colorable graph in a Fary grid fashion by using $O(n^2)$ volume, so that all three dimensions are $O(n^{3/2})$. This would guarantee a better bound for the size of the drawing and for the maximum distance between the vertices.

3.2 4-colorable graphs

In order to draw in $O(n^2)$ volume a 4-colorable graph, we use again straight-lines lying on four different parallel planes ($z = 0, 1, 2, 3$), such that their projections on the xy plane form a convex quadrilateral (see fig. 5).

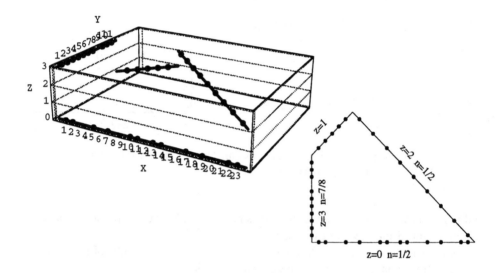

Fig. 5. Drawing of a 4-colorable graph in $O(n^2)$ volume and projection of the same drawing on the xy plane.

We call $\mathcal{R}, \mathcal{G}, \mathcal{B}$ and $\mathcal{Y}$ the four straight-lines, lying on the planes $z = 0, 3, 1, 2$ and containing red, green, blue and yellow vertices, respectively.

Assume $b, \lfloor \frac{8}{7}g \rfloor$ be odd numbers. If they are not, round them to the next integer odd value.

The parametric equations of the 4 lines are:

line $\mathcal{R}$: $(x = p, y = 0, z = 0)$
line $\mathcal{G}$: $(x = 0, y = q, z = 3)$
line $\mathcal{B}$: $(x = s, y = s + \frac{8}{7}g, z = 1)$
line $\mathcal{Y}$: $(x = t + b, y = \frac{8}{7}g + b - t, z = 2)$

Position the vertices at the following coordinates:

r-vertices: $(x = p, y = 0, z = 0)$, where $p = 1..2r$ and $p \bmod 3 \neq 0, p \bmod 4 \neq 0$
g-vertices: $(x = 0, y = q, z = 3)$, where $q = 1..\frac{8}{7}g$ and $q \bmod 8 \neq 0$
b-vertices: $(x = s, y = s + \frac{8}{7}g, z = 1)$, where $s = 1..b$

y-vertices: $(x = t + b, y = \frac{8}{7}g + b - t, z = 2)$, where $t = 1..2y$ and $q \bmod 2 = 0$

Theorem 5. *Given a 4-colorable graph having n vertices, the described method is a Fary grid drawing of $O(n^2)$ volume.*

Proof. In order to prove that the stated drawing is Fary grid we must show that no crossing of edges can arise. Given a couple of edges, they can connect either 2, or 3 or 4 different colors (see fig. 6). Therefore there are three possible kinds of crossings we need to avoid:

Fig. 6. possible crossings in a 4-colorable graph.

involving 2 colors: By placing the lines with different directions on different planes, we obtain mutually skew lines and then we are assured that no crossings can arise between two edges involving only 2 colors.

involving 3 colors: Using a technique similar to the proof of theorem 4, by removing some vertices along the lines, we assure that no crossings can arise between two edges involving 3 colors.

It can be shown that, on $\mathcal{G}$ we must forbid one grid-crossing out of 8, on $\mathcal{B}$ all the positions remain available, on $\mathcal{Y}$ one out of 2 can be filled up, finally one out of 3 plus one out of 4 (i.e. $(4 + 3 - 1)/12 = 1/2$) grid-crossings cannot be used on $\mathcal{R}$ (fig. 5).

involving 4 colors: We must check for crossings originated by two edges with their four endpoints all belonging to different colors. In view of the shape of the quadrilateral, the only way to have such a crossing is between edges connecting colors lying on opposite sides of the quadrilateral, that is, between an edge connecting an r-vertex to a b-vertex and an edge connecting a g-vertex to a y-vertex. These crossings can be avoided by placing the four lines $\mathcal{R}$, $\mathcal{G}$, $\mathcal{B}$ and $\mathcal{Y}$ on the planes at levels $z = 0, 3, 1$ and 2, respectively. In this way, all the edges connecting an r-vertex to a b-vertex and a g-vertex to a y-vertex lie in different half-spaces (separated, for example, by the plane $z = 3/2$).

Now we justify the assumption that $b, \lfloor \frac{8}{7} \rfloor$ are odd.

Let us call L_i the number of gridpoints of line i, where $i \in \{\mathcal{R}, \mathcal{B}, \mathcal{G}, \mathcal{Y}\}$. In view of the shape of the quadrilateral, the following relations hold:

$L_{\mathcal{R}} = L_{\mathcal{G}} + 2L_{\mathcal{B}}$ and $L_{\mathcal{Y}} = L_{\mathcal{G}} + L_{\mathcal{B}}$.

From them, it may be inferred that:

- if $L_{\mathcal{G}}$ is even then $L_{\mathcal{R}}$ is even
- if $L_{\mathcal{G}}$ and $L_{\mathcal{B}}$ have the same parity than $L_{\mathcal{Y}}$ is even

therefore, by choosing both $L_{\mathcal{G}}$ and $L_{\mathcal{B}}$ even, we obtain all four even-length lines. This is desirable, otherwise all the grid-crossings on one line would be forbidden in order to avoid crossings involving 3 colors.

Since the height of the rectangular hull is 3, its volume depends only on the area of the xy projection of the rectangular hull. The worst case arises when all the four lines have $O(n)$ length, therefore the area of the quadrilateral is $O(n^2)$.

□

Notice that it is not immediate to extend this construction to 5-colorable graphs for two main reasons. First of all, too many positions should be 'forbidden'; secondly, the 'trick' used to guarantee no crossings involving 4 colors cannot be used anymore. Actually, it is not clear if there is a way to draw 5-colorable graphs in a Fary grid fashion using $O(n^2)$ volume or not.

4 Conclusions and Open Problems

In this work we have described some results dealing with the volume of three-dimensional drawing of graphs. Namely, we have established a lower bound of $\Omega(n^{3/2})$ volume for each k-colorable graph and an upper bound of $\Omega(n^2)$ volume for 2-, 3- and 4-colorable graphs. Observe that we have made no efforts to compute a drawing with optimal multiplying constant since our aim was to improve the previous bounds for these classes of graphs. It may be that a different arrangment of colors on the skew-lines allows to 'forbid' less positions and therefore to get a smaller constant. Furthermore, it would be interesting to compute a more precise volume by means of a more general bounding convex prism instead of the rectangular hull. Moreover, there are a lot of open problems to solve and of questions to answer. Among them, we will thick those ones encountered in the paper:

- we conjecture that complete bipartite graphs $K_{m,n}$ with $m, n \geq 3$ cannot be drawn in less than $\Omega(n^2)$ volume. This would imply that all k-colorable graphs need at least the same volume.
  How is it possible to close the gap between the lower bound and the upper bound?
- Is it possible to draw in a Fary grid fashion 2-, 3- and 4-colorable graphs such that the size is minimum, that is $O(n^{2/3})$?
- Is it possible to extend our construction based on skew lines to k-colorable graphs, $k \geq 5$? If the answer is no, is there a way to prove that these graphs need more than $\Omega(n^2)$ volume?
- Do there exist other classes of graphs needing less than $O(n^3)$ volume?

References

1. R.F.Cohen, P.Eades, T.Lin and F.Ruskey: *Three-dimensional graph drawing*, In Proc.GD '94, Lecture Notes in Comp. Sci. 894, pp. 1-11, 1994.
2. I.Fary: *On straight lines representations of planar graphs*, Acta Sci. Math. Szeged, 11, pp. 229-233 , 1948.
3. G.Frank, D.Hui and C.Ware: *Visualizing object oriented software in three dimensions*, In Proc.CASCON, 1993.
4. J.Mackinley, G.Robertson and S. Card: *Cone trees: Animated 3d visualization of hierarchical information*, In Proc.SIGCHI Conf. on Human Factors in Computing, pp. 189-194, 1991.
5. P.Reid: *Dynamic Interactive Display of Complex Data Structures*, In Graphics Tools for Software Engineers, pp. 62–70, Cambridge, 1989.
6. O.Tversky, S.Snibbe and R.Zeleznik: *Cone trees in the uga graphics system: suggestions of a more robust visualization tool*, Technical Report CS-93-07, Brown University, 1993.

Optimizing Area and Aspect Ratio in Straight-Line Orthogonal Tree Drawings*

TIMOTHY CHAN**
Center for Geometric Computing
Dept. of Computer Science
Johns Hopkins Univ.
tchan@cs.jhu.edu

MICHAEL T. GOODRICH***
Center for Geometric Computing
Dept. of Computer Science
Johns Hopkins Univ.
goodrich@cs.jhu.edu

S. RAO KOSARAJU†
Center for Geometric Computing
Dept. of Computer Science
Johns Hopkins Univ.
kosaraju@cs.jhu.edu

ROBERTO TAMASSIA‡
Center for Geometric Computing
Dept. of Computer Science
Brown Univ.
rt@cs.brown.edu

Abstract. We investigate the problem of drawing an arbitrary n-node binary tree orthogonally in an integer grid using straight-line edges. We show that one can simultaneously achieve good area bounds while also allowing the aspect ratio to be chosen as being $O(1)$ or sometimes even an arbitrary parameter. In addition, we show that one can also achieve an additional desirable aesthetic criterion, which we call "subtree separation." We investigate both upward and non-upward drawings, achieving area bounds of $O(n \log n)$ and $O(n \log \log n)$, respectively, and we show that, at least in the case of upward drawings, our area bound is optimal to within constant factors.

1   Introduction

Binary trees are, of course, very common structures in many application areas, so obtaining good drawings of binary trees is an important com-

---

* This work is a consequence of the participation of Drs. Goodrich and Tamassia in the 1996 International Workshop on 3D Graph Drawing at Bellairs Research Inst. of McGill University.

** This research was performed while the author was visiting the Center for Geometric Computing at Johns Hopkins University, and it was supported in part by by ARO under grant DAAH04-96-1-0013.

*** This research supported by NSF under Grants CCR-9300079 and CCR-9625289, and by ARO under grant DAAH04-96-1-0013.

† This research supported by NSF under Grant CCR-9508545 and by ARO under grant DAAH04-96-1-0013.

‡ This research supported by NSF under Grant CCR-9423847 and by ARO under grant DAAH04-96-1-0013.

ponent in a wide variety of visualization tasks. Nevertheless, there are a number of interesting issues regarding binary-tree drawings that are still unresolved, including those related to drawings that optimize the easily-motivated aesthetic criterion of using straight line segments to display edges while also optimizing the area and aspect ratio of the drawing.

Optimizing the area of a drawing is important, because a drawing typically needs to be displayed on a medium of limited area and resolution, such as a terminal window on a workstation screen. Formally, we define the *area* of a drawing to be the area of a smallest rectangle enclosing the drawing. Of course, this assumes a reasonable rule for defining the resolution of a drawing, such as that used in *grid* drawings, where all vertices are placed at integer grid points and edges are drawn as polygonal chains that bend only at integer grid points. Additionally, one may wish to restrict the drawing further to be an *orthogonal* drawing, which is a drawing where the polygonal chains representing edges must be composed of only vertical and horizontal segments. When drawn on a rastered device such as a laser printer or computer monitor, such drawings avoid the aliasing effect caused by the "staircased" drawing of edges that are neither vertical nor horizontal.

An optimization parameter that is perhaps equal in importance to area for a drawing, however, is the *aspect ratio* of a drawing's enclosing rectangle, i.e., the ratio of the width and height of the rectangle. A drawing that is, for example, tall and narrow would be difficult to display nicely on a printed page or in a screen window even if the area is reasonably small (although it might fit quite nicely on a cash-register tape). Ideally, the aspect ratio should be a parameter that could be chosen from a large range of values, or, failing that, it should at least be allowed to be that of a "well proportioned" rectangle (e.g., 1, 5/3, 8.5/11, or $(1 + \sqrt{5})/2$).

Another aesthetic criterion that may be desirable in some applications is that a tree drawing be *upward*. That is, that the tree be drawn so that no child is placed higher (in the y-direction) than its parent. This criterion is desirable, for example, if the tree represents an inherently hierarchical relationship, such as the organizational structure of a large business.

1.1 Previous related research

There has been a fair amount of research involving area and aspect ratio tradeoffs of tree drawings (e.g., see the annotated bibliography of Di Battista *et al.* [4]). We summarize the previous bounds for planar *polyline* grid drawings, for example, where edges are drawn as polygonal chains that bend only at integer grid points, in Table 1.

Class	Drawing Type	Area	Aspect Ratio(s)	Source
degree-$O(1)$ rooted tree	upward	$\Theta(n)$	$[1/n^{\alpha}, n^{\alpha}]$	[5]
binary tree	upward orthogonal	$\Theta(n \log \log n)$	$O(n \log \log n / \log^2 n)$	[5]
degree-4 tree	orthogonal	$\Theta(n)$	$O(1)$	[6, 7, 11]
degree-4 tree	leaves-on-hull orthogonal	$\Theta(n \log n)$	$O(1)$	[1]

Table 1. Summary of some area/aspect ratio results for planar polyline grid drawings of trees. We use α to denote an arbitrary constant such that $0 \le \alpha < 1$.

Notice that each of the area bounds for polyline drawings are tight, to within constant factors, even for upward orthogonal drawings. The related issues for straight-line drawings are not as well-understood, however. We summarize relevant previous results for this class of drawings in Table 2.

Class	Drawing Type	Area	Aspect Ratio(s)	Source
rooted tree	upward layered grid	$O(n^2)$	$O(1)$	[8]
rooted tree	upward grid	$O(n \log n)$	$O(n/\log n)$	[2, 9]
rooted tree	strictly upward grid	$\Theta(n \log n)$	$O(n/\log n)$	[2]
complete or Fibonacci tree	strictly upward grid	$\Theta(n)$	$O(1)$	[2, 10]
AVL tree	strictly upward grid	$\Theta(n)$	$[(\log^{\beta} n)/n, n/(\log^{\beta} n)]$	[3]
balanced tree of height $O(\log n)$	upward grid	$O(n \log n)$	$O(n/\log n)$	[2, 9]

Table 2. Summary of previous area/aspect ratio results for planar straight-line grid drawings. We use β to denote an arbitrary constant such that $\beta > 1$.

We are not aware, for example, of any non-trivial previous work on straight-line orthogonal grid drawings of arbitrary binary trees. This seems to be a fairly serious omission, since straight-line edges are easier for the eye to follow than polyline edges, and orthogonal drawings automatically avoid small angles between edges, which can also cause confusion, and they also avoid aliasing edges drawn on a rasterized device.

1.2 Subtree separation

There is, in fact, an additional desirable aesthetic property for drawings of binary trees. We say that a region R in the plane is *rectilinearly convex* if the intersection of R and any vertical or horizontal line is connected. For any set S of integer grid points and edges, define the *rectilinear convex hull* of S to be the smallest rectilinearly-convex region containing S. Let $T[v]$ denote the subtree of a tree T that is rooted at $v \in T$ and contains all the descendents of v in T, i.e., $T[v]$ is the subtree of T induced by v. If, for any disjoint induced subtrees $T[v]$ and $T[w]$ in a binary tree T, the rectilinear convex hulls of $T[v]$ and $T[w]$ are disjoint in a drawing D of T, then we say that D achieves *subtree separation*. This property is desired for binary tree drawings, because it allows the eye to quickly distinguish between different parts of the tree. It also allows for multi-resolutional renderings of a drawing D, so that, for example, if D has too many nodes to all simultaneously fit in a screen window, then D can be rendered up to the resolution of the screen, with some induced subtrees rendered as filled-in rectilinearly-convex regions. Of course, it might not always be possible to achieve subtree separation while also optimizing for other aesthetic criteria. For example, many of the drawings produced by the algorithms of Garg *et al.* [5] do not achieve subtree separation. But it is certainly desirable to achieve this property whenever possible.

1.3 Our results

In this paper we present a general approach, based upon a simple "recursive winding" paradigm, for drawing arbitrary binary trees in small area with good aspect ratio, for both upward and non-upward straight-line orthogonal drawing criteria. Intuitively, our recursive winding paradigm draws a binary tree T by laying down a small chain of nodes monotonically in the x-direction leading to a distinguished node, v_k, and then "winding" by recursively laying out $T[v']$ and $T[v'']$ in the other direction, where v' and v'' denote the children of v_k. We apply this approach in both upward and non-upward drawing frameworks, and we also show that the area bound on our upward orthogonal straight-line drawings is optimal to within constant factors. Specifically, we establish the following results regarding a planar grid drawing, D, of an arbitrary n-node binary tree, T:

- D can be made to realize an upward orthogonal straight-line drawing of T with $O(n \log n)$ area and $O(1)$ aspect ratio. Moreover, D can be made to achieve subtree separation, and it can be produced in $O(n)$ time.

- There are n-node trees that require $\Omega(n \log n)$ area in any upward orthogonal straight-line drawing that achieves an aspect ratio in the range $[1/n^\alpha, n^\alpha]$, for any fixed α with $0 \le \alpha < 1$.

- D can be made to realize a (non-upward) orthogonal straight-line drawing of T with $O(n \log \log n)$ area and aspect ratio in the range $[(\log n)/n, n/\log n]$. Moreover, this D can be made to achieve subtree separation, and it can be produced in $O(n)$ time.

In Section 2 we describe our method for producing upward orthogonal straight-line tree drawings with $O(1)$ aspect ratio. We give our lower bound showing that our $O(n \log n)$ area bound in this case is optimal in Section 3. In Section 4 we describe how one can produce non-upward orthogonal straight-line tree drawings with arbitrary aspect ratios in area that is $O(n \log \log n)$, and we conclude in Section 5.

2 Upward Orthogonal Straight-Line Drawings

We begin by presenting our recursive winding paradigm and show how it can be applied to prove the following:

Theorem 1. *Given any binary tree T with n nodes, there is a planar upward straight-line orthogonal grid drawing of T with height and width $O(\sqrt{n \log n})$. Such a drawing can be constructed in $O(n)$ time, and it achieves subtree separation.*

Without loss of generality, assume that each internal node has degree 2. Given an internal node v, let $left(v)$ and $right(v)$ denote the left child and the right child of v respectively. Let $T[v]$ again denote the subtree of T rooted at v, and let $N[v]$ be the number of leaves in $T[v]$. Arrange the tree so that $N[left(v)] \leq N[right(v)]$ at every node v. This preprocessing requires only linear time. We first review the following lemma:

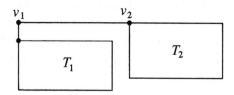

Fig. 1. Drawing of a binary tree with $O(\log n)$ height and $O(n)$ width.

Lemma 2 [2, 9]. *If T has n leaves, then*

(i) *there is a planar upward orthogonal straight-line grid drawing of T with height at most $\lfloor \log_2 n \rfloor$ and width $n-1$, and*

(ii) *there is a planar upward orthogonal straight-line grid drawing of T with width at most $\lfloor \log_2 n \rfloor$ and height $n-1$.*

In both drawings, the root is placed at the upper left-hand corner and the construction time is $O(n)$.

Proof: The construction can be described as follows. If $n = 1$, the drawing is trivial. Suppose $n > 1$ and v_0 is the root of T. Letting $T_1 = T[left(v_0)]$ and $T_2 = T[right(v_0)]$, we can draw T as shown in Figure 1, where the subtrees T_1 and T_2 are drawn recursively. Since $N[left(v_0)] \leq N[right(v_0)]$, it is not difficult to see that the height of the drawing is bounded by $\log_2 n$. Part (ii) can be proven using a similar construction. □

Before we proceed to prove Theorem 1, we first analyze a certain recurrence relation.

Lemma 3. *Suppose $A > 1$ and f is a function such that*

- *if $n \leq A$, then $f(n) \leq 1$; and*

- *if $n > A$, then $f(n) \leq f(n') + f(n'') + 1$ for some $n', n'' \leq n - A$ with $n' + n'' \leq n$.*

Then $f(n) < 4n/A - 1$ for all $n > A$.

Proof: The proof is by induction. Suppose the theorem is true for n' and n''. If both $n', n'' \leq A$, then $f(n) \leq 3 < 4n/A - 1$. If $n' \leq A$ and $n'' > A$, then

$$f(n) \leq f(n'') + 2 < 4n''/A + 1 \leq 4(n-A)/A + 1 < 4n/A - 1.$$

Finally, if both $n', n'' > A$, then

$$f(n) \leq f(n') + f(n'') + 1 < 4n'/A + 4n''/A - 1 \leq 4n/A - 1.$$

□

We now use a recursive winding scheme to prove Theorem 1. Let $H(n)$, $W(n)$, and $T(n)$ denote the height, width, and construction time for drawing a tree with n leaves. Fix a parameter $A > 1$ to be determined later. If $n \leq A$, then we use the scheme in Lemma 2. This provides the base case:

$$H(n) \leq \log_2 n, \quad W(n) \leq A, \quad \text{and } T(n) = O(A) \quad \text{if } n \leq A.$$

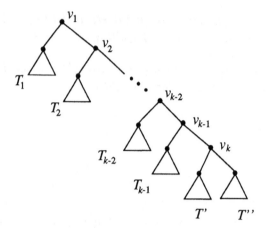

Fig. 2. The binary tree T.

Suppose $n > A$. Define a sequence $\{v_i\}$ of vertices as follows: v_1 is the root and $v_{i+1} = right(v_i)$ for $i = 1, 2 \ldots$ Let $k \geq 1$ be an index with $N[v_k] > n - A$ and $N[v_{k+1}] \leq n - A$; such an index can be found in $O(k)$ time, since $N[v_1], N[v_2], \ldots$ is a strictly decreasing sequence of integers. Let $T_i = T[left(v_i)]$ and $n_i = N[left(v_i)]$ for $i = 1, \ldots, k-1$. Let $T' = T[left(v_k)]$, $T'' = T[right(v_k)]$, $n' = N[left(v_k)]$, and $n'' = N[right(v_k)]$. Note that $n' \leq n''$, since T is "right heavy." (See Figure 2.) Note also the following properties:

1. $n_1 + \cdots + n_{k-1} = n - N[v_k] < A$, and

2. $\max\{n', n''\} = N[v_{k+1}] \leq n - A$.

Now, consider the planar upward orthogonal straight-line drawing of T in Figure 3(a), (b), or (c), depending on whether $k = 1$, $k = 2$, or $k > 2$. The root v_1 is placed at the upper left-hand corner. In (a), the subtrees T' and T'' are drawn recursively below v_1. In (b), the subtree T_1 is drawn according to Lemma 2(i), while the subtrees T' and T'' are drawn recursively below this. In (c), the subtrees $T_1, \ldots, T_{k-2}$ are drawn from left to right according to Lemma 2(i); the subtree T_{k-1} is drawn down the right side according to Lemma 2(ii); and the subtrees T' and T'' are drawn recursively below $T_1, \ldots, T_{k-2}$ and then reflected so that their roots are placed at upper right-hand corners (this is the "recursive winding").

In any case, the drawing can be made with the following bounds on the height, width, and construction time:

$$H(n) \leq \max\{H(n') + H(n'') + \log_2 A + 3, \ n_{k-1} - 1\}$$

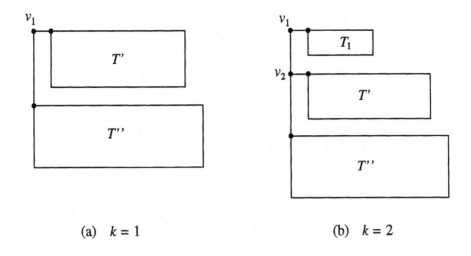

(a) $k = 1$ (b) $k = 2$

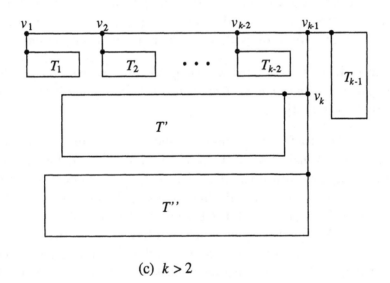

(c) $k > 2$

Fig. 3. Drawing of the binary tree T with $O(\sqrt{n \log n})$ height and width.

$$W(n) \leq \max\{W(n') + 1, \ W(n''), \ n_1 + \cdots + n_{k-2}\} + \log_2 n_{k-1} + 1$$
$$T(n) \leq T(n') + T(n'') + O(n_1 + \cdots + n_{k-1} + 1).$$

By property 1, we can write the recurrences as

$$H(n) \leq \max\{H(n') + H(n'') + O(\log A), \ A\}$$
$$W(n) \leq \max\{W(n'), \ W(n''), \ A\} + O(\log A)$$

$$T(n) \leq T(n') + T(n'') + O(A).$$

By property 2, we can see that $W(n) = O(\lceil n/A \rceil \log A + A)$. Using in addition the fact that $n' + n'' \leq n$, we can also conclude that $H(n) = O(\lceil n/A \rceil \log A + A)$ and $T(n) = O(\lceil n/A \rceil A)$ by a direct application of Lemma 3. Theorem 1 then follows by setting $A = \sqrt{n \log_2 n}$.

Moreover, a simple induction argument based upon this proof can be used to show that this construction results in a drawing with subtree separation.

3 A Lower Bound on Upward Orthogonal Straight-Line Drawings

In this section we show that there is a family of n-node trees, for each $n \geq 2$, such that any planar upward straight-line orthogonal grid drawing of such a tree with width W requires $\Omega(n \log n)$ area, for any $\log n \leq W \leq n/\log n$. We begin with a simple, but important, lemma.

Lemma 4. *If T is an n-node complete binary tree, then any planar straight-line orthogonal grid drawing of T must have width and height at least $\lfloor (\log n)/2 \rfloor$.*

Proof: The proof is based upon an induction argument similar to that used by Crescenzi *et al.* [2] to prove a similar bound for upward (non-orthogonal) grid drawings of T. Let T be an n-node complete binary tree and let D be a straight-line orthogonal drawing of T. We prove that height$(T) \geq (\log n)/2$; the bound for the width follows by a similar argument. Any internal node v in T can have at most two of its adjacent nodes in its same row. If it has one neighbor w in a different row, then the height of the subdrawing for v must be at least one greater than the subdrawing for w, and the subtree of T rooted at w must have size at least half of that of the subtree of T rooted v. On the other hand, if v has two neighbors on the same row as v in D, and one of these nodes, call it w, is not a leaf, and w must have an adjacent node u in T on a different row, and the size of the and the subtree of T rooted at u must have size at least half of that of the subtree of T rooted w. The proof is implied by applying this argument repeatedly (through an induction argument) starting at the root of T. $\qquad\Box$

This lemma is for non-upward orthogonal drawings, but the application we give here is for upward orthogonal drawings.

Theorem 5. *There is an n-node binary tree T that requires $\Omega(n \log n)$ area in any planar straight-line orthogonal upward grid drawing D with width at most n^α, for any fixed constant $\alpha > 0$.*

Proof: The proof is based upon a non-trivial adaptation of a "chain pinning" argument of Garg *et al.* [5]. Let $\alpha > 0$ be a fixed constant, and let T be a tree defined by a chain of $n/2$ nodes $C = (v_1, v_2, \ldots, v_{n/2})$ such that every node v_i with i being a multiple of $n^{\alpha/2}$ has a complete binary tree of size $n^{\alpha/2}$ rooted at an adjacent node (not on C). Assume that v_1 is defined as the root of T and partition C into $n^{1-\alpha}/4$ subchains with $2n^\alpha$ vertices each. We claim that each subchain S so defined has height that is $\Omega(\log n)$. This claim implies the lemma, since D is upward, so all that remains is to establish this claim.

Let S be a subchain of C as defined above. If the vertical distance between the two endpoints of S is at least $\log n$, then we are done. So, suppose the vertical distance between these two nodes is less than $\log n$. In traversing from endpoint to endpoint of S we alternately march to the left and to the right, possibly with vertical drops mixed into these "left-flowing" and "right-flowing" subchains of S. Since there are less than $\log n$ such chains, there must be two subchains S' and S'' of S that completely cross the same vertical strip of width $2n^\alpha / \log n$. Assuming, without loss of generality, that S' is above S'', there are at least $n^{\alpha/2} / \log n > \log n$ subtrees of size $2n^\alpha$ each hanging from S'. Let v be a node in S' adjacent to one of these subtrees. Suppose the root of this subtree is on the same vertical level of D as v. Then S' must drop down at least one vertical unit continuing (to the right or left) at this lower level. Since there are at least $\log n$ subtrees like this adjacent to S', this implies that either the subchain S must have height at least $\log n$ or one of these subtrees is positioned directly below the node v on C to which it is adjacent. But if the root of this subtree is directly below v in D, then, by Lemma 4, the subchain S'' must cross the vertical line through v at distance $(\alpha/2) \log n$ below v. This establishes the claim, and the theorem. $\qquad \square$

Having established tight asymptotic bounds on the area of nice-aspect-ratio upward straight-line orthogonal drawings of binary trees, we next consider non-upward drawings of such trees.

4 Non-Upward Drawings

Let us now consider non-upward straight-line orthogonal grid drawings of a binary tree T in small area with arbitrary aspect ratio, again applying

73

the recursive winding paradigm. Our approach is similar to that of Section 2, in that we select the chain of vertices $(v_1, v_2, \ldots, v_k)$ according to the same rule as in that section. Likewise, if $k \leq 2$, then we recursively lay out T as in that construction (see Figure 3(a) and (b)). If $k > 2$, however, then we use the construction shown in Figure 4. The subtrees $T_1, \ldots, T_{k-1}$ are drawn according to Lemma 2(i), with the drawing of T_{k-1} rotated 180 degrees. As in the previous construction, the subtrees T' and T'' are drawn recursively and then reflected so that their roots are placed at upper right-hand corners. Also, as in the previous construction, the size of T' is at most that of T'', since T is right-heavy.

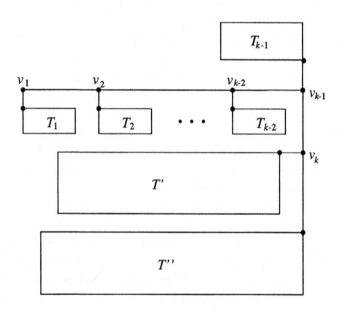

Fig. 4. Recursive strategy for non-upward drawings.

This all implies that the drawing can be made with the following bounds on the height, width, and construction time:

$$H(n) \leq H(n') + H(n'') + 2\log_2 A + 3$$
$$W(n) \leq \max\{W(n') + 1, \ W(n''), \ n_1 + \cdots + n_{k-1}\}$$
$$T(n) \leq T(n') + T(n'') + O(n_1 + \cdots + n_{k-1} + 1).$$

We can re-write the recurrences as

$$H(n) \leq H(n') + H(n'') + O(\log A)$$

$$W(n) \leq \max\{W(n') + 1, \ W(n''), \ A\}$$
$$T(n) \leq T(n') + T(n'') + O(A).$$

Using the fact that $n' \leq n''$, this implies that $W(n) = O(A + \log n)$ and $H(n) = O(\lceil n/A \rceil \log A)$. We can also conclude that $T(n) = O(\lceil n/A \rceil A)$. So, for example, we can take $A = \log n$, and achieve a drawing with width $O(\log n)$ and height $O((n/\log n) \log \log n)$, i.e., the area of this drawing is $O(n \log \log n)$. Of course, the aspect ratio of this drawing is fairly poor.

Still, we can now substitute this method as the base case of our method, rather than using Lemma 2. This allows us to achieve a better aspect ratio while keeping the area to be $O(n \log \log n)$. In particular, this results in the following new recurrence relations characterizing the height and width of the resulting drawing:

$$H(n) \leq H(n') + H(n'') + O(\log A)$$
$$W(n) \leq \max\{W(n') + 1, \ W(n''), \ O((A/\log A) \log \log A)\}.$$

This, in turn, implies that $W(n) = O((A/\log A) \log \log A + \log n)$ while we still have $H(n) = O(\lceil n/A \rceil \log A)$. Thus, we can achieve arbitrary aspect ratio (less than $n/\log n$) while maintaining the area at $O(n \log \log n)$.

This implies the following theorem:

Theorem 6. *Given any binary tree T with n nodes, there is a planar straight-line orthogonal grid drawing of T with area $O(n \log \log n)$. Such a drawing can be constructed in $O(n)$ time so as to achieve subtree separation and any aspect ration in the range $[(\log n)/n, n/\log n]$.*

Proof: As described above, the drawing of T does not necessarily achieve subtree separation. We can easily modify the method, however, to achieve this desired property. The only modification is that in the case $k > 2$ we must be sure that the leftmost point of the drawing of T_{k-1} is to the right of the rightmost point of the drawing of T_{k-2}. Since this modification does not necessitate any changes to the recurrence relations, it achieves the same height and width bounds. □

By Lemma 4 this range of aspect ratios is the largest possible, to within constant factors.

5 Conclusion

We have investigated several issues related to space-efficient planar straight-line orthogonal grid drawings of arbitrary binary trees. In the case of upward drawings we have established tight upper and lower bounds of $\Theta(n \log n)$ on the area needed, with good aspect ratios, and in the non-upward case we have given a method that achieves area $O(n \log \log n)$ with good aspect ratios. Some interesting problems that remain open include the following:

- Can one achieve arbitrary aspect ratios for planar upward straight-line orthogonal grid drawings while maintaining $O(n \log n)$ area?

- Are there binary trees that require $\Omega(n \log \log n)$ area for any (non-upward) planar straight-line orthogonal grid drawing?

Acknowledgements

We would like to thank Stina Bridgeman, Marek Chrobak, and Sue Whitesides for several stimulating discussions and e-mail exchanges relating to the topics of this paper.

References

1. R. P. Brent and H. T. Kung. On the area of binary tree layouts. *Inform. Process. Lett.*, 11:521–534, 1980.
2. P. Crescenzi, G. Di Battista, and A. Piperno. A note on optimal area algorithms for upward drawings of binary trees. *Comput. Geom. Theory Appl.*, 2:187–200, 1992.
3. P. Crescenzi and A. Piperno. Optimal-area upward drawings of AVL trees. In R. Tamassia and I. G. Tollis, editors, *Graph Drawing (Proc. GD '94)*, volume 894 of *Lecture Notes in Computer Science*, pages 307–317. Springer-Verlag, 1995.
4. G. Di Battista, P. Eades, R. Tamassia, and I. G. Tollis. Algorithms for drawing graphs: an annotated bibliography. *Comput. Geom. Theory Appl.*, 4:235–282, 1994.
5. A. Garg, M. T. Goodrich, and R. Tamassia. Area-efficient upward tree drawings. In *Proc. 9th Annu. ACM Sympos. Comput. Geom.*, pages 359–368, 1993.
6. C. E. Leiserson. Area-efficient graph layouts (for VLSI). In *Proc. 21st Annu. IEEE Sympos. Found. Comput. Sci.*, pages 270–281, 1980.
7. C. E. Leiserson. *Area-efficient graph layouts (for VLSI)*. ACM Doctoral Dissertation Award Series. MIT Press, Cambridge, MA, 1983.
8. E. Reingold and J. Tilford. Tidier drawing of trees. *IEEE Trans. Softw. Eng.*, SE-7(2):223–228, 1981.
9. Y. Shiloach. *Arrangements of Planar Graphs on the Planar Lattice*. PhD thesis, Weizmann Institute of Science, 1976.
10. L. Trevisan. A note on minimum-area upward drawing of complete and Fibonacci trees. *Information Processing Letters*, 57(5):231–236, 1996.
11. L. Valiant. Universality considerations in VLSI circuits. *IEEE Trans. Comput.*, C-30(2):135–140, 1981.

Drawing Directed Acyclic Graphs:
An Experimental Study*

Giuseppe Di Battista[1], Ashim Garg[2], Giuseppe Liotta[2], Armando Parise[3],
Roberto Tamassia[2], Emanuele Tassinari[3], Francesco Vargiu[4], Luca Vismara[2]

[1] Dipartimento di Discipline Scientifiche, Sezione Informatica
Università degli Studi di Roma Tre, Italy
dibattista@iasi.rm.cnr.it
[2] Center for Geometric Computing, Department of Computer Science
Brown University, USA
{ag,gl,rt,lv}@cs.brown.edu
[3] Dipartimento di Informatica e Sistemistica
Università degli Studi di Roma "La Sapienza", Italy
{parise,tassinar}@dis.uniroma1.it
[4] Autorità per l'Informatica nella Pubblica Amministrazione
Italy
vargiu@aipa.it

Abstract. In this paper we consider the class of directed acyclic graphs
(DAGs), and present the results of an experimental study on four drawing algorithms specifically developed for DAGs. Our study is conducted
on two large test suites of DAGs and yields more than 30 charts comparing the performance of the drawing algorithms with respect to several
quality measures, including area, crossings, bends, and aspect ratio. The
algorithms exhibit various trade-offs with respect to the quality measures, and none of them clearly outperforms the others.

1 Introduction

Over a decade of research in the graph drawing area, motivated by applications
to information visualization, has produced a wealth of drawing algorithms. As-
sessing the practical performance of some of these algorithms has been recently
the goal of experimental comparative studies.

In [18] the performance of four planar straight-line drawing algorithms is
compared. The standard deviations in angle size, edge length, and face area are
used to compare the quality of the planar straight-line drawings produced. Since
the experiments are limited to randomly generated maximal planar graphs, this

* Research supported in part by the National Science Foundation under grant CCR–
9423847, by the U.S. Army Research Office under grant DAAH04-96-1-0013, by
ESPRIT LTR of the European Community under Project no. 20244 (ALCOM-IT),
by the NATO-CNR Advanced Fellowships Programme, and by a gift from Tom
Sawyer Software.

work gives only partial insight on the performance of the algorithms on general planar graphs.

Himsolt [17] presents a comparative study of twelve graph drawing algorithms. The algorithms selected are based on various approaches (e.g., force-directed, layering, and planarization) and use a variety of graphic standards (e.g., orthogonal, straight-line, polyline). Only three algorithms draw general graphs, while the others are specialized for trees, planar graphs, Petri nets, and graph grammars. The experiments are conducted with the graph drawing system GraphEd [17]. Many examples of drawings constructed by the algorithms are shown, and various objective and subjective evaluations on the aesthetic quality of the drawings produced are given. However, statistics are provided only on the edge length, and few details on the experimental setting are provided. The charts on the edge length have marked oscillations, due to the small size of the test suite (about 100 graphs). This work provides an excellent overview and comparison of the main features of some popular drawing algorithms. However, it does not give detailed statistical results on their performance.

Di Battista et al. [7, 8] present an extensive experimental study comparing four general-purpose graph drawing algorithms. The four algorithms take as input general undirected graphs and construct orthogonal grid drawings. The test graphs are generated from a core set of 112 graphs used in "real-life" software engineering and database applications. The experiments provide a detailed quantitative evaluation of the performance of the four algorithms, and show that they exhibit trade-offs between "aesthetic" properties (e.g., crossings, bends, edge length) and running time. The observed practical behavior of the algorithms is consistent with their theoretical properties.

Brandenburg, Himsolt, and Rohrer [2] compare five "force-directed" algorithms for constructing straight-line drawings of general undirected graphs. The algorithms are tested on a a wide collection of examples and with different settings of the force parameters. The quality measures evaluated are crossings, edge length, vertex distribution, and running time. They also identify trade-offs between the running time and the aesthetic quality of the drawings produced.

Jünger and Mutzel [19] investigate crossing minimization strategies for straight-line drawings of 2-layer graphs, and compare the performance of eight popular heuristics for this problem.

In this paper we consider the important class of directed acyclic graphs (DAGs), and compare the performance of four drawing algorithms specifically developed for them. DAGs are commonly used to model hierarchical structures such as PERT diagrams in project planning, class hierarchies in software engineering, and is-a relationships in knowledge representation systems.

The contributions of this work can be summarized as follows:

- We have developed a general experimental setting for comparing the practical performance of graph drawing algorithms for DAGs. Our setting consists of (i) two large test suites of DAGs, one obtained from the collection of directed graphs submitted to the e-mail graph drawing service at Bell Labs [21], and the other randomly generated by a program that simulates a

PERT project planner; (*ii*) a set of quality measures for drawings of DAGs derived from [7].

- Within our experimental setting, we have performed a comparative study of four popular drawing algorithms for DAGs: two of them are based on the layering paradigm [20, 23], while the other two are based on the grid paradigm [10, 11].
- Our comparison highlights how more than ten years of research in this field have produced a complex landscape. Namely, the four algorithms exhibit various trade-offs with respect to the quality measures, and none of them clearly outperforms the others. The sometimes surprising findings of our investigations include:
  - Some algorithms construct very compact drawings at the expense of a relaxed resolution rule that does not consider crossing-crossing and vertex-crossing distances. Other algorithms produce drawings that distribute vertices and crossings with great regularity at the expenses of a larger area requirement.
  - Concerning bends, an algorithm with good theoretical worst-case bounds performs in practice worse than algorithms for which no theoretical bounds are available.
  - Concerning crossings, grid-based algorithms tend to perform worse than layering-based algorithms, where part of the geometry of the drawing is decided at the very first step.
  - The performance of a drawing algorithm on planar DAGs is not a good predictor of the performance of the same algorithm on nonplanar DAGs
  - Algorithms with a topological foundation tend to distribute the bends and the lengths of the edges more evenly.
- Our analysis of the performance of the four algorithms has motivated us to develop a new hybrid strategy for drawing DAGs, which uses a layering-based method to perform the initial planarization and a grid-based method to compute the final drawing. The new strategy performs quite well in practice.

The rest of the paper is organized as follows. The four algorithms being compared are described in Section 2. Details on the experimental setting are given in Section 3. In Section 4, we summarize our experimental results in 14 charts, and perform a comparative analysis of the performance of the four algorithms. The new hybrid strategy is described and its performance is discussed in Section 5. Open problems are addressed in Section 6. Due to space limitations we could not include all the charts, which appear in a longer version of this paper [6].

2   The Drawing Algorithms Under Evaluation

We have tested four different algorithms for producing drawing of DAGs. Our main graph drawing tool has been *GDW* [3], a system for prototyping and testing graph drawing algorithms.

The four algorithms can be classified into two categories on the basis of their approach to constructing drawings.

Layering-Based: These algorithms construct *layered drawings*, i.e., drawings where the vertices and edge-bends are placed at integer coordinates on a set of horizontal layers, and each edge is drawn as a curve monotonically increasing in the y-direction. Note that in such drawings, even though vertices and edge bends are placed at integer coordinates, the edge crossings can be arbitrarily close to each other or to the vertices and edge bends. These algorithms accept as input directed graphs without any particular restriction (the input directed graph can be planar or not, acyclic or cyclic). For constructing drawings, they generally follow the methodology of Sugiyama et al. [23], which consists of the following three steps:

Step 1 Assign vertices to layers heuristically optimizing some criteria, such as the total edge length.

Step 2 Reduce the crossings among edges by permuting the order of vertices on each layer.

Step 3 Reduce the number of bends by readjusting the position of vertices on each layer.

Because of their generality and conceptual simplicity, these algorithms are very popular among the designers of practical graph drawing systems. Several layering-based algorithms have been designed [5]. The above steps have also been investigated separately, and various heuristics have been proposed for each of them.

In this paper, we have evaluated and compared the performance of two layering based algorithms: *Dot* and *Layers*.

Dot is a highly optimized algorithm, developed by Koutsofios and North [20] as a successor to *Dag* [13, 14]. *Dot* first constructs a polyline layered drawing of the input directed graph and then, as a final step, converts the polygonal chains representing the edges into smooth curves using splines. An implementation of *Dot* is available at `ftp://ftp.research.att.com/dist/drawdag/`, and this is the implementation we used. However, since all other algorithms considered in this study represent edges as polygonal chains, we decided to analyze the polyline drawing produced by *Dot* and not the final drawing with curved lines.

Layers is the original algorithm by Sugiyama, Tagawa and Toda [23]. For our study we have used the implementation of *Layers* available in *GDW* [3].

Grid-Based: These algorithms accept, as input, a planar *st*-graph, i.e., a planar DAG with exactly one source and one sink, and construct an *upward grid drawing* of it. In an *upward grid drawing* the vertices, the edge-bends and the edge-crossings are all placed at integer coordinates, and each edge is drawn as a curve monotonically increasing in the y-direction. Although the requirement of having just one source and one sink may appear too restrictive, such directed graphs occur in several practical applications, such as activity planning (where they are called PERT graphs), network flows, etc. These algorithms are also called *numbering-based* algorithms because they typically construct a numbering of the vertices and faces of the planar *st*-graph, and compute the coordinates of the vertices and bends using this numbering.

The grid-based algorithms have two advantages: first, their performance on planar planar st-graphs has been theoretically analyzed, and second, their running times are usually low. The disadvantage is that a nonplanar DAG needs to be converted into a planar st-graph, before it can be drawn using these algorithms. This is done by introducing a fictitious vertex for each crossing between two edges. These fictitious vertices are assigned a position on the grid, but are not represented in the final drawing. The simple planarization method we have used for our study is the one described in [9]. The grid-based algorithms that we evaluated and compared fall under two categories:

Visibility Representation-Based: These algorithms use a two-step process for constructing drawings [10]. In the first step, they construct a *visibility representation* of the input planar st-graph. (In a visibility representation, vertices and edges are represented as horizontal and vertical line-segments, respectively; two vertices are connected by an edge if and only if they are visible to each other.) In the second step, they construct a polyline drawing of the planar st-graph from the visibility representation; this is done by replacing each vertex-segment with a point and by approximating each edge-segment with a polygonal line containing at most two bends.

The visibility representation is constructed using two numberings [12, 24]: a topological numbering of the vertices of the planar st-graph, and a topological numbering (in the dual graph) of the faces of the planar st-graph. A *topological numbering* of a DAG is such that for every directed edge (u, v), v is assigned a higher number than u.

We have evaluated the performance of an algorithm, called *Visibility*, which follows this approach, and of three variations of it, called *Barycentric Visibility*, *Long Edge Visibility*, and *Median Visibility*. For our study we have used the implementations of these algorithms available in *GDW* [3]. The differences between Algorithms *Visibility*, *Barycentric Visibility*, *Long Edge Visibility*, and *Median Visibility* are in the strategy they use for substituting the vertex- and edge-segments of the intermediate visibility representation with the points and polygonal chains of the final drawing; they put the vertex in the middle point of the vertex-segment, in the barycenter of the endpoints of the incident edge-segments, on the endpoint of the longest incident edge-segment, and on the median of the endpoints of the incident edge-segments, respectively.

Poset-Based: These algorithms view planar st-graphs as covering graphs of partially ordered sets (*posets*). They exploit the relationship between the upward planarity of DAGs and the order-theoretic properties of planar lattices (see, e.g., [16, 22]).

In our study we have evaluated the performance of one poset-based algorithm: the dominance drawing algorithm of [11]. This algorithm computes two topological numberings of the vertices of the input planar st-graph; one numbering gives the x-coordinates of the vertices and bends, and the other gives the y-coordinates. These numberings are obtained by scanning the outgoing edges of each vertex of the planar st-graph in the left-to-right and the right-to-left order respectively. For this reason, this algorithm is also known as the *left-right* algorithm. In this paper we have referred to this algorithm as *Lattice*. For our study we have used the implementation of *Lattice* available in *GDW* [3].

3 Experimental Setting

The following quality measures of a drawing of a DAG have been considered:

Area: area of the smallest rectangle with horizontal and vertical sides covering the drawing;

Cross: total number of edge-crossings;

TotalBends: total number of edge-bends;

TotalEdgeLen: total edge length;

MaxEdgeBends: maximum number of bends on any edge;

MaxEdgeLen: maximum length of any edge;

UnifBends: standard deviation of the number of edge-bends;

UnifLen: standard deviation of the edge length;

ScreenRatio: deviation from the optimal aspect ratio, computed as the difference between the width/height ratio of the best of the two possible orientations (portrait and landscape) of the drawing and the standard 4/3 ratio of a computer screen.

ResFactor: Inverse of the minimum distance between two vertices, or two edge-crossings, or an edge-crossing and a vertex.

It is widely accepted (see, e.g., [5]) that small values of the above measures are related to the perceived aesthetic appeal and visual effectiveness of the drawing.

The issue of *resolution* of a drawing has been extensively studied, motivated by the finite resolution of physical rendering devices. Several papers have been published about the resolution and the area of drawings of graphs; (see, e.g., [1, 4, 11, 15]). The resolution of a drawing is defined as the minimum distance between two vertices. The grid-based algorithms consider edge-bends and edge-crossings as "dummy" vertices for computing the resolution. The layering-based algorithms, however, do not consider the edge-crossings as dummy vertices for computing the resolution. Since the measures Area, TotalEdgeLen and MaxEdgeLen of a drawing depend on its resolution, two drawings can be compared for these measures only if they have the same resolution. ResFactor allows us to scale a drawing D_1 produced by a layering-based algorithm so that it has the same resolution as that of a drawing D_2 produced by a grid-based algorithm; the scaling factor is equal to R_1/R_2, where R_1 and R_2 are the value of ResFactor for D_1 and D_2 respectively.

The experimental study was performed on two different sets of DAGs, both with a strong connection to "real-life" applications. We considered two typical contexts where DAGs play a fundamental role, namely software engineering and project planning.

The first set of test DAGs are what we call the *North DAGs*. They are obtained from a collection of directed graphs [21], that North collected at AT&T Bell Labs by running for two years *Draw DAG*, an e-mail graph drawing service that accepts directed graphs formatted as e-mail messages and returns messages with the corresponding drawings [20].

Originally, the North DAGs consisted of 5114 directed graphs, whose number of vertices varied in the range $1\ldots7602$. However, the density of the directed graphs with a number of vertices that did not fall in the range $10\ldots100$ was very low (see also the statistics in [21]); since such directed graphs represent a very sparse statistical population we decided to discard them. Then we noted that many directed graphs were isomorphic; since the vertices of the directed graphs have labels associated with them, the problem is tractable. For each isomorphism class, we kept only one representative directed graph. Also, we deleted the directed graphs where subgraphs were specified as clusters, to be drawn in their own distinct rectangular region of the layout, because constrained algorithms are beyond the scope of this study. This filtering left us with 1277 directed graphs.

Still, 491 directed graphs were not connected and this was a problem for running algorithms implemented in *GDW* (they assume input directed graphs to be connected). Instead of discarding the directed graphs, we followed a more practical approach, by randomly adding a minimum set of directed edges that makes each directed graph connected. Finally, we made the directed graph acyclic, where necessary, by applying some heuristics for inverting the direction of a small subset of edges.

We then ran a first set of experiments and produced the statistics by grouping the DAGs by number of vertices. Although the comparison among the algorithms looked consistent (the produced plots never oddly overlapped), each single plot was not satisfactory, because it showed peaks and valleys. We went back to study the test suite and observed that grouping them by number of vertices was not the best approach. In fact, the North DAGs come from very heterogeneous sources, mainly representing different phases of various software engineering projects; as a result, directed graphs with more or less the same number of vertices may be either very dense or very sparse.

Since most of the analyzed quality measures strongly depend on the number of edges of the DAG (e.g. area, number of bends, and number of crossings), we decided that a better approach was to group the DAGs by number of edges. After some tests, we clustered the DAGs into nine groups, each with at least 40 DAGs, so that the number of edges in the DAGs belonging to group i, $1 \leq i \leq 9$, is in the range $10i\ldots10i+9$ (see Fig. 3). The resulting test suite consists of 1158 DAGs, each with edges in the range $10\ldots99$.

The second set of test DAGs are what we call the *Pert DAGs*. Although such DAGs have been randomly generated by one of the facilities of *GDW*, their construction is based on refinement operations typical of project planning.

First, we generated a set of skeleton planar DAGs consisting of a small number of vertices to simulate the initial models of the projects. This was done by randomizing an ear-composition for each DAG. Second, we performed a random sequence of typical planning-refinement steps, i.e., expanding an edge into a series and/or a parallel component and inserting new edges between existing vertices. The inserted edges represent precedences between activities that were not captured by the starting skeleton projects.

The resulting test suite contains 813 DAGs with edges in the range $10 \ldots 150$ and vertices in the range $10 \ldots 100$. As for the North DAGs, we grouped the Pert DAGs by number of edges, so that the number of edges in the DAGs belonging to group i, $1 \leq i \leq 14$, is in the range $10i \ldots 10i + 9$ (see Fig. 3).

The Pert DAGs are generally denser than the North DAGs and they are single-source single-sink. As shown in the next section, there are some quality measures for which the relative performance of the algorithms is different for the North DAGs and the Pert DAGs. Also, the plots obtained for the Pert DAGs are in general smoother, reflecting the relative uniformity of the statistical population.

4 Analysis of the Experimental Results

Algorithms *Dot, Layers, Visibility* (with its variations *Barycentric Visibility, Long Edge Visibility, Median Visibility*), and *Lattice* were executed on every North DAG and every Pert DAG, and the data for all ten quality measures were collected. Because of the different nature of the two test suites, we compared the performance of the algorithms for the North DAGs and the Pert DAGs separately. In addition, since the quality measures Area, TotalEdgeLen, and MaxEdgeLen depend upon the resolution of the drawings, we compared the layering-based and grid based algorithms separately for these three quality measures. The other seven quality measures do not depend upon the resolution, so we compared all four algorithms together for them. This gave us a total of 28 comparison charts. Figures 1–4 display the comparison charts showing the average values for the quality measures Area, ResFactor, Cross, TotalBends, and ScreenRatio; the left column of these figures contains the charts for the North DAGs, and the right column contains the ones for the Pert DAGs. The x-axis of each chart shows the number of edges. The average is computed over each group of DAGs with number of edges in the range $10 \ldots 19, 20 \ldots 29$, etc.

We started the experimental analysis by comparing the behavior of *Visibility* and its variations *Barycentric Visibility, Long Edge Visibility*, and *Median Visibility*. As a result, we found that the behavior of the visibility representation-based algorithms is almost identical for all the quality measures, with Algorithm *Visibility* performing slightly better than the others. In order to improve the readability of the other charts and to simplify the presentation of the experimental results, we have used Algorithm *Visibility* as the representative visibility representation-based algorithm.

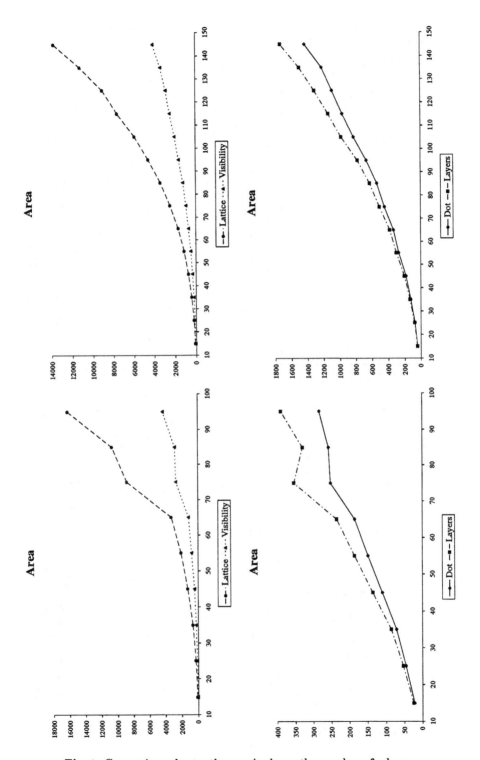

Fig. 1. Comparison charts: the x-axis shows the number of edges.

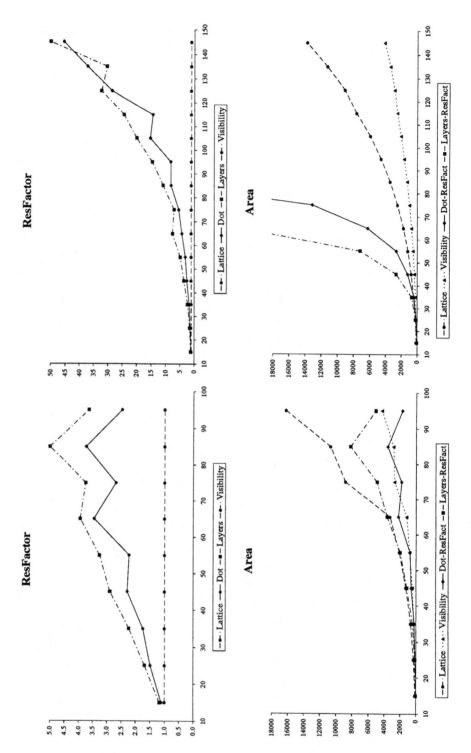

Fig. 2. Comparison charts: the x-axis shows the number of edges.

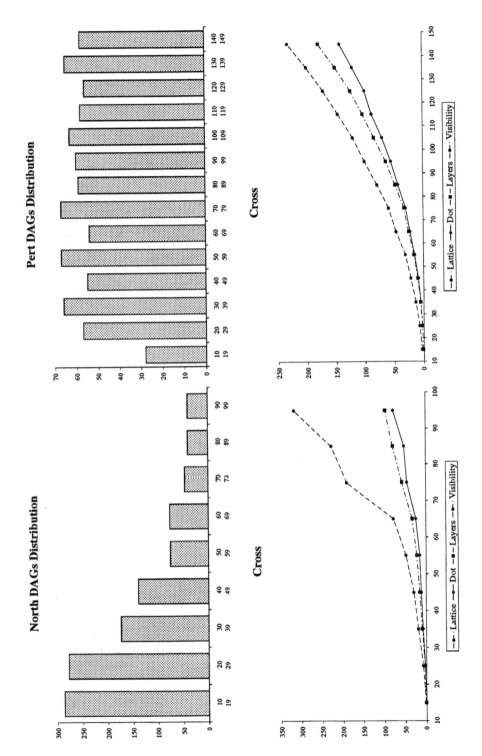

Fig. 3. Comparison charts: the x-axis shows the number of edges.

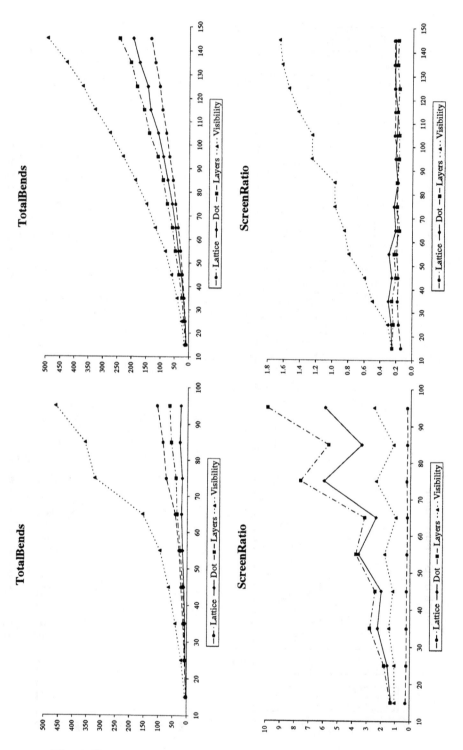

Fig. 4. Comparison charts: the x-axis shows the number of edges.

The analysis of the performance of the four algorithms for each quality measure, and for each set of input DAGs is summarized below:

Area: (see Fig. 1) *Dot* performs better than *Layers*, and *Lattice* performs better than *Visibility* for both the North DAGs and the Pert DAGs. While for the North DAGs, the plots grow linearly for #edges in the range 10 . . . 60, for the Pert DAGs they show quadratic growth in the entire range. Also observe that the difference in the performance of the two grid-based algorithms is significant for DAGs with more than 75 edges, whereas the two layering-based algorithms perform about the same in the entire range.

ResFactor: (see Fig. 2) Not surprisingly, ResFactor is equal to one for the grid-based algorithms in the entire range. On the other hand, the layering-based algorithms tend to have a non-constant ResFactor. This reflects the fact that they do not take edge-crossings into consideration for defining the resolution. The bottom charts of Fig. 2 show a comparative study of the area of the drawings produced by the four algorithms. Unlike the bottom charts of Fig. 1, the plots for the two layering-based algorithms now take ResFactor into account. Note that they are comparable with the plots of the grid-based algorithms.

Cross: (see Fig. 3) Since *Lattice* and *Visibility* use the same planarizer, the drawings produced by them have the same number of edge-crossings. All the algorithms have quadratic behavior for both sets of DAGs. *Dot* has the best performance among the four. The difference between the performance of the layering-based and the grid-based algorithms reduces considerably for the Pert DAGs. Also observe that the slope of the plots is steeper for the Pert DAGs. This reflects the fact that the Pert DAGs are in general denser than the North DAGs and that the number of edge-crossings tend to increase with the ratio #edges/#vertices.

TotalBends: (see Fig. 4) The performance of *Visibility* is unsatisfactory. For the North DAGs, the plots of the other three algorithms grow almost linearly for #edges up to 65. After that, *Dot* is clearly the best. The experimentation with the Pert DAGs produced a surprising result. Namely, *Lattice* outperforms the layering-based algorithms while *Visibility* has still the worst behavior. As for measure Cross, the slope of the plots is steeper for the Pert DAGs. Note, however, that the behavior of *Lattice* seems to be quite independent from the density of the input DAG, at least for DAGs with up to 75 edges.

ScreenRatio: (see Fig. 4) *Lattice* seems to be the algorithm of choice with respect to this quality measure. All the algorithms have a better performance for the Pert DAGs. We believe that this a consequence of the relative density of the Pert DAGs; the drawings tend to spread in both the x- and y-dimension.

TotalEdgeLen and MaxEdgeLen: These two measures are dependent on ResFactor. Therefore, we compared the performance of the layering-based and grid-based algorithms separately. *Dot* performs better than *Layers*, and *Visibility* performs better than *Lattice* for both the North DAGs and the Pert DAGs.

MaxEdgeBends: Quite interestingly, the plots grow linearly for the Pert DAGs for all four algorithms. While *Dot* has the best performance for the North DAGs, *Lattice* is the best for the Pert DAGs. The overall performance of the algorithms is much better for the North DAGs than for the Pert DAGs. Again, we believe that this is because the North DAGs are in general sparser.

5   Cross-Fertilization of Grid- and Layering-Based Algorithms

The analysis of the experimental results of *Dot*, *Layers*, *Visibility*, and *Lattice* clearly shows that the layering-based algorithms (*Dot* and *Layers*) produce drawings with fewer crossings than the grid-based algorithms (*Visibility* and *Lattice*). This indicates that the crossing reduction step of the layering-based algorithms is more effective than the simple planarization strategy [9] used in *Visibility* and *Lattice*. On the other hand, *Visibility* and *Lattice* perform well with respect to other quality measures (see Section 4).

The above considerations suggest the development of a hybrid strategy that substitutes the original planarization step of *Visibility* and *Lattice* with the crossing reduction step of *Layers* (we choose *Layers* over *Dot* for simplicity of implementation). More specifically, we first execute the crossing reduction step of *Layers* and then visit the resulting drawing, replacing each crossing with a fictitious vertex. This planarizes the input graph. Finally, we execute the remaining algorithmic steps of *Visibility* and *Lattice*. The new drawing algorithms so obtained will be called *VisibilityLayers* and *LatticeLayers*, respectively.

Algorithms *VisibilityLayers* and *LatticeLayers* always perform better than their "parent algorithms" *Visibility* and *Lattice*, respectively. In particular, we observe the following:

Area: The improvement of *VisibilityLayers* and *LatticeLayers* over *Visibility* and *Lattice* is especially significant for the North DAGs, where it ranges between 30% and 50%.

TotalBends: Again, the improvement is especially significant (about 50%) for the North DAGs. Also, while *Layers* is always better than *Lattice*, we have that *LatticeLayers* is slightly better than *Layers* for the North DAGs with more than 70 edges.

MaxEdgeBends: Analogous considerations to those for TotalBends apply. Also, the improvement of *VisibilityLayers* over *Visibility* is substantial.

The analysis of the other quality measures shows similar trends.

We conclude this section by observing that the performance of the grid-based algorithms (*Visibility* and *Lattice*) is strongly influenced by the number of crossings introduced in the planarization step.

6   Open Problems

Our experiments lead to many interesting theoretical and practical questions:

- The angular resolution of a drawing is the magnitude of the smallest angle between any two edges incident on a vertex. The readability of a drawing can be improved by increasing its angular resolution. Unfortunately, not much is known either theoretically or empirically about the angular resolution of drawings of directed graphs. This issue is worth exploring.
- *Dot* in a final step converts the polylines into Bezier curves using splines. This has a dramatic impact on the quality of the drawing. Similarly, we believe that the performance of several algorithms, such as *Visibility*, can be improved by a postprocessing "beautification" step. For example, it would be interesting to study bend-stretching techniques [25] that reduce the bends by doing local transformations.
- Similarly, the role of the preprocessing step should also be studied. In particular, the performance of grid-based algorithms can be improved by using a more sophisticated planarizer.

Acknowledgements

We are grateful to Stephen North for many discussions about *Dot* and the collection of directed graphs. We thank Petra Mutzel for answering our questions about the statistics on such graphs. Finally we thank Paola Vocca for her encouragement during the writing of this paper.

References

1. P. Bertolazzi, R. F. Cohen, G. Di Battista, R. Tamassia, and I. G. Tollis. How to draw a series-parallel digraph. *Internat. J. Comput. Geom. Appl.*, 4:385–402, 1994.
2. F. J. Brandenburg, M. Himsolt, and C. Rohrer. An experimental comparison of force-directed and randomized graph drawing algorithms. In F. J. Brandenburg, editor, *Graph Drawing (Proc. GD '95)*, volume 1027 of *Lecture Notes in Computer Science*, pages 76–87. Springer-Verlag, 1996.
3. L. Buti, G. Di Battista, G. Liotta, E. Tassinari, F. Vargiu, and L. Vismara. GD-Workbench: A system for prototyping and testing graph drawing algorithms. In F. J. Brandenburg, editor, *Graph Drawing (Proc. GD '95)*, volume 1027 of *LNCS*, pages 111–122. Springer-Verlag, 1996.
4. M. Chrobak, M. T. Goodrich, and R. Tamassia. Convex drawings of graphs in two and three dimensions. In *Proc. 12th Annu. ACM Sympos. Comput. Geom.*, pages 319–328, 1996.
5. G. Di Battista, P. Eades, R. Tamassia, and I. G. Tollis. Algorithms for drawing graphs: An annotated bibliography. *Comput. Geom. Theory Appl.*, 4:235–282, 1994.
6. G. Di Battista, A. Garg, G. Liotta, A. Parise, R. Tamassia, E. Tassinari, F. Vargiu, and L. Vismara. Drawing directed graphs: An experimental study. Technical Report CS-96-24, Center for Geometric Computing, Dept. Computer Science, Brown Univ., 1996. `ftp://ftp.cs.brown.edu/pub/techreports/96/cs96-24.ps.Z`.

7. G. Di Battista, A. Garg, G. Liotta, R. Tamassia, E. Tassinari, and F. Vargiu. An experimental comparison of three graph drawing algorithms. In *Proc. 11th Annu. ACM Sympos. Comput. Geom.*, pages 306–315, 1995.

8. G. Di Battista, A. Garg, G. Liotta, R. Tamassia, E. Tassinari, and F. Vargiu. An experimental comparison of four graph drawing algorithms. *Comput. Geom. Theory Appl.*, 1996 (to appear). http://www.cs.brown.edu/cgc/papers/dglttv-ecfgd-96.ps.gz.

9. G. Di Battista, E. Pietrosanti, R. Tamassia, and I. G. Tollis. Automatic layout of PERT diagrams with XPERT. In *Proc. IEEE Workshop on Visual Languages (VL'89)*, pages 171–176, 1989.

10. G. Di Battista and R. Tamassia. Algorithms for plane representations of acyclic digraphs. *Theoret. Comput. Sci.*, 61:175–198, 1988.

11. G. Di Battista, R. Tamassia, and I. G. Tollis. Area requirement and symmetry display of planar upward drawings. *Discrete Comput. Geom.*, 7:381–401, 1992.

12. G. Di Battista, R. Tamassia, and I. G. Tollis. Constrained visibility representations of graphs. *Inform. Process. Lett.*, 41:1–7, 1992.

13. E. R. Gansner, E. Koutsofios, S. C. North, and K. P. Vo. A technique for drawing directed graphs. *IEEE Trans. Softw. Eng.*, 19:214–230, 1993.

14. E. R. Gansner, S. C. North, and K. P. Vo. DAG – A program that draws directed graphs. *Softw. – Pract. Exp.*, 18(11):1047–1062, 1988.

15. A. Garg and R. Tamassia. Planar drawings and angular resolution: Algorithms and bounds. In J. van Leeuwen, editor, *Algorithms (Proc. ESA '94)*, volume 855 of *Lecture Notes in Computer Science*, pages 12–23. Springer-Verlag, 1994.

16. A. Garg and R. Tamassia. Upward planarity testing. *Order*, 12:109–133, 1995.

17. M. Himsolt. Comparing and evaluating layout algorithms within GraphEd. *J. Visual Lang. Comput.*, 6(3), 1995. (Special Issue on Graph Visualization, I. F. Cruz and P. Eades, editors).

18. S. Jones, P. Eades, A. Moran, N. Ward, G. Delott, and R. Tamassia. A note on planar graph drawing algorithms. Technical Report 216, Department of Computer Science, University of Queensland, 1991.

19. M. Jünger and P. Mutzel. Exact and heuristic algorithms for 2-layer straightline crossing minimization. In F. J. Brandenburg, editor, *Graph Drawing (Proc. GD '95)*, volume 1027 of *Lecture Notes in Computer Science*, pages 337–348. Springer-Verlag, 1996.

20. E. Koutsofios and S. North. Drawing graphs with *dot*, 1993. *dot* user's manual. ftp://ftp.research.att.com/dist/drawdag/.

21. S. North. 5114 directed graphs, 1995. Manuscript. ftp://ftp.research.att.com/dist/drawdag/.

22. I. Rival. Reading, drawing, and order. In I. G. Rosenberg and G. Sabidussi, editors, *Algebras and Orders*, pages 359–404. Kluwer Academic Publishers, 1993.

23. K. Sugiyama, S. Tagawa, and M. Toda. Methods for visual understanding of hierarchical systems. *IEEE Trans. Syst. Man Cybern.*, SMC-11(2):109–125, 1981.

24. R. Tamassia and I. G. Tollis. A unified approach to visibility representations of planar graphs. *Discrete Comput. Geom.*, 1(4):321–341, 1986.

25. R. Tamassia and I. G. Tollis. Planar grid embedding in linear time. *IEEE Trans. Circuits Syst.*, CAS-36(9):1230–1234, 1989.

Circular Layout in the Graph Layout Toolkit

Uğur Doğrusöz, Brendan Madden, Patrick Madden

Tom Sawyer Software, 804 Hearst Avenue, Berkeley, CA 94710
info@tomsawyer.com
http://www.tomsawyer.com

Abstract. The *Graph Layout Toolkit* is a family of portable, automated, graph layout libraries designed for integration into graphical user interface application programs. The *Circular Library* is one of the four styles currently available with the Graph Layout Toolkit. It produces layouts that emphasize natural group structures inherent in a graph's topology, and is well suited for the layout of ring and star network topologies. It clusters (groups) the nodes of a graph by group IDs, by IP addresses, and by biconnectivity or node degree, and allows the user to specify a range for the size of each cluster. The Library positions the nodes of a cluster on a radiating circle, and employs heuristics to reduce the crossings not only between edges incident to nodes of the same cluster but also between edges that connect different clusters.

1   Introduction

Graph layout is the automatic positioning of the nodes and the edges of a graph in order to produce an aesthetically pleasing drawing of the graph that is easy to comprehend. This is very important for visualization tools in numerous areas such as project management, software development, database design, and network management.

Many graph layout and editing systems have been developed in the past. Please refer to [DETT95] for an overview of such systems. The *Graph Layout Toolkit* [MMPH95, GLT96a, GLT96b] is a family of graph layout libraries with ANSI C++ and C APIs that facilitate easy integration with graphical user interface programs for development of graph visualization and editing tools.

Graph layout comes in different flavors, each being more suitable for a different area. The Graph Layout Toolkit offers four different layout styles: *Circular*, *Hierarchical*, *Orthogonal*, and *Symmetric*.

The Circular Library uses robust techniques such as clustering by biconnectivity and methods for minimizing the cut when a cluster needs to split. In addition, it can provide application oriented partitioning. It also shows good performance on the important aesthetic layout criterion of low number of crossings. In the following sections, we describe the methods used by the Circular Library. In particular, we focus on clustering, positioning, and crossing minimization techniques.

2 The Circular Library

The layout algorithm of the Circular Library is primarily designed for the layout of ring and star network topologies. It is an advanced version of the one developed by Kar, Madden, and Gilbert [KMG89]. It functions by partitioning the nodes of a graph into logical groups (clusters) based on a number of flexible grouping methods. These clusters are placed on radiating circles based on their logical interconnection. Part of the graph of clusters is laid out on a circle. A virtual cluster is used to manage this cluster of clusters. The remaining parts of the cluster graph (subgraphs that form trees), on the other hand, are laid out using the Hierarchical Library, and attach to the virtual cluster from their roots.

The clustering is either performed with a generic method, such as the biconnectivity of the graph or the degrees of the nodes, or can be performed with IP addresses, IP subnet masks or another domain specific technique. Domain specific techniques clearly should not the primary focus for the development of graph drawing techniques, however, it is noteworthy that domain specific clustering can be easily provided in certain circumstances. The Library also supports manually configured clustering with the help of group IDs.

The algorithm minimizes cluster to cluster crossings as well as crossings within each cluster. In addition, it employs tree balancing routines, and has ring and star detection and placement techniques within each cluster. Figure 1 shows two sample drawings produced by the Circular Library.

3 Definitions

In this section, we give some basic definitions.

A *cluster* C is a group of nodes to be laid out together. The Circular Library lays the nodes of a cluster around a circle. We refer to the radius of the circle around which the nodes of a cluster C is positioned as the *radius of C*.

The cluster graph G^C of a graph G is a graph in which each node is uniquely associated with a cluster in G, and there is an edge in between two nodes of G^C representing clusters C_1 and C_2 of G if and only if there is at least one pair of nodes $n_1 \in C_1$ and $n_2 \in C_2$ that are adjacent in G.

For a given graph $G = (N, E)$, *reduction of trees* refers to the process of recursively removing degree one nodes along with their incident edges until no degree one node is left in the graph. All such reduced nodes form a forest. These trees can be later inserted back into the original graph by a reverse procedure. We call this *expansion of trees*.

Clusters in a cluster graph can be of two types: *sub site* clusters are those that are removed as a result of the reduction of trees of the cluster graph, whereas *main site* clusters are the ones that remain in the cluster graph after the reduction of the trees. We will refer to each reduced tree of the cluster graph as a *sub site tree*. In addition, the cluster formed from the main site clusters and their interconnections is called the *virtual* cluster. Circular positioning of the clusters in the virtual cluster forms the *backbone* of the drawing.

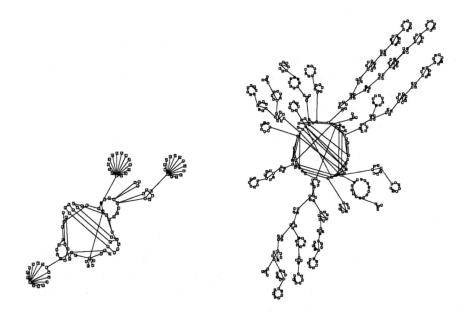

Fig. 1. Sample drawings produced by the Circular Library.

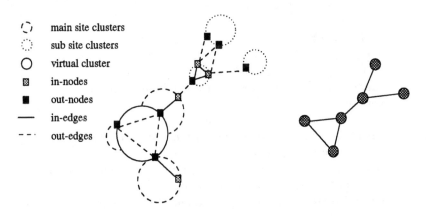

Fig. 2. A circular drawing of a graph with *only a portion* of the set of nodes and edges shown as an illustration of the definitions (left), and the corresponding cluster graph (right).

The nodes (edges) of a cluster are categorized into two: *in* and *out*-nodes (edges). An *in*-node (edge) of a cluster C is one that is connected to (connects) *only* nodes in cluster C. An *out*-node (edge) of C, on the other hand, is one that is connected to (incident to) at least one node in a cluster other than C.

A graph $G = (N, E)$ is said to be *biconnected* if there are at least two distinct paths between any pair of nodes u, $v \in N$. We use n (n_C) and e (e_C) to denote the number of nodes and edges in a graph (cluster), respectively.

The *ports* of a node help define a circular ordering for the edges that are incident to the node. These edges, when drawn, are clipped according to their port assignments.

4 Algorithm

An outline of the algorithm used in the Circular Library follows. The first phase of the algorithm partitions the nodes of a graph into clusters, each of which is to be laid out as a circle. It also gathers information to be used for the positioning of the nodes within each cluster as well as relative positioning of the clusters during the second and final phases of the algorithm.

Partition:
(1)   cluster nodes
(2)   **foreach** cluster C **do**
(3)   calculate in and out-nodes and edges in C
(4)   calculate radius of C
(5)   **end foreach**
(6)   **repeat**
(7)   merge neighboring clusters that are too small
(8)   **until** each cluster is larger than the minimum given
(9)   **repeat**
(10)   split clusters that are too large
(11)   **until** each cluster is smaller than the maximum given
(12)   construct cluster graph
(13)   reduce trees of cluster graph (remaining nodes represent main
            site clusters and form a virtual cluster)
(14)   expand trees of cluster graph (expanded nodes represent sub site
            clusters)

Position (draft):
(15)   **foreach** cluster C **do**
(16)   reduce trees of the subgraph associated with cluster C only
(17)   **end foreach**
(18)   order, position, and swap main site clusters on the virtual cluster
(19)   **foreach** main site cluster C **do**
(20)   position and swap out-nodes

(21)   order, position, and swap in-nodes
(22)   **end foreach**
(23)   **foreach** sub site cluster C, in a pre-order traversal order **do**
(24)   position and swap out-nodes
(25)   order, position, and swap in-nodes
(26)   assign ports for the inter-cluster edges in between C and its
               children clusters
(27)   **end foreach**
(28)   position sub site clusters with a layout of sub site tree using
           Hierarchical Library with port support

         Position (final):
(29)   **foreach** cluster C **do**
(30)   expand the reduced trees of C
(31)   **end foreach**
(32)   **foreach** sub site cluster C **do**
(33)   position in-nodes
(34)   position out-nodes
(35)   **end foreach**
(36)   rotate the drawing according to the aspect ratio of the window in
           which the graph is to be drawn.

In the following sections, we discuss some parts of the algorithm in more detail that we believe deserve more attention.

4.1 Clustering

The Circular Library clusters the nodes of a given graph in three stages. First, the *group IDs* of the nodes, if any, are used; that is, two nodes with the same group ID are put in the same cluster. This gives the user a chance to manually group nodes. Then, the remaining unclustered nodes are clustered by *IP addresses* if desired. IP addresses are a very effective way to logically group network devices.[1] Finally, any unclustered node is clustered with *biconnectivity*, the default clustering method. The user has the option of using a clustering based on *node degrees* instead of biconnectivity, however. In this paper, we will concentrate on clustering by biconnectivity.

We believe that the most natural way of capturing the structure of graphs encountered in networking environments is to first find the biconnected components of the graph to be laid out. By clustering only nodes that are in the same biconnected component together, we make sure that there are at least two distinct paths in between each node pair. In a network map, this implies that the failure of no single device in a cluster will leave the entire network disconnected. This will naturally reveal the weaknesses in the design of a network topology.

---

[1] IP is the most widely used network protocol today; support for it was specifically requested by several of our customers.

In addition, such clustering will be reasonably "stable" over modifications to the topology of the network. For instance, the overal clustering of the graph will not change over operations such as addition of a link in between two nodes that belong to the same biconnected component cluster, or deletion of a link within a cluster that remains biconnected after the deletion. Similarly, in many cases the addition or deletion of a node or edge does not change the block tree of biconnected graphs, or causes only a minor change to the block tree, which helps with drawing stability.

Furthermore, the Circular Library allows users to control the sizes of clusters via several tailoring options. The user can specify the minimum and the maximum number of nodes that each cluster may have. This is especially useful when the underlying graph has large biconnected components.

The Library merges clusters that are too small (i.e., smaller than the minimum chosen by the user) with neighboring clusters. The merging continues until the minimum size requirement is met by all clusters.

Then, a heuristic algorithm derived from Kernighan-Lin's algorithm [KL70] is employed in order to split the clusters that are larger than the maximum specified. This algorithm minimizes the cut (number of inter-cluster connections) created by the split operation. Similarly, this process is repeated until all the clusters are within the specified range.[2]

4.2   Positioning and Crossing Minimization

In this section, we discuss the ideas for positioning clusters along with the nodes within each cluster. In addition, the methods applied to minimize the crossings inside each cluster and the crossings in between clusters are explained.

"Ordering", "positioning", and "swapping" (for crossing minimization) are the main operations used throughout the entire layout algorithm of the Circular Library. The in-nodes of a cluster are ordered such that we have a good initial positioning around the cluster's circle for crossing minimization. This is significant because the minimization algorithm works locally and settles for a local minimum. The ordering is determined based on a method that tries to find a depth-first search tree with a maximal depth. The ordering of the out-nodes of the same cluster is also taken into account to reduce the number of crossings for the edges in between in and out-nodes.

Once we have an ordering for the in-nodes, we position them around a circle spread out according to a spacing option set by the user. After that, we minimize the crossings inside a cluster by repeatedly swapping adjacent pairs of nodes as long as the swap results in a lower number of crossings. This process is repeated until there is a pass with no swaps. The number of crossings is calculated *only* for the in-edges of the two in-nodes considered. The out-nodes in a cluster are positioned and swapped similarly.

---

[2] Notice however that the algorithm might not always yield clusters within the desired range due to the fact that a cluster may not split if it would result in clusters that are smaller than desired.

After the clustering phase is finished, the second phase of the algorithm first reduces trees in each cluster independently. The benefit of this is two-fold. First, it speeds up the draft positioning phase since operations in this phase are performed on subgraphs with reduced trees. Secondly, it ensures that each tree is treated as a single entity and its expansion does *not* introduce any crossings.

Then, the algorithm orders, positions, and swaps the main site clusters around a main circle (backbone). For this, we use the concept of a virtual cluster whose nodes and edges correspond to the main site clusters, and the interconnections among the main site clusters, respectively. After this the number of crossings of the interconnections among the main site clusters will be minimal.

After that, the Library applies the same operations to each main site cluster followed by the application of these operations to the sub site clusters in a pre-order traversal of the sub sites in the corresponding sub site tree.

The ordering of the in-nodes of each cluster help reduce the number of crossings. This is due to the fact that the longer the depth-first search tree used for ordering in-nodes of a cluster, the better chances are for a lower number of edge crossings within the cluster initially. Furthermore this depth-first search traversal yields deterministic ring drawing when a cluster is comprised of a chain of degree two nodes. In most cases, this will result in a better minima for the final number of crossings than a random initial positioning.

In addition, the order in which each sub site cluster is processed is crucial to the overall success of crossing reduction because the crossing minimization in a particular cluster assumes the stability of the positions of the nodes in its parent clusters. Another important factor in crossing reduction is the usage of ports during the layout of the cluster graph with the Hierarchical Library. In the Hierarchical Library, each node can have ports at the bottom and top (or left and right depending on the orientation of the drawing) of the rectangle representing the drawing of the node. Port support is used to reflect the order in which in-nodes connect to children clusters.

The final phase of the algorithm positions the clusters according to the coordinates calculated with a hierarchical layout of the cluster graph after the reduced trees are expanded back into each cluster in a way that does *not* introduce any crossings.

4.3 Time Complexity

Lines (1)-(8), (12)-(14), (15)-(17), (29)-(31), and (36) of the algorithm take linear time in the number of edges of the graph to execute. A variant of the KL algorithm in lines (9)-(11) is of $O(n_C \cdot log n_C)$ for each cluster C. The major operations in lines (18)-(27) and lines (32)-(35) are "order", "position", and "swap". These operations take $O(e_C)$, $O(n_C)$, and $O(n_C^2)$ time for each cluster C, respectively. The hierarchical layout algorithm used in line (28) is loosely based on [STT81] and the time it takes to execute is normally negligible since it lays out sub site trees of relatively small sizes. Overall, the layout algorithm of the Circular Library is of $O(n^2 + e)$ in the worst case. However, the algorithm performs much better on average since most these operations are applied to parts of the graph in a divide-and-conquer fashion. Please refer to Figure 3 for some statistical results on the performance of the Library's layout algorithm.

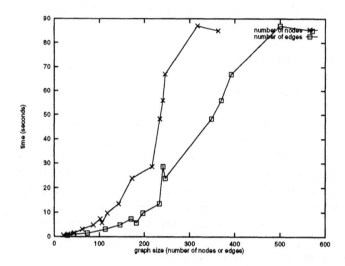

Fig. 3. The performance of the layout algorithm of the Circular Library on randomly created graphs. The tests were run on a Sun Sparc 5 workstation.

5 Conclusion

In this paper, we described some of the interesting algorithms employed by the Circular Library of the Graph Layout Toolkit, a family of portable graph layout libraries designed for integration into graphical user interface application programs.

The Library cluster by biconnectivity by default. After that we apply a variant of Kernighan-Lin's heuristic algorithm for clustering the nodes of a graph which further splits biconnected component clusters if necessary. It applies heuristic algorithms to minimize the crossings inside each cluster as well as the crossings between edges connecting two clusters.

The algorithms employed by the Circular Library take $O(n^2 + e)$ time to execute in the worst case. The average time complexity of the Library's layout algorithm, however, is much lower according to the statistical data gathered on random graphs.

One possible modification for our layout algorithm that we are currently looking into is support for multiple backbones and/or layout of the cluster graph with a spring embedder based algorithm. Both approaches seem to demand much more complicated crossing minimization techniques. Proper handling of crossings among inter-cluster edges may yield better drawings of cluster graphs.

Acknowledgements: The authors wish to thank Dr. Ioannis Tollis for his helpful suggestions, Therese Biedl for providing us with the software for biconnectivity, and Dr. Michael Doorley for revising an earlier version of this paper.

References

[DETT95] Di Battista, G., Eades, P., Tamassia, R., Tollis, I.G.: *Algorithms for drawing graphs: An annotated bibliography*. In Computational Geometry: Theory and Applications, 4, (1994), 235–282.

[GLT96a] Tom Sawyer Software: *Graph Layout Toolkit User's Guide*, Berkeley, CA, (1992 - 1996).

[GLT96b] Tom Sawyer Software: *Graph Layout Toolkit Reference Manual*, Berkeley, CA, (1992 - 1996).

[KMG89] Kar, G., Madden, B.P., Gilbert, R.S.: *Heuristic Layout Algorithms for Network Management Presentation Services*. IEEE Network November (1988) 29–36.

[KL70] Tollis, I.G., Xia, C.: *Drawing Telecommunications Networks*. Proc. Graph Drawing '94, Lecture Notes in Computer Science 894, Springer Verlag, (1994), 206–217.

[KL70] Kernighan, B.W., Lin, S.: *An Efficient Heuristic for Partitioning Graphs*. Bell Systems Technical Journal, **49**, (1970), 291–307.

[MMPH95] Madden, B., Madden, P., Powers, S., Himsolt, M.: *Portable Graph Layout and Editing*. Proc. Graph Drawing '95, Lecture Notes in Computer Science 1027, Springer Verlag, (1995), 385–395.

[STT81] Sugiyama, K., Tagawa, S., Toda, M.: *Methods for visual understanding of hierarchical systems*. IEEE Transactions on Systems, Man and Cybernetics, **11**, (1981) 109–125.

Multilevel Visualization of Clustered Graphs*

Peter Eades and Qing-Wen Feng

Department of Computer Science and Software Engineering,
University of Newcastle, NSW 2308, Australia
Email: {eades, qwfeng}@cs.newcastle.edu.au

Abstract. Clustered graphs are graphs with recursive clustering structures over the vertices. This type of structure appears in many systems.
Examples include CASE tools, management information systems, VLSI
design tools, and reverse engineering systems. Existing layout algorithms
represent the clustering structure as recursively nested regions in the
plane. However, as the structure becomes more and more complex, two
dimensional plane representations tend to be insufficient. In this paper,
firstly, we describe some two dimensional plane drawing algorithms for
clustered graphs; then we show how to extend two dimensional plane
drawings to three dimensional multilevel drawings. We consider two conventions: straight-line convex drawings and orthogonal rectangular drawings; and we show some examples.

1 Introduction

Graph drawing algorithms are widely used in graphical user interfaces of software systems. As the amount of information that we want to visualize becomes
larger, we need more structure on top of the classical graph model. Graphs with
recursive clustering structures over the vertices are called *clustered graphs* (see
Fig. 1). This type of structure appears in many systems. Examples include CASE
tools [16], management information systems [8], and VLSI design tools [7].

In two dimensional representations, the clustering structure is represented by
region inclusions, i.e. a cluster is represented by a simple region that contains
the drawing of all the vertices which belong to that cluster (see Fig. 2). For
such drawings, some heuristic methods have been developed by Sugiyama and
Misue [13, 10], by North [11], and by Madden et al. [12, 9]. Algorithms for planar
straight-line convex drawings have been developed by Eades, Feng and Lin [6, 4].
An algorithm for planar orthogonal rectangular drawings is presented by Eades
and Feng in [3]. However, as the clustering structure becomes more and more
complex, two dimensional representations tend to be insufficient. A common
strategy for visualizing large graphs with recursive clusterings is to visualize
the graph at multiple abstraction levels. A natural method for such multiple
level representations is a three dimensional drawing with each level drawn on a
plane at different z-coordinate; and with the clustering structure drawn as a tree
in three dimensions. This type of representation not only facilitates visualizing

* This work was supported by a research grant from the Australian Research Council.

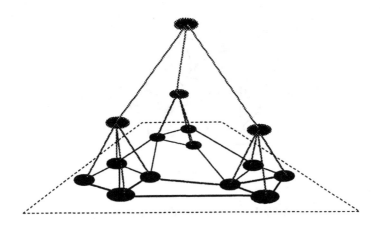

Fig. 1. An Example of a Clustered Graph

the graph at different depth of abstractions, but also keeps the track of the abstractions from one level to another. This is useful in preserving the mental map between abstraction levels.

In this paper, firstly, we describe some two dimensional drawing algorithms for clustered graphs; then we show how to extend two dimensional plane drawings to three dimensional multilevel drawings. We consider two conventions: straight-line convex drawings and orthogonal rectangular drawings; and we show some examples.

2 Terminology

A *clustered graph* $C = (G, T)$ consists of an undirected graph G and a rooted tree T such that the leaves of T are exactly the vertices of G. Each node ν of T represents a *cluster* $V(\nu)$ of the vertices of G that are leaves of the subtree rooted at ν. Note that tree T describes an inclusion relation between clusters. The *height* of a cluster ν, denoted by $h(\nu)$, is defined as the depth of the subtree of T rooted at ν. The *span* of an edge (ν_1, ν_2) of T is $|h(\nu_1) - h(\nu_2)|$. If the span of an edge of T is greater than one, we say it is *long*. In the rest of the paper, we assume every edge of T has a span of one. We consider long edges of T as a sequence of edges, each has a span of one.

For a clustered graph $C = (G, T)$, its *view at level i* is a graph $G_i = (V_i, E_i)$, where V_i consists of the set of nodes of height i in T. There is an edge (μ, ν) in E_i if there is an edge (u, v) of G where u belongs to cluster μ, and v belongs to cluster ν; in other words, edge (μ, ν) of E_i is the abstraction of all edges between cluster μ and cluster ν in G.

In a *plane drawing* of a clustered graph $C = (G, T)$, graph G is drawn as points and curves in the plane as usual. For each node ν of T, the cluster is drawn as a simple closed region R that contains the drawing of $G(\nu)$, such that:

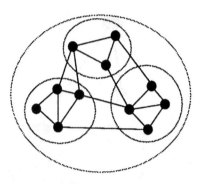

Fig. 2. A 2D Representation of a Clustered Graph

- the regions for all sub-clusters of R are completely contained in the interior of R;
- the regions for all other clusters are completely contained in the exterior of R;
- if there is an edge e between two vertices of $V(\nu)$, then the drawing of e is completely contained in R.

We say that the drawing of edge e and region R have an *edge-region crossing* if the drawing of e crosses the boundary of R more than once. A plane drawing of a clustered graph is *c-planar* if there are no edge crossings or edge-region crossings. If a clustered graph C has a c-planar drawing, then we say it is *c-planar* (see Fig. 2). An edge is said to be *incident* to a cluster $V(\nu)$ if one end of the edge is a vertex of that cluster but the other end is not in $V(\nu)$. An *embedding* of a clustered graph consists of the circular ordering of edges around each cluster which are incident to that cluster.

In a *plane drawing of a view* (see Fig.3), each node is drawn as a simple region in the plane, each edge is drawn as a curve between the region boundaries of its two ends. A plane drawing of a view is *c-planar* if there are no edge crossings or edge-region crossings.

A *multilevel drawing* (see Fig. 4) of a clustered graph $C = (G, T)$ consists of:

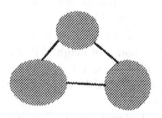

Fig. 3. The Drawing of a View

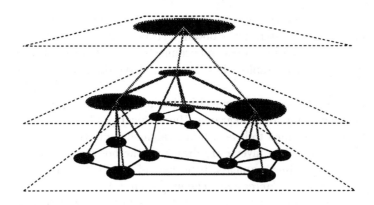

Fig. 4. A Multilevel Drawing

- A sequence of plane drawings of views from the leaf level (level 0) to the root level. The view at level i is drawn on the plane $z = i$.
- A three dimensional drawing of tree T, with each node ν of height i drawn as a point on the plane $z = i$, and within the region of ν in the drawing of the view at that level.

A multilevel drawing of a clustered graph is *c-planar* if the plane drawings of views at all levels are c-planar.

3 Plane Drawings

In this section we describe algorithms which produce c-planar plane drawings of clustered graphs. From these plane drawings, c-planar multilevel drawings can be constructed. We consider two conventions: straight-line convex drawings and orthogonal rectangular drawings.

3.1 Straight-line Convex Drawings

One of the basic graph drawing convention consists of representing edges as straight-line segments. In a straight-line convex drawing of a clustered graph $C = (G, T)$, edges of G are drawn as straight-line segments, regions for clusters are drawn as convex polygons. We use two approaches for such drawings.

An approach based on Tutte's algorithm. This approach from [6] applies a well known algorithm of Tutte [15], which creates a straight-line planar drawing of a triconnected planar graph G such that every face is a convex polygon. To apply Tutte's algorithm, we construct a *skeleton* $\Gamma(\nu)$ for each cluster ν. The skeleton $\Gamma(\nu)$ is the subgraph of $G(\nu)$ consisting of the vertices and edges on the outer faces of the child clusters of ν. Intuitively, a child cluster μ is represented by the outer face of $G(\mu)$ in the skeleton $\Gamma(\nu)$. We recursively apply

Tutte's algorithm to every skeleton graph, and compute a convex polygon for the outer face of each cluster, hence obtain a straight-line convex drawing. However, since Tutte's algorithm works on triconnected planar graphs, this approach is restricted to clustered graphs whose skeletons have the required connectivity property.

An approach based on hierarchical drawings. This approach uses the technique of drawing hierarchical graphs. Hierarchical graphs are directed graphs where vertices are assigned to layers. In a straight-line drawing of a hierarchical graph, vertices assigned to layer i are drawn on the horizontal line $y = i$, arcs are drawn as straight-line segments. If no pair of nonincident arcs intersect in the drawing, we say it is *hierarchical planar (h-planar)*.

In this approach, we transform a clustered graph to a hierarchical graph by computing an *st numbering*[1] of the vertices of G, such that the vertices which belong to the same cluster are numbered consecutively. We call this numbering *c-st numbering*. We use this numbering as a layer assignment to transform a clustered graph to a hierarchical graph, then apply the algorithm presented in [4] to produce a h-planar straight-line drawing.

The *c-st* numbering ensures that each cluster occupies consecutive layers in the drawing. For every cluster, we draw a convex hull of its vertices. It can be shown that in this drawing, there are no edge crossings; and there are no edges that cross the region (the convex hull) of a cluster where they do not belong. Note that if we draw regions as rectangles instead of convex hulls, edge-region crossings are still possible. In fact, by this algorithm, vertices of every cluster are bounded inside a trapezoid region which is formed by two horizontal lines for the highest layer and lowest layer of the cluster, and two straight lines (but not necessarily vertical) on the left and right of the the cluster.

3.2 Orthogonal Rectangular Drawings

In this section, we consider a drawing convention known as *orthogonal rectangular drawings*. In an orthogonal rectangular drawing of a clustered graph $C = (G, T)$, edges of G are drawn as sequences of horizontal and vertical segments, vertices of G are drawn on grid points and regions for clusters are drawn as rectangles. We use a method in [3] which produce such drawings with $O(n^2)$ area, and with constant number of bends on every edge.

Roughly speaking, this method works as follows. First, we transform a clustered graph to a planar *st*-graph[2], taking into account the clustering structure.

---

[1] Given any edge (s, t) in a biconnected graph G with n vertices, a *st* numbering for G is defined as follows. The vertices of G are numbered from 1 to n so that vertex s receives number 1, vertex t receives number n, and any vertex except s and t is adjacent both to a lower-numbered and a higher-numbered vertex. Vertices s and t are called the *source* and the *sink* respectively. Such a numbering is an *st* numbering for G. An *st* numbering of a biconnected graph can be computed in linear time [5].

[2] A planar *st*-graph [1] is a planar directed graph with one source s and one sink t; and both source and sink above can be embedded on the boundary of the same face, say the external face.

Then we produce a visibility representation of the planar st-graph. Finally, we use orthogonalization method to produce our orthogonal rectangular drawing from the visibility representation.

Here again, we compute a c-st numbering of G. Then we apply a direction for each edge of G according to the c-st numbering, and therefore obtain a planar st-graph. We use the technique in [2] of producing visibility representations of planar st-graphs. To obtain a rectangle for each cluster ν, we add 4 dummy vertices, each represents one side of a rectangle. We also add some dummy edges to obtain the two vertical sides of a rectangle. Then, using the algorithm in [2], we obtain a visibility representation of the graph. Finally, we construct an orthogonal rectangular drawing from the visibility drawing using some local operations similar to [14].

4   Multilevel Drawings

In this section we discuss methods of producing multilevel drawings of clustered graphs. We take the two dimensional plane drawings produced by the algorithms described in the previous section, and we show how to construct three dimensional multilevel drawings from the plane drawings.

To extend plane drawings of clustered graphs to multilevel drawings, we need to consider the following issues:

- Construct the drawing of the view at every level.
- Construct the drawing of the inclusion tree.

To construct the drawing of a view graph, we need to construct the regions for each node of the view, and route every edge between the boundaries of the regions of its two ends. For every node ν of a view at level i, we simply use its representation in the two dimensional plane drawing, and translate them to the plane where $z = i$. Note that every edge (μ, ν) in the view of level i is the abstraction of all the edges that connect between vertices of cluster μ and cluster ν. Therefore, an edge (μ, ν) in the view graph may correspond to multiple edges in G. We choose one edge (u, v) between cluster μ and cluster ν as a representative edge, and derive the drawing of edge (μ, ν) in the view from the drawing of edge (u, v). Suppose that in the two dimensional plane drawing, cluster μ and ν are drawn as regions $R(\mu)$ and $R(\nu)$ respectively; the drawing of edge (u, v) crosses the boundaries of $R(\mu)$ and $R(\nu)$ at points x and y respectively (see Fig. 5). To construct the drawing of edge (μ, ν) in the view, we use the segment between x and y and translate it to the plane where $z = i$. It can been shown that if the two dimensional plane drawing is c-planar, i.e. with no edge crossings or edge-region crossings, then the drawing derived for each view also has no edge crossings or edge-region crossings. It can also be shown that the derived drawing for each view preserves the convention of the two dimensional plane drawing.

To form the drawing of the inclusion tree T, we need to decide the position of every node, and route the edges between the nodes. Note that a node ν of

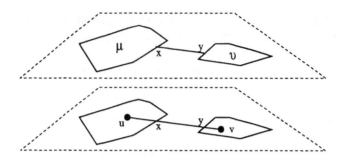

Fig. 5. Forming an Edge in the View

level i has to be positioned on the plane $z = i$ and in the corresponding region of the view. Here, we compute the position of each node recursively from bottom to top of T, as follows:

> FOR $i = 0$ to h (the depth of tree T) DO
> - If $i = 0$, then for each node of level 0 (leaf node), we simply place it at the position where it is drawn in the two dimensional plane drawing.
> - For every node ν of level i, we compute the average of the xy-coordinates of its children (at level $i - 1$), and use them as the xy-coordinates for ν.
>
> END

It is easily shown that by this method, every node ν is positioned within the corresponding region in the drawing of the view.

To route the edges of T, we simply draw a straight-line segment between the two nodes. Since we have replaced long edges of T by a sequence of edges, crossings between the edges of T cannot occur. Note that we use the average xy-coordinates of the children as the coordinates of a node. This will put a node right above most of its children and therefore let the edges between a node and most of its children drawn at a large angle to the xy-plane. If a node has only one child, then the edge is strictly vertical. Consequentially, by this method, a long edge of span k is drawn as a line with only one bend. The first segment is strictly vertical and spans $k - 1$ levels. The second segment spans one level.

5 Examples

In this section, we show some examples of drawings produced by our method.

Figure 6 shows a straight-line convex drawing produced using the approach based on Tutte's algorithm. Figure 7 shows the same drawing of Figure 6, but has a viewing direction almost orthogonal to the z axis; this shows the inclusion tree. Figure 8 shows a straight-line convex drawing produced using the approach based on hierarchical drawings. Figure 9 shows the same drawing of Figure 8, but focuse on a certain part. This reveals the detail of some substructures. Figure 10 shows an orthogonal rectangular drawing we produced. Figure 11 shows a view inside the orthogonal rectangular drawing of Figure 10; this emphasizes a single level of the abstraction hierarchy.

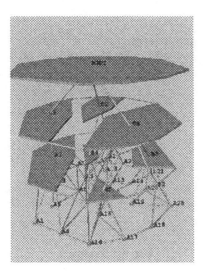

Fig. 6. Example 1

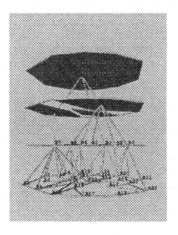

Fig. 7. Example 2

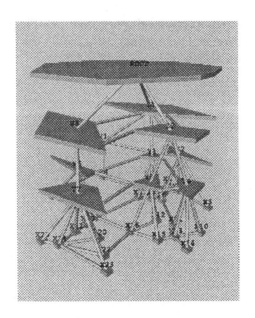

Fig. 8. Example 3

Fig. 9. Example 4

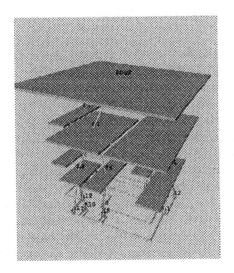

Fig. 10. Example 5

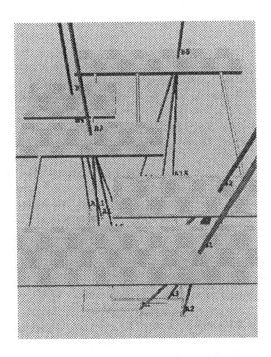

Fig. 11. Example 6

6 Conclusion and Future Work

This paper represents the first attempt to investigate methods for visualizing clustered graphs at multiple abstraction levels and in three dimensions. Particularly, we have considered two drawing conventions: straight-line convex drawings and orthogonal rectangular drawings. We have described some algorithms for two dimensional plane drawings and have shown how to extend them to multilevel three dimensional drawings.

In this paper, every view that we consider is at a specific abstraction level. However, in many applications, we need to visualize a graph at an arbitrary cross-section. For example, sometimes we need to visualize a view with some portions in very detail, and other portions in abstract. This seems an interesting topic for our future research. Further, it will be interesting to investigate the methods of making smooth changes between views based on three dimensional drawings. This would be helpful to some mental map issues in human computer interface design.

Although some of the methods described in this paper may look naive and straightforward, we hope, with the increasing interest in compound structure visualization, more and more results could come forward.

References

1. G. Di Battista and R. Tamassia. Algorithms for plane representations of acyclic digraphs. *Theoretical Computer Science*, 61:175–198, 1988.
2. G. Di Battista, R. Tamassia, and I.G. Tollis. Constrained visibility representations of graphs. *Information Processing Letters*, 41:1–7, 1992.
3. Peter Eades and Qing-Wen Feng. Orthogonal grid drawing of clustered graphs. Technical Report 96-04, Department of Computer Science, The University of Newcastle, Australia, 1996.
4. Peter Eades, Qing-Wen Feng, and Xuemin Lin. Straight-line drawing algorithms for hierarchical graphs and clustered graphs. Technical Report 96-02, Department of Computer Science, The University of Newcastle, Australia, 1996.
5. S. Even and R. E. Tarjan. Computing an st-numbering. *Theoretical Computer Science*, 2:339–344, 1976.
6. Qing-Wen Feng, Robert F. Cohen, and Peter Eades. How to draw a planar clustered graph. In *COCOON'95*, volume 959 of *Lecture Notes in Computer Science*, pages 21–31. Springer-Verlag, 1995.
7. D. Harel. On visual formalisms. *Communications of the ACM*, 31(5):514–530, 1988.
8. J. Kawakita. The KJ method – a scientific approach to problem solving. Technical report, Kawakita Research Institute, Tokyo, 1975.
9. Brendan Madden, Patrick Madden, Steve Powers, and Michael Himsolt. Portable graph layout and editing. In *GD'95*, volume 1027 of *Lecture Notes in Computer Science*, pages 385–395. Springer-Verlag.
10. K. Misue and K. Sugiyama. An overview of diagram based idea organizer: D-abductor. Technical Report IIAS-RR-93-3E, ISIS, Fujitsu Laboratories, 1993.

11. Stephen C. North. Drawing ranked digraphs with recursive clusters. In *Proc. ALCOM Workshop on Graph Drawing '93*, September 1993.
12. Tom Sawyer Software. Graph layout toolkit. available from bmadden@TomSawyer.COM.
13. K. Sugiyama and K. Misue. Visualization of structural information: Automatic drawing of compound digraphs. *IEEE Transactions on Systems, Man and Cybernetics*, 21(4):876–892, 1991.
14. R. Tamassia, G. Di Battista, and C. Batini. Automatic graph drawing and readability of diagrams. *IEEE Transactions on Systems, Man and Cybernetics*, SMC-18(1):61–79, 1988.
15. W. T. Tutte. How to draw a graph. *Proceedings of the London Mathematical Society*, 3(13):743–768, 1963.
16. C. Williams, J. Rasure, and C. Hansen. The state of the art of visual languages for visualization. In *Visualization 92*, pages 202 – 209, 1992.

Straight-Line Drawing Algorithms for Hierarchical Graphs and Clustered Graphs

Peter Eades[1]   Qing-Wen Feng[1]   Xuemin Lin[2]

[1] Department of Computer Science and Software Engineering, University of Newcastle, NSW 2308, Australia. Email: {eades,qwfeng}@cs.newcastle.edu.au
[2] Department of Computer Science, University of Western Australia, Nedlands, WA 6009, Australia. Email: lxue@cs.uwa.edu.au

(extended abstract)

Abstract. Hierarchical graphs and clustered graphs are useful nonclassical graph models for structured relational information. Hierarchical graphs are graphs with layering structures; clustered graphs are graphs with recursive clustering structures. Both have applications in CASE tools, software visualization, VLSI design, etc. Drawing algorithms for hierarchical graphs have been well investigated. However, the problem of straight-line representation has not been addressed. In this paper, we answer the question: does every planar hierarchical graph admit a planar straight-line hierarchical drawing? We present an algorithm that constructs such drawings in $O(n^2)$ time. Also, we answer a basic question for clustered graphs, i.e. does every planar clustered graph admit a planar straight-line drawing with clusters drawn as convex polygons? A method for such drawings is provided in this paper.

1   Introduction

Graph drawing algorithms are widely used in graphical user interfaces of many software systems. Examples include CASE tools, reverse engineering systems and software design systems. A good picture is worth a thousand words, while a poor drawing can be misleading.

As the information that we want to visualize becomes more and more complicated, we need more structure on top of the classical graph model. Hierarchical graphs are directed graphs with layering structures (see Fig. 1). They appear in applications where hierarchical structures are involved e.g. PERT networks and organization charts [19, 10]. Clustered graphs are graphs with clustering structures (see Fig. 2) which appear in many structured diagrams [18, 11, 12].

A hierarchical graph is conventionally drawn with vertices of a layer on the same horizontal line, and arcs as curves monotonic in y direction (see Fig. 1). A hierarchical graph is *hierarchical planar (h-planar)* if it admits a drawing without edge crossings. For a clustered graph, the clustering structure is represented by a closed curve that defines a region. The region contains the drawing of all the vertices which belong to that cluster (see Fig. 2). A clustered graph is *compound*

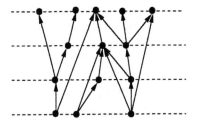

Fig. 1. An Example of a Hierarchical Graph

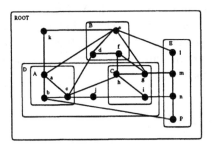

Fig. 2. An Example of a Clustered Graph

planar (c-planar) if it admits a drawing with no edge crossings or edge-region crossings.

One of the basic graph drawing convention consists of representing edges as straight-line segments. For classical graphs, it has been shown independently by Fáry [7], Stein [17], and Wagner [21] that every planar graph admits a straight-line drawing without edge crossings. Tutte [20] proved that every 3-connected planar graph admits a planar straight-line drawing where all the face boundaries are drawn as convex polygons.

In this paper, we answer the question: does every planar hierarchical graph admit a planar straight-line hierarchical drawing? Although many results have been obtained on drawing hierarchical graphs [19, 10, 3], the basic problem of planar straight-line drawings has not been studied. It has been shown by Di Battista and Tamassia [1] that every planar st-graph admits an *upward drawing* i.e. a drawing where all arcs are drawn as straight-line segments pointing upward. However, the problem for hierarchical graphs is different. We have more constraints: vertices of the same layer should be drawn on the same horizontal line; layers should be equal distance apart. A method to construct straight-line drawings of hierarchical graphs has been presented by Eades, Lin and Tamassia [5]. However, it is restricted to a special class of hierarchical graphs known as *collapsible free* graphs.

The second question answered in this paper is for clustered graphs: does every planar clustered graph admit a planar straight-line drawing with clusters drawn as convex polygons? Although an algorithm for straight-line drawing of clustered graphs has been presented by Feng, Cohen and Eades [8], it does not apply to all clustered graphs. If the graph induced by a cluster is not biconnected, then its external facial cycle is not a simple cycle. In this case, we cannot use the drawing of its external facial cycle as the region boundary, since it cannot form a simple region. This question that we answer in this paper was posed as an open problem in [8].

The rest of the paper is organized as follows. In section 2, we present some terminology for hierarchical graphs. In section 3, we prove that every planar hierarchical graph admit a planar straight-line drawing. An $O(n^2)$ time algorithm that produces such drawings is presented. We introduce the model of clustered graphs in section 4. In section 5, we show that every planar clustered graph admits a planar straight-line convex cluster drawing. This is accomplished by transforming clustered graphs into hierarchical graphs. Section 6 concludes with some remarks and discussion.

2   Hierarchical Graphs

Hierarchical graphs are directed graphs where vertices are assigned to layers. As described in [5], a *hierarchical graph* $H = (V, A, \lambda, k)$ consists of a directed graph (V, A), a positive integer k, and, for each vertex u, an integer $\lambda(u) \in 1, 2, ..., k$, with the property that if $(u, v) \in A$, then $\lambda(u) < \lambda(v)$. For $1 \leq i \leq k$, the set $\{u : \lambda(u) = i\}$ is the ith *layer* of H and is denoted by L_i. The *span* of an arc (u, v) is $\lambda(v) - \lambda(u)$. An arc of span greater than one is *long*, and a hierarchical graph with no long arcs is *proper*.

A hierarchical graph is conventionally drawn with layer L_i on the horizontal line $y = i$, and arcs as curves monotonic in y direction. If no pair of nonincident arcs intersect in the drawing, we say it is a *hierarchical planar (h-planar)* drawing. Note that a nonproper hierarchical graph can be transformed into a proper hierarchical graph by adding dummy vertices on long arcs. It can be shown that a nonproper hierarchical graph is h-planar if and only if the corresponding proper hierarchical graph is h-planar. A *hierarchical planar embedding* of a proper hierarchical graph is defined by the ordering of vertices on each layer of the graph. Note that every such embedding has an unique external face. It is easily shown that every proper h-planar graph admits a *straight-line hierarchical drawing*, that is, a drawing where arcs are drawn as straight-line segments. However, for nonproper hierarchical graphs, the problem is not trivial, since no bends are allowed on long arcs.

We call a planar embedded graph a *plane graph*. If a hierarchical plane graph has only one source s and one sink t, we call it a *hierarchical-st plane graph*. We will show that every hierarchical plane graph can be extended to a hierarchical-st plane graph. The following lemma gives some basic properties which are useful to our algorithm.

Lemma 1. *Let H be a hierarchical-st plane graph, then: (a) Every biconnected component of H is also a hierarchical-st plane graph. (b) If B_1 and B_2 are two biconnected components of H, and u is a vertex of B_1 and u is not a cut vertex, then either for all vertices v of B_2, $\lambda(u) < \lambda(v)$, or for all vertices v of B_2, $\lambda(u) > \lambda(v)$. (c) H has a planar straight-line hierarchical drawing if and only if each of its biconnected component has a planar straight-line hierarchical drawing.*

With the above lemma, we can assume that we are given a hierarchical-st plane graph that is biconnected, which implies its external face is bounded by a simple cycle if the graph is not just a single edge. Since hierarchical graphs are directed graphs, the terms "cycle" and "path" that we use in the rest of the paper are all for the underlying undirected graphs. To denote a cycle of a plane graph, we use the sequence of vertices on the cycle in clockwise order. We say a path is *monotonic* if the directions of the edges (arcs) do not change along the path.

The following lemma is also very useful for to algorithm.

Lemma 2. *Let H be a hierarchical-st plane graph which is biconnected, and has external facial cycle $C = (v_1, \ldots, v_k, v_1)$. Suppose that $\mathcal{P} = (v_i, x_1, \ldots, x_l, v_j)$ is a monotonic path from v_i to v_j in H, and the vertices of $\mathcal{P}$ are not on C except v_i and v_j. Let H_1 and H_2 be the two subgraphs bounded inside by cycles $C_1 = (v_1, \ldots, v_i, x_1, \ldots, x_l, v_j, \ldots, v_k, v_1)$ and $C_2 = (v_i, \ldots, v_j, x_l, \ldots, x_1, v_i)$ inclusive. Then H_1 and H_2 are hierarchical-st plane graphs and are biconnected (see Fig.3).*

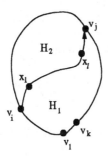

Fig. 3. Illustration of Lemma 2

3   Straight-Line Hierarchical Drawings

In this section, we show that given an n-vertex hierarchical plane graph, we can compute a planar straight-line hierarchical drawing in $O(n^2)$ time.

We apply a divide and conquer approach: divide the hierarchical graph into subgraphs, compute the drawings of the subgraphs, and obtain a drawing of the graph by combining the drawings of the subgraphs. The key part of this approach is to find a suitable partition.

Our method works on *triangular hierarchical-st plane graphs*. In a triangular hierarchical-st plane graph, the boundary of every nonexternal face consists of exactly three edges. We prove that every hierarchical plane graph can be extended to a triangular hierarchical-st plane graph which admits a straight-line drawing with a prescribed polygon as its external face. We provide a straight-line drawing algorithm based on our proof.

In our method, we are given a prescribed polygon as the external face of the drawing. Note that there can be vertices of the external facial cycle which are not drawn as apexes of the polygon. This can give some problems if the external facial cycle has a chord (i.e. an edge between two nonconsecutive vertices). To deal with this problem, we need some terminology. Let H be a hierarchical-st plane graph with source s and sink t; let cycle C be the boundary of its external face; let polygon P be a straight-line hierarchical drawing of cycle C. We say that P is *feasible* for H if the following conditions hold:

- P is a convex polygon.
- If cycle C has a chord (x, y), then on each of the two paths of cycle C between x and y, there exists a vertex v which is drawn as an apex of polygon P.

In our divide and conquer approach, we distinguish two situations. If the external facial cycle has a chord, we simply divide the graph into two parts with the chord. Otherwise, we find a vertex not on the external facial cycle, such that there are three monotonic paths that connect the vertex with the external facial cycle. Therefore, by using Lemma 2 twice, we divide the graph into three parts. The following lemma is useful in finding such vertex in the graph. We need some more terminology. For a hierarchical graph H with vertices u and v, an *st-component for* (u, v) is the union of all subgraphs of H for which u is the unique source, and v is the unique sink. In other words, the st-component for (u, v) is the maximal subgraph with a single source u and a single sink v.

Lemma 3. *Let H be a triangular hierarchical-st plane graph with single source s and single sink t. Suppose that the external facial cycle C of H has no chords. Let v be a vertex on cycle C other than s or t. Suppose that $H_{st}(v)$ is the st-component of the hierarchical graph $H - v$ for vertex pair (s, t). Then there exists a vertex w incident to v in H and not on cycle C, such that $w \in H_{st}(v)$ and w has the following properties: (1) Vertex w is on the external face of $H_{st}(v)$. (2) vertex w is not a cut vertex of $H_{st}(v)$. (3) Suppose that $H_{main}(v)$ denotes the biconnected component of $H_{st}(v)$ that contains w. Then the external facial cycle of $H_{main}(v)$ consists of two paths: path $(x, \ldots, y)$ which belongs to C, and path $(y, \ldots, w, \ldots, x)$ which does not belong to C; and path $(x, \ldots, w, \ldots, y)$ is monotonic (see Fig.4).*

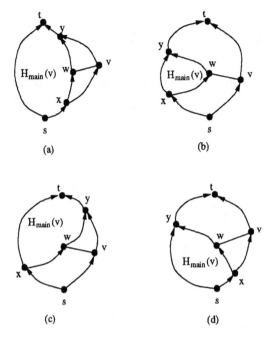

Fig. 4. Possible Partitions of H

Sketch of Proof: Fig 5(a)-(d) shows all possible situations of the path formed by the vertices incident to v. We show in each situation there exists such vertex w incident to v in H and not on cycle C, such that $w \in H_{st}(v)$. We show that the situation illustrated in Fig. 5(e) would not occur. The proofs for *Property 1* and *Property 2* of such vertex w are immediate. For *Property 3*, Fig. 4 illustrates all possible situations of the external face of $H_{main}(v)$, and we show that the property holds for all these situations. □

Theorem 4. *Suppose that H is a triangular hierarchical-st plane graph, and polygon P is a straight-line hierarchical drawing of its external facial cycle C. If P is feasible for H, then there exists a planar straight-line hierarchical drawing of H with external face P.*

Proof. We prove by induction on the number n of vertices of H. The basis of the induction, $n = 3$ is immediate. Now, assume that the theorem holds for graphs with less than n vertices. We distinguish two cases:

$\quad$ *Case 1:* The external facial cycle C of H has a chord (x, y). By Lemma 2, chord (x, y) divides H into two subgraphs H_1 and H_2. We draw a straight line segment between x and y, which divides P into two polygons P_1 and P_2. It can be verified that P_1 and P_2 are feasible for H_1 and H_2. Since both H_1 and H_2 have less than n vertices, by induction, there exist straight-line hierarchical drawings of H_1 and H_2 with external faces P_1 and P_2. Hence, by combining the

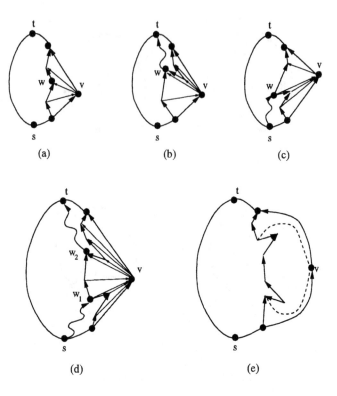

Fig. 5. Illustration of the Proof of Lemma 3

two drawings, we obtain a straight-line hierarchical drawing of H with external face P.

Case 2: The external facial cycle C of H has no chords. There exists a vertex v other than the source s or sink t on the external face, such that v is drawn as an apex of P. Otherwise P would not be a convex polygon. By Lemma 3, there exists a vertex w incident to v but not on cycle C such that $w \in H_{st}(v)$ and w has those properties stated in the lemma . Hence, we have a monotonic path $(x, \ldots, w, \ldots, y)$ inside H, and also an edge (w, v) inside H (see Fig. 4). Using Lemma 2 twice, they divide H into three parts:

- $H_{main}(v)$ bounded by cycle $(x, \ldots, y, \ldots, w, \ldots x)$;
- $H_{asso1}(v)$ bounded by cycle $(x, \ldots, w, v, \ldots x)$;
- $H_{asso2}(v)$ bounded by cycle $(y, \ldots, v, w, \ldots y)$.

Now we need to adjust this partition such that a feasible polygon can be computed for each part. Note that path $(x, \ldots, w)$ has no chords in $H_{asso1}(v)$, otherwise it would not belong to the external face of $H_{main}(v)$. Similarly, path $(w, \ldots, y)$ has no chords in $H_{asso2}(v)$.

Now consider chords in $H_{main}(v)$. We reduce $H_{main}(v)$ and accordingly, extend $H_{asso1}(v)$ or $H_{asso2}(v)$ to eliminate such chords. If path

$(x, \ldots, u_1, \ldots, u_2, \ldots, w)$ has a chord (u_1, u_2), we modify the path to $(x, \ldots, u_1, u_2, \ldots, w)$. Also, we modify path $(w, \ldots, y)$ in similar way if it has a chord. Graphs $H_{main}(v)$, $H_{asso1}(v)$ and $H_{asso2}(v)$ change accordingly when we modify the paths. After the modification, paths $(x, \ldots, w)$ and $(w, \ldots, y)$ have no chords in $H_{main}(v)$, $H_{asso1}(v)$ and $H_{asso2}(v)$.

It can be verified that $H_{main}(v)$, $H_{asso1}(v)$ and $H_{asso2}(v)$ are triangular hierarchical-st plane graphs.

Let $H_{frame}(v)$ be the graph that consists of only the external faces of $H_{main}(v)$, $H_{asso1}(v)$ and $H_{asso2}(v)$. Now $H_{frame}(v)$ is a hierarchical-st plane graph with the same external face as H. Hence polygon P is also a hierarchical planar drawing of the external face of $H_{frame}(v)$. We need to find a position for w such that the drawing of the three internal faces of $H_{frame}(v)$ are convex polygons, and therefore feasible for $H_{main}(v)$, $H_{asso1}(v)$ and $H_{asso2}(v)$. We compute the x coordinate of vertex w using the following equation derived from [5]:

$$x(a) = \frac{\frac{1}{2d_a^+}\sum_{b \to a} \frac{x(b)}{l} + \frac{1}{2d_a^-}\sum_{a \to b} \frac{x(b)}{l}}{\frac{1}{2d_a^+}\sum_{b \to a} \frac{1}{l} + \frac{1}{2d_a^-}\sum_{a \to b} \frac{1}{l}}. \tag{1}$$

Here, $x(a)$ and $x(b)$ denote the x coordinates of vertices a and b respectively; d_a^+ denotes the indegree of a; d_a^- denotes the outdegree of a; and $l = |\lambda(a) - \lambda(b)|$.

This formula computes the x coordinate of vertex w as a weighted barycenter of its neighbors x, y and v. (Vertices with degree 2 are not considered here.) We place other internal vertices of $H_{frame}(v)$ (those with degree 2) onto the line segments from x to w and from w to y at appropriate horizontal lines. It can be shown that the drawings P_0, P_1 and P_2 of the three internal faces of $H_{frame}(v)$ are convex polygons [5]. Note that edges on path $(x, \ldots, w)$ are drawn on the same line, so are the edges on path $(w, \ldots, y)$. However, since there are no chords on these paths, P_0, P_1 and P_2 are feasible for $H_{main}(v)$, $H_{asso1}(v)$ and $H_{asso2}(v)$ respectively. As each of $H_{main}(v)$, $H_{asso1}(v)$ and H_{asso2} has less than n vertices, by induction, there exist straight-line hierarchical drawings of $H_{main}(v)$, $H_{asso1}(v)$ and H_{asso2} with external faces P_0, P_1 and P_2. Hence, by combining these drawings, we obtain a straight-line hierarchical drawing of H with external face P.

The algorithm to compute a planar straight-line hierarchical drawing is based on the proof of Theorem 4. The input of the algorithm is a hierarchical plane graph H; the output is a planar straight-line hierarchical drawing of H. The algorithm consists of two phases: *Preprocessing* and *Drawing*. In the preprocessing phase, we extend the hierarchical plane graph to a triangular-st plane graph. The drawing phase is a recursive procedure that actually constructs the drawing of the graph. Now we describe them in more detail.

Preprocessing. We extend the hierarchical plane graph to a triangular hierarchical-st plane graph in three steps. (1) Extend the hierarchical plane graph such that all the sources and sinks lie on the bottom layer and top layer. We can use a method similar to those in [1, 15] which performs two sweeps from bottom to top and from top to bottom to eliminate the sources and sinks in

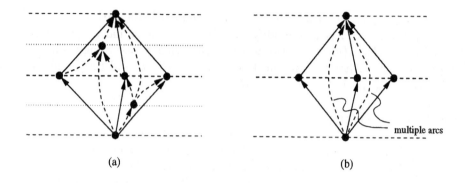

(a)   (b)

Fig. 6. Triangulating the Hierarchical Graph

between. (2) Add one more vertex s below the bottom layer and connect it to all the sources; then add one more vertex t above the top layer and connect all the sinks to it. Therefore, a hierarchical-st plane graph is obtained. (3) Extend the hierarchical-st plane graph to a triangular one as follows: insert a layer between every two consecutive layers (This ensures that original layers still to be evenly distributed.); add a "star" structure inside each face (see Fig. 6(a)), and place the center of the each star on an inserted layer. After this, every internal face is bounded by exactly three edges. Note that this operation does not increase the size of the graph by more than a constant. This triangulation method is a little unusual, but necessary. Fig. 6(b) shows that if we do not add new vertices, multiple arcs can be produced. Further, we cannot allow dummy nodes on the arcs because this may introduce bends. Also note that no arcs are allowed between two vertices of the same layer.

Drawing. The drawing phase is realized with a recursive procedure which is based on the proof of Theorem 4. Firstly, it is easy to find a feasible polygon P for an input graph H. Then we call the procedure and obtain a drawing of H.

Procedure *Straight-line_Hierarchical_Draw(H, $\mathcal{E}$, P, Γ)*
{H is a triangular hierarchical-st plane graph with planar embedding $\mathcal{E}$; P is a polygon feasible for H. Γ is a planar straight-line hierarchical drawing of H returned by the procedure.}

(1) If H has three vertices, then draw H as P. Let $\Gamma = P$, exit.
(2) Check H for possible chords of the external facial cycle C.
(3) If C has a chord (x, y), then:
    (3.1) divide H into H_1 and H_2 with chord (x, y); draw straight-line segment between x and y in P; divide P into P_1 and P_2;
    (3.2) call Straight-line_Hierarchical_Draw(H_1, P_1, Γ_1);
            call Straight-line_Hierarchical_Draw(H_2, P_2, Γ_2);
    (3.3) Let $\Gamma = \Gamma_1 \cup \Gamma_2$, exit.
(4) If C has no chords, then:

(4.1) choose vertex v on C; find the st-component $H_{st}(v)$ of $H - v$ for the source and sink pair (s, t); choose vertex w that is incident to v but not on C, and on the external face of $H_{st}(v)$.

(4.2) find the biconnected component $H_{main}(v)$ of $H_{st}(v)$ that contains w; modify the two paths $(w, \ldots, x)$ and $(y, \ldots, w)$ on the external face of $H_{st}(v)$ to avoid chords;

(4.3) construct $H_{frame}(v)$ and compute its drawing using equation 1; hence obtain polygons P_0, P_1 and P_2 of the three internal faces of $H_{frame}(v)$;

(4.4) divide H into $H_{main}(v)$, $H_{asso1}(v)$ and $H_{asso2}(v)$ with paths $(w, \ldots, x)$, $(y, \ldots, w)$ and edge (w, v);

(4.5) call Straight-line_Hierarchical_Draw($H_{main}(v)$, P_0, Γ_0);
call Straight-line_Hierarchical_Draw($H_{asso1}(v)$, P_1, Γ_1);
call Straight-line_Hierarchical_Draw($H_{asso2}(v)$, P_2, Γ_2);

(4.6) Let $\Gamma = \Gamma_0 \cup \Gamma_1 \cup \Gamma_2$, exit.

In the preprocessing phase, each of the three steps takes linear time.

In the drawing phase, we maintain an edge list and a face list for each hierarchical-st plane subgraph through the procedure Straight-line_Hierarchical_Draw. With this data structure, we can check for chords of a cycle C (or path) of a graph in linear time; we can divide graph H with a chord of its external facial cycle or a path inside it in linear time. An st-component for vertex pair (u, v) can be found in linear time by performing depth-first search from u in one direction, and from v in the opposite direction. Also, the biconnected components of a graph can be found in linear time [2]. In the procedure call of the drawing phase, every vertex is processed at most $O(n)$ times by step (2) or step (4.2). Every vertex is processed also at most $O(n)$ times by step (3.1) or steps (4.2) and (4.4). Every vertex is processed once by step (4.3) for computing its x coordinate. Consequentially, the drawing phase costs $O(n^2)$ time.

Note that each edge appears in at most two subgraphs through the procedure. Therefore, our algorithm requires linear space.

The following theorem summarizes the performance of the algorithm.

Theorem 5. *Let H be a hierarchical plane graph with n vertices. The above algorithm constructs a planar straight-line hierarchical drawing for H in $O(n^2)$ time and $O(n)$ space.*

Based on our results for hierarchical graphs, we consider the straight-line drawing problem for clustered graphs in the following sections.

4   Clustered Graphs

A *clustered graph* $C = (G, T)$ consists of an undirected graph G and a rooted tree T such that the leaves of T are exactly the vertices of G. Each node ν of

T represents a *cluster* $V(\nu)$ of the vertices of G that are leaves of the subtree rooted at ν. Note that tree T describes an inclusion relation between clusters.

In a *drawing* of a clustered graph $C = (G, T)$, graph G is drawn as points and curves as usual. For each node ν of T, the cluster is drawn as a simple closed region R that contains the drawing of $G(\nu)$, such that:

- the regions for all sub-clusters of R are completely contained in the interior of R;
- the regions for all other clusters are completely contained in the exterior of R;
- if there is an edge e between two vertices of $V(\nu)$, then the drawing of e is completely contained in R.

We say that the drawing of edge e and region R have an *edge-region crossing* if the drawing of e crosses the boundary of R more than once. A drawing of a clustered graph is *c-planar* if there are no edge crossings or edge-region crossings. If a clustered graph C has a c-planar drawing then we say that it is *c-planar* (see Fig. 2).

An edge is said to be *incident* to a cluster $V(\nu)$ if one end of the edge is a vertex of that cluster but the other end is not in $V(\nu)$. An *embedding* of a clustered graph consists of the circular ordering of edges around each cluster which are incident to that cluster. A clustered graph $C = (G, T)$ is a *connected clustered graph* if each cluster induces a connected subgraph of G. The following results from [9] characterize c-planarity in a way which can be exploited by our drawing algorithm.

Theorem 6. *A clustered graph $C = (G, T)$ is c-planar if and only if it is a sub-clustered graph of a connected and c-planar clustered graph.*

From Theorem 6, we can assume that we are given a connected clustered graph when drawing a c-planar clustered graph. According to [13], a c-planar embedding of a connected clustered graph can be found in linear time. In the rest of the paper, we assume there are no degenerated clusters, that is, every nonleaf node of T has at least two children.

5 Straight-Line Convex Cluster Drawings

One of the fundamental questions in planar clustered graph drawing is: does every c-planar clustered graph admit a planar drawing such that edges are drawn as straight-line segments and clusters are drawn as convex polygons? In this section, we answer this question based on our results for hierarchical graphs. We transform a clustered graph into a hierarchical graph, and construct a straight-line convex cluster drawing on top of the straight-line hierarchical drawing.

By Theorem 6, we assume that we are given a c-planar connected clustered graph $C = (G, T)$ with a c-planar embedding. Roughly speaking, our algorithm works as follows. First, we triangulate G (including triangulating the external

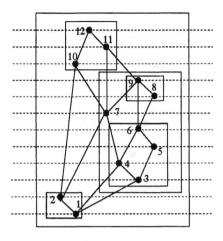

Fig. 7. Clustered Graph → Hierarchical Graph

face) [16]; then compute an *st* numbering [3] of the vertices of G such that the vertices that belong to the same cluster are numbered consecutively. We call this numbering *c-st numbering*. This numbering gives us a layer assignment of the vertices of G. Hence, the clustered graph is transformed to a hierarchical graph (see Fig. 7), and each cluster has consecutive layers. Because of this property, we show that a straight-line convex cluster drawing can be constructed from the straight-line hierarchical drawing.

The critical part of this method is the construction of the *c-st* numbering. To ensure that the vertices of the same cluster are numbered consecutively, we need to compute an ordering of the child clusters for every parent cluster ν. To do this, we construct a graph $F(\nu)$ from $G(\nu)$ by shrinking each child cluster of ν to a vertex while preserving the embedding. First of all, we add a dummy node on every edge of G; this prevents edges from collapsing into one edge when shrinking. We use a top down approach , ordering the children of the root first.

For the root node γ of T, the graph $F(\gamma)$ is constructed as follows. We choose an edge e of G that does not belong to any other cluster except the root cluster. Since we are given a connected clustered graph, such an edge exists. We choose s and t to be the two ends of this edge. Then shrink every child cluster of the root into a single vertex and preserve the planar embedding in the meantime. The resulting graph is $F(\gamma)$. Every vertex of $F(\gamma)$ represents a child cluster of the root cluster. Since st numberings are constructed on biconnected graphs, we need the following lemma:

---

[3] Given any edge (s, t) in a biconnected graph G with n vertices, a st numbering for G is defined as follows. The vertices of G are numbered from 1 to n so that vertex s receives number 1, vertex t receives number n, and any vertex except s and t is adjacent both to a lower-numbered and a higher-numbered vertex. Vertices s and t are called the *source* and the *sink* respectively. Such a numbering is an st numbering for G. An st numbering of a biconnected graph can be computed in linear time [6].

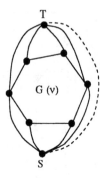

Fig. 8. Illustration of Computing c-st Numbering

Lemma 7. *Suppose that $C = (G, T)$ is a c-planar clustered graph, γ is the root of T, and G is triangulated. Then the graph $F(\gamma)$ is biconnected.*

Sketch of Proof: Since G has been triangulated and we preserve the embedding in the shrinking operation, it can be shown that every face in the resulting graph $F(\gamma)$ is bounded by a simple cycle. Therefore $F(\gamma)$ is biconnected. □

Since $F(\gamma)$ is biconnected, we can order the children of the root Γ by computing an *st* numbering, choosing the vertex that represents the cluster where s belongs as the source, and the vertex that represents the cluster where t belongs as the sink.

We proceed top down from the root. For a nonroot node ν, we construct a graph $F(\nu)$ in a similar but slightly more complex way; $F(\nu)$ depends on the place of ν in the ordering of ν and its siblings. For each child cluster μ of ν, we shrink graph $G(\mu)$ into one vertex while preserving the planar embedding. For those edges that connect cluster ν with clusters which are ordered before ν (note that this order is computed recursively as mentioned above), we connect them to a single vertex S. For those edges that connect cluster ν with clusters which are ordered after ν, we connect them to a single vertex T (see Fig. 8). Finally, we connect S and T in the external face of $G(\nu)$, hence forming $F(\nu)$. Here, if vertex $s(t)$ belongs to cluster ν, we simply choose the vertex which represents the child cluster that contains $s(t)$ as $S(T)$. We need the following lemma; its proof is similar to that of Lemma 7.

Lemma 8. *Suppose that $C = (G, T)$ is a c-planar clustered graph, and G is triangulated. For every node ν of T, the graph $F(\nu)$ is biconnected.*

□

With the lemma, we order every vertex of $F(\nu)$ by computing an *st* numbering, choosing vertex S as the source, and vertex T as the sink.

Now, each cluster ν is assigned a number of order within its parent. Therefore, a recursive hierarchy of orders is formed. We expand it lexicographically into a linear order and hence form an ordering of the all vertices of G. It can be verified that this order gives us an st numbering on the vertices of G such that the vertices that belong to the same cluster are numbered consecutively.

With this c-st numbering, we transform a clustered graph into hierarchical graph by assigning the layer of each vertex with its c-st number. Then, apply the straight-line hierarchical drawing algorithm described in section 3, hence, obtain a planar straight-line hierarchical drawing of G. The c-st numbering ensures that each cluster occupies consecutive layers in the drawing. For every cluster, we draw a convex hull of the vertices of the cluster. In this drawing, there are no edge crossings; there are no edges that cross the region (the convex hull) of a cluster where they do not belong. Since we are given a connected clustered graph, each cluster forms a connected subgraph of G. If the drawing of an edge crosses the convex hull of a cluster where it does not belong, then there would be an edge crossing. This forms a contradiction. Note that if we draw regions as rectangles instead of convex hulls, edge-region crossings are still possible.

Since an st numbering can be constructed in linear time [6], the computation of c-st numbering takes linear time in terms of the size of the graph. An algorithm to compute a convex hull of a set of m points requires $O(m \log m)$ time [15]. By Theorem 5, our method takes $O(n^2)$ time.

The following theorem summarizes our result on planar straight-line convex cluster drawings.

Theorem 9. *Let $C = (G, T)$ be a c-planar clustered graph with n vertices. A planar straight-line convex cluster drawing of C can be constructed in $O(n^2)$ time.*

6 Conclusion and Remarks

In this paper, we answer one of the basic questions for hierarchical graphs that has not be investigated before. We show that every hierarchical planar graph admits a planar straight-line hierarchical drawing, and present an algorithm that produces such drawings in $O(n^2)$ time. With this result, we answer a similar basic question for clustered graphs that has been posed as an open problem in [8]. We show that every c-planar clustered graph admits a planar straight-line convex cluster drawing. A method to construct such drawings is provided.

The algorithms that we present in this paper take quadratic time. From the computational point of view, it is interesting to know whether the time complexity can be improved, say, to $O(n \log n)$; and whether $O(n \log n)$ is optimal for this type of problems.

The drawings produced by our algorithms may require exponential area. This is justified by the area lower bounds for these drawing conventions presented in [14, 8]. Relaxing the straight-line constraints can give us polynomial area bounds [4].

For future work on clustered graphs, other drawing conventions such as polyline rectangular cluster drawings, and also nonplanar drawings will be studied.

For clustered graphs, we only ensure that clusters are drawn as convex polygons, while it is desirable to represent clusters as more regular bodies such as circles and rectangles. This also forms an interesting topic for our future research.

References

1. G. Di Battista and R. Tamassia. Algorithms for plane representations of acyclic digraphs. *Theoretical Computer Science*, 61:175–198, 1988.
2. J.A. Bondy and U.S.R. Murty. *Graph Theory with Applications*. North-Holland, New York, N.Y., 1976.
3. P. Eades and K. Sugiyama. How to draw a directed graph. *Journal of Information Processing*, 424–437, 1991.
4. Peter Eades and Qing-Wen Feng. Orthogonal grid drawing of clustered graphs. Technical Report 96-04, Department of Computer Science, The University of Newcastle, Australia, 1996.
5. Peter D. Eades, Xuemin Lin, and Roberto Tamassia. An algorithm for drawing a hierarchical graph. *International Journal of Computational Geometry and Applications*, 1995.
6. S. Even and R. E. Tarjan. Computing an st-numbering. *Theoretical Computer Science*, 2:339–344, 1976.
7. I. Fary. On straight lines representation of planar graphs. *Acta Sci. Math. Szeged.*, 11:229–233, 1948.
8. Qing-Wen Feng, Robert F. Cohen, and Peter Eades. How to draw a planar clustered graph. In *COCOON'95*, volume 959 of *Lecture Notes in Computer Science*, pages 21–31. Springer-Verlag, 1995.
9. Qing-Wen Feng, Robert F. Cohen, and Peter Eades. Planarity for clustered graphs. In *ESA'95*, volume 979 of *Lecture Notes in Computer Science*, pages 213–226. Springer-Verlag, 1995.
10. E.R. Gansner, S.C. North, and K.P. Vo. Dag – a program that draws directed graphs. *Software – Practice and Experience*, 18(11):1047–1062, 1988.
11. D. Harel. On visual formalisms. *Communications of the ACM*, 31(5):514–530, 1988.
12. Wei Lai. *Building Interactive Digram Applications*. PhD thesis, Department of Computer Science, University of Newcastle, Callaghan, New South Wales, Australia, 2308, June 1993.
13. Thomas Lengauer. Hierarchical planarity testing algorithms. *Journal of ACM*, 36:474–509, 1989.
14. Xuemin Lin. *Analysis of Algorithms for Drawing Graphs*. PhD thesis, Department of Computer Science, University of Queensland, Australia, 1992.
15. Franco P. Preparata and Michael I. Shamos. *Computational geometry: an introduction*. Springer-Verlag, New York, 1985.
16. R. Read. Methods for computer display and manipulation of graphs and the corresponding algorithms. Technical Report 86-12, Faculty of Mathematics, Univ. of Waterloo, July 1986.
17. S.K. Stein. Convex maps. *Proceedings American Mathematical Society*, 2:464–466, 1951.
18. K. Sugiyama and K. Misue. Visualization of structural information: Automatic drawing of compound digraphs. *IEEE Transactions on Systems, Man and Cybernetics*, 21(4):876–892, 1991.

19. K. Sugiyama, S. Tagawa, and M. Toda. Methods for visual understanding of hierarchical systems. *IEEE Transactions on Systems, Man and Cybernetics*, SMC-11(2):109–125, 1981.

20. W.T. Tutte. How to draw a graph. *Proceedings London Mathematical Society*, 3(13):743–768, 1963.

21. K. Wagner. Bemerkungen zum vierfarbenproblem. *Jber. Deutsch. Math.-Verein*, 46:26–32, 1936.

Graph-Drawing Contest Report

Peter Eades[1], Joe Marks[2], and Stephen North[3]

[1] Department of Computer Science
University of Newcastle
University Drive – Callaghan
NSW 2308, Australia
E-mail: eades@cs.newcastle.edu.au
[2] MERL–A Mitsubishi Electric Research Laboratory
201 Broadway
Cambridge, MA 02139, U.S.A.
E-mail: marks@merl.com
[3] AT&T Research
600 Mountain Avenue
Murray Hill, NJ 07974, U.S.A.
E-mail: north@research.att.com

Abstract. This report describes the the Third Annual Graph Drawing Contest, held in conjunction with the 1996 Graph Drawing Symposium in Berkeley, California. The purpose of the contest is to monitor and challenge the current state of the art in graph-drawing technology.

1 Introduction

Text descriptions of the four graphs for the 1996 contest can be found on the World Wide Web at URL `www.research.att.com/conf/gd96/contest.html`. Graph A represents a finite automaton used in a natural-language processing system. Graph B represents the calls made between a set of telephone numbers. Graph C is an artificial graph that was designed as a special challenge for standard algorithms. Graph D represents the structure and content of a fragment of the World Wide Web. An effective graph drawing had to communicate not only the edge connections between vertices, but also any vertex- or edge-attribute values peculiar to the graph. Thus the main judging criterion was one of information visualization.

Approximately 35 graphs were submitted by the contest deadline. The winners were selected by a panel of judges, and are shown below.

2 Winning submissions and honorable mentions

2.1 Graph A

This directed graph contains 1,096 nodes and 1,691 edges. Each node is either a terminal or a nonterminal node, and each edge is labeled with a single character. It depicts part of a finite automaton used in a natural-language processing system. Our intention was to award separate prizes for the best overall drawing of this graph and for the best distorted drawing that emphasized a particular node.

However, only four submissions (two each for the regular and distorted views) were received, and the judges felt that none of them was good enough to win. But because of the challenge offered by this graph, the judges awarded special honorable mentions to Gilles Paris (paris@ireq.ca) of IREQ Institut de recherche d'Hydro-Quebec, Canada, and to Falk Schreiber and Carsten Friedrich ([schreibe, friedric]@fmi.uni-passau.de) of Universität Passau, Germany. Paris submitted three-dimensional color drawings of the graph. Schreiber and Friedrich did not draw the graph explicitly, but instead decompiled it by listing all the words that could be spelled out by traversing the graph's edges.

2.2 Graph B

This graph contains 111 nodes and 193 edges. It was extracted from a large telephone-call database by a utility that finds connected components of graphs in external storage. Graphs like this are used by the police in the investigation of telephone fraud and other criminal activities. For obvious reasons, random numbers were substituted for real numbers. However, the area codes are actual area codes for the United States and Canada.

The winning drawing for Graph B, shown in Figure 1, was submitted by Ulrich Fößmeier and Michael Kaufmann ([foessmei, mk]@informatik.uni-tuebingen.-de) of Universität Tübingen. An initial drawing was generated by an algorithm for finding partially layered representations of planar bipartite graphs.[4] The final version of the drawing was refined manually.

An honorable mention for this graph was awarded to François Bertault (Francois.Bertault@loria.fr) of CRIN/INRIA-Lorraine, France. His drawing is shown in Figure 2. The layout algorithm used was a spring method. Node positions were adjusted manually, and nodes are color-coded according to the area code of the corresponding telephone number.[5]

An honorable mention was also awarded to Vladimir Batagelj and Andrej Mrvar ([vladimir.batagelj, andrej.mrvar]@uni-lj.si) from the University of Ljubljana, Slovenia, for the drawing in Figure 3. This layout was obtained by a program that positions vertices on a rectangular net so as to minimize edge crossings. Nodes were repositioned manually. Color coding and a key (not shown) associate nodes with area codes and telephone numbers, respectively.

2.3 Graph C

Unlike the other graphs, Graph C was contrived without reference to a real-world application. It contains 65 nodes and 125 edges. The winning drawing was

---

[4] Graph B was in fact the inspiration for developing this algorithm.

[5] To obtain a color hard copy or a PostScript version of this report, please contact Joe Marks (marks@merl.com).

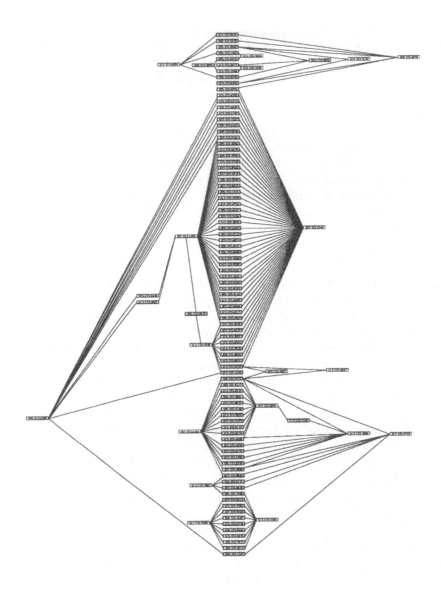

Fig. 1. Winner, Graph B.

submitted by Vladimir Batagelj and Andrej Mrvar ([vladimir.batagelj, andrej.-mrvar]@uni-lj.si) from the University of Ljubljana, Slovenia. It appears in Figure 4. The graph was first partitioned into two parts automatically, and then each part was drawn using an energy-minimization approach. Some manual editing of the planar portion of the graph was also done.

The three honorable mentions (Figures 5-7) have approximately the same visual structure as the winning drawing. Figure 5 is the work of Falk Schreiber and

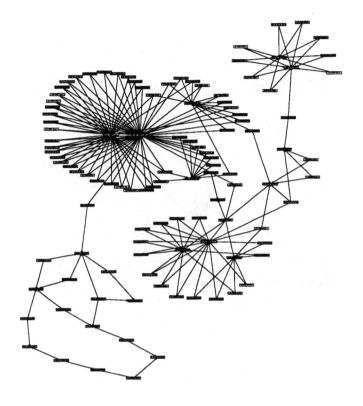

Fig. 2. Honorable mention, Graph B (original in color).

Carsten Friedrich ([schreibe, friedric]@fmi.uni-passau.de) of Universität Passau, Germany. The layout results from a spring method. Figure 6 is due to Günter Rote (rote@opt.math.tu-graz.ac.at) from the Technische Universität Graz, Austria. The layout techniques used to produce the drawing were not described in the submission. Lastly, Figure 7 was submitted by François Bertault (Francois.-Bertault@loria.fr) of CRIN/INRIA-Lorraine, France. A spring algorithm was used to separate the graph into two components. The layout of the planar component was found by first computing a planar embedding, and then applying a spring method that conserves planarity. The grid component was also handled by the spring method. Finally, the curved edges were added by hand.

2.4   Graph D

This directed graph contains 180 nodes and 229 edges. It represents some of AT&T's WWW sites and their contents. Each node represents either a URL, a text label, or an image; the node type can be inferred from the node's text label. So although the graph is relatively small, the node-attribute data make for a challenging visualization task.

The two best submissions for this graph took basically the same approach, which is to allow the user to view subsets of the graph interactively. Figure 8

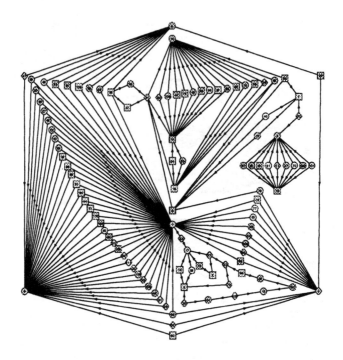

Fig. 3. Honorable mention, Graph B (original in color).

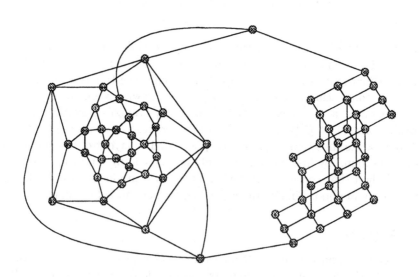

Fig. 4. Winner, Graph C (original in color).

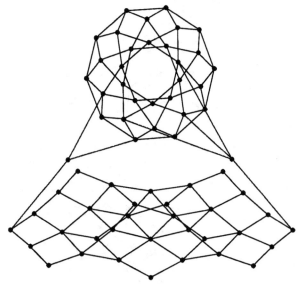

Fig. 5. Honorable mention, Graph C (original in color).

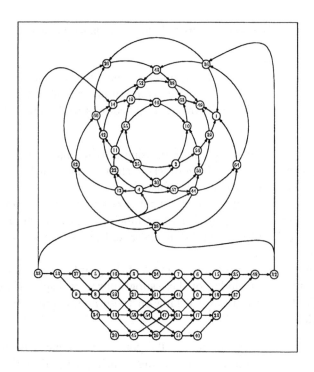

Fig. 6. Honorable mention, Graph C.

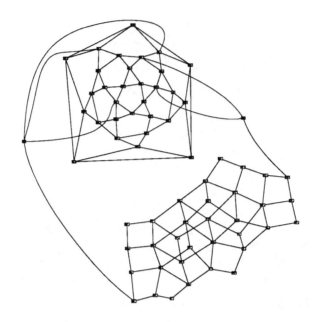

Fig. 7. Honorable mention, Graph C (original in color).

contains a screen snapshot from the winning system, developed by Falk Schreiber and Carsten Friedrich ([schreibe, friedric]@fmi.uni-passau.de) of Universität Passau, Germany. They made the following modifications to the graph before computing layouts:

1. Node clusters were identified initially using a spring method.
2. Nodes that were referenced from different clusters and which had no successor were duplicated in each of the clusters.
3. Nodes whose labels had the prefix "`http://www.att.com/`" were displayed as an AT&T icon, which eased the text-labeling task considerably.
4. Some nodes with exactly one predecessor and one successor were replaced with a labeled edge.
5. Clusters that were connected to the rest of the graph via just one node were made into subgraphs. Nodes for these subgraphs were displayed large AT&T icons in the top-level graph. Clicking on these nodes causes the subgraphs to be displayed.

The subgraphs were drawn automatically (for the most part) by a Sugiyama algorithm. The main graph was drawn using a spring method, with subsequent modification by hand.

The drawing in Figure 9 was submitted by Thomas Kamps, Jörg Kleinz, and Thomas Reichenberger ([kamps, kleinz, reichen]@darmstadt.gmd.de) from IPSI, GMD Darmstadt, Germany. They also made similar structural changes to the graph to enable it to be visualized and explored interactively. All layouts were

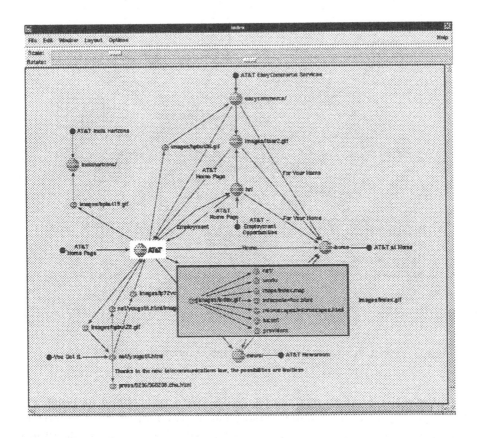

Fig. 8. Winner, Graph D.

computed using a spring method, with the exception of the drawing of the curved edges and the label abbreviation, which were done by hand. A screen shot of their system is shown in Figure 9.

3 Observations and Conclusions

Our first observation is that Graph A proved to be too challenging. It has 5 to 15 times as many nodes as the other graphs, and it also has labeled edges, which are not handled well by most current graph-drawing systems. We had hoped that the graph would serve to showcase the capabilities of distorted-view graph drawing, but no entries of this kind were submitted. Nevertheless, the graph may serve well as a near-term challenge for the next generation of graph-drawing software and may return in future contests.

The widespread use of spring methods among the better submissions is our next observation. A majority of the winning or honorable-mention drawings

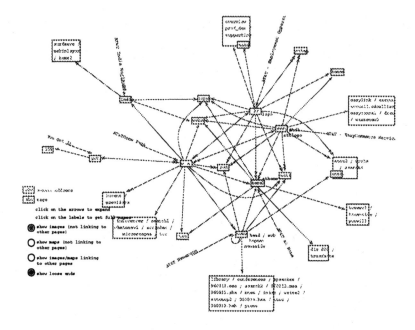

Fig. 9. Honorable mention, Graph D (original in color).

made use of the spring concept. Spring methods were used not only for producing final layouts, but also for initial exploration of graph structure. The insights gained from this exploration sometimes suggested other, non-spring algorithms for producing the final layout. The widespread use of the spring method as an investigatory layout technique seems to be new, and worthy of note.

A third observation is the lack of success achieved by orthogonal-edge drawings and three-dimensional drawings. In the former case, this year's results may be anomalous: in the two previous contests, orthogonal-edge drawings have figured prominently among the prize winners [1, 2]. However, no three-dimensional drawing has ever been awarded a prize in any of the three contests to date (we are not counting the special honorable mention given to a three-dimensional drawing of this year's Graph A – see above). The sample – a total of 11 graphs – may be too small to draw any strong conclusions,[6] but what evidence there is suggests that static three-dimensional graph drawings are not very effective at all.

As in previous years, the winners and honorable mentions often combined automatic layout and manual fine-tuning, which the rules allow. We suspect that the extent of manual modification and the editing tools used to do it vary greatly from one submission to the next, but we have had no good way of accurately

---

[6] To be fair, we note that contest judging is done from static page-size drawings (color or grayscale, as appropriate), which certainly does not capture the full effect of using an interactive three-dimensional viewer.

classifying and quantifying post factum editing.

Lastly, we note how well interaction was used to visualize Graph D. Three-dimensional graph drawing may be a very good idea if the third dimension is temporal, not spatial! A graph that would be near impossible to explore and comprehend as a single drawing was made quite accessible in two well-designed interactive contexts that use a discrete "expand/contract" metaphor for navigating the graph.

The two final observations lead to our main conclusion, which is that future graph-drawing contests need to encourage and better accommodate interactive graph-drawing systems. Existing two-dimensional methods can be made more effective in well-designed interactive systems; interactivity may be essential for making three-dimensional graph drawing useful. We are investigating the possibility of introducing a separate category of video submissions next year as a way to foster research into interactive graph drawing.

4 Acknowledgments

Sponsorship for this contest was provided by AT&T Research, Fujitsu, MERL–A Mitsubishi Electric Research Laboratory, and Tom Sawyer Software. Takao Nishizeki, Yanni Tollis, and Sue Whitesides assisted with the judging. Graph A was contributed by Emmanuel Roche and Yves Schabes of MERL, and Graphs B and D was contributed by Eleftherios Koutsofios of AT&T. Patrick Garvan from the University of Newcastle helped to devise Graph C.

References

1. P. Eades and J. Marks. Graph-drawing contest report. In R. Tamassia and I. G. Tollis, editors, *Lecture Notes in Computer Science: 894 (Proceedings of the DIMACS International Workshop on Graph Drawing '94)*, pages 143–146, Berlin Heidelberg, October 1994. Springer-Verlag.
2. P. Eades and J. Marks. Graph-drawing contest report. In F. J. Brandenburg, editor, *Lecture Notes in Computer Science: 1027 (Proceedings of the Symposium on Graph Drawing GD '95)*, pages 224–233, Berlin Heidelberg, September 1995. Springer-Verlag.

Two Algorithms for Three Dimensional Orthogonal Graph Drawing

Peter Eades[1] and Antonios Symvonis[2] and Sue Whitesides[3]

[1] Dept. of Computer Science, University of Newcastle, N.S.W. 2308, Australia;
eades@cs.newcastle.edu.au; supported by ARC
[2] Basser Dept. of Computer Science, University of Sydney, N.S.W. 2006, Australia
symvonis@cs.su.oz.au; supported by an ARC Institutional grant
[3] School of Computer Science, McGill University, Montreal, Canada H3A 2A7;
sue@cs.mcgill.ca; supported by FCAR and NSERC

Abstract. We use basic results from graph theory to design two algorithms for constructing 3-dimensional, intersection-free orthogonal grid drawings of n vertex graphs of maximum degree 6. Our first algorithm gives drawings bounded by an $O(\sqrt{n}) \times O(\sqrt{n}) \times O(\sqrt{n})$ box; each edge route contains at most 7 bends. The best previous result generated edge routes containing up to 16 bends per route. Our second algorithm gives drawings having at most 3 bends per edge route. The drawings lie in an $O(n) \times O(n) \times O(n)$ bounding box. Together, the two algorithms initiate the study of bend/bounding box trade-off issues for 3-dimensional grid drawings.

1 Introduction

The 3-dimensional orthogonal grid consists of *grid points* whose coordinates are all integers, together with the axis-parallel *grid lines* determined by these points. A *3-dimensional orthogonal grid drawing* of a graph G places the vertices of G at grid points and routes the edges of G along sequences of contiguous segments contained in the grid lines. Edge routes are allowed to contain bends but are not allowed to cross or to overlap, i.e., no internal point, not necessarily a grid point, of one edge route may lie in any other edge route. Throughout this paper, *grid* refers to the 3-dimensional orthogonal grid, and *grid drawing* refers to the type of 3-dimensional orthogonal grid drawing just described. Note that because each grid point lies at the intersection of three grid lines, any graph that admits a grid drawing necessarily has maximum vertex degree at most 6.

Figure 1 shows a grid drawing of a graph. This particular drawing lies in a $2 \times 3 \times 2$ bounding box. The edge route joining the two extremal vertices in the Z-direction lies along the top, back and bottom faces of the box and contains 2 bends. The edge route joining the two extremal vertices in the X-direction also contains 2 bends, but passes through the interior of the box.

While the graph drawing literature has extensively investigated 2-dimensional grid drawings of graphs (see [7]), 3-dimensional grid drawing has been little studied. Our research is motivated in part by recent interest in exploring the utility of 3-dimensional drawings of graphs for visualization purposes. It should also be

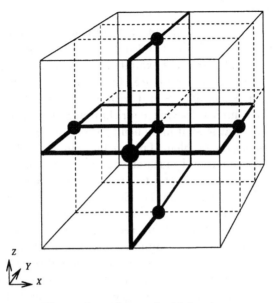

Fig. 1. *An orthogonal grid drawing*

noted that since VLSI technology now permits the stacking of many layers, this work may be relevant to that application area as well.

This paper offers two algorithms for obtaining grid drawings. Both algorithms use basic graph theory methods to preprocess the input graph by colouring its edges; then the algorithms route edges according to their colour class. One algorithm produces drawings with a small number of bends per edge, while the other algorithm produces more compact drawings, but at the cost of an increased number of bends per edge. This raises for 3-dimensional grid drawing the same kind of bend versus bounding box trade-off issues that have been studied in the 2-dimensional case. Our algorithmic results establish upper bounds for the number of bends per route on the one hand, and for various measures of the bounding box on the other hand.

Since there are many measures of bounding box compactness (e.g., volume, maximum dimension, sum of dimensions, length of long diagonal) we simply give the dimensions of the bounding box of the drawings produced by our algorithms, and we do not define compactness precisely. Note that volume is generally not the most appropriate measure of bounding box suitability for 3-dimensional drawings to be displayed in projection on screen or paper for visualization purposes.

Our first algorithm takes as input any n-vertex graph of maximum degree at most 6 and produces a grid drawing bounded by a box of dimensions $O(\sqrt{n}) \times O(\sqrt{n}) \times O(\sqrt{n})$. Each edge route has length $O(\sqrt{n})$ and contains at most 7 bends. This improves the best previously known result [9] of 16 bends maximum per edge route, described further below. Komolgorov and Bardzin ([16]; see also [9]) showed that no algorithm can produce asymptotically more compact drawings; hence we refer to our first algorithm as the *compact drawing* algorithm.

Our second algorithm produces drawings having only 3 bends maximum per edge route. However, the bounding box of the drawing increases to one of dimensions $O(n) \times O(n) \times O(n)$. We refer to this algorithm as the *3-bends* algorithm. The maximum of 3 bends per route exhibited by the 3-bends algorithm may be best possible; we conjecture the existence of a graph of maximum degree at most 6 that cannot be embedded in the grid with a maximum of 2 bends per edge route. However, the compact drawing algorithm suggests that the bounding box can be reduced in size, at least if more bends per route are admitted.

For two dimensions, the problem of producing compact orthogonal grid drawings with few bends has received much attention. Several methods for obtaining crossing-free drawings in area $O(n) \times O(n)$ with $O(1)$ bends per edge are available, e.g. [3, 4, 11, 15, 21, 23, 24, 25, 17]. Some recent research aims at reducing the number of bends while allowing crossings between edges, e.g. [5].

Not surprisingly, problems that are seemingly computationally intractable arise in both 2-dimensional and 3-dimensional grid drawing. For example, in two dimensions, minimizing the number of bends is NP-complete [12], but can be solved in polynomial time for any fixed planar embedding [22]. Eades, Stirk and Whitesides [9] have shown how to generalize various 2-dimensional NP-completeness results such as minimizing the number of bends, the volume of the drawing, and the maximum individual edge route length [12, 8, 2] to 3 dimensions.

As cited above, [9] provided an algorithm, based on the technique of Komolgorov and Bardzin [16], for obtaining 3-dimensional orthogonal grid drawings. The algorithm produces drawings for n-vertex, maximum degree at most 6 graphs in which each edge route has at most 16 bends and has $O(\sqrt{n})$ length. The drawing lies in an $O(\sqrt{n}) \times O(\sqrt{n}) \times O(\sqrt{n})$ bounding box. Thus the compact drawing algorithm we present here reduces the number of bends per edge route to a maximum of 7 while still achieving the bounding box dimensions and maximum edge route length obtained by [9].

Biedl (private communication) has shown that drawings with similar bounds on the edge length and bounding box dimensions can be obtained by using the techniques of 3-dimensional VLSI layout from the early 1980's [18, 19, 20].

The rest of this paper is organized as follows. Section 2 gives the simple graph theoretic methods that our two algorithms use to preprocess the input graph. Sections 3 and 4 present and analyse the compact drawing algorithm and the 3-bends algorithm, respectively. Section 5 concludes.

2 Preliminaries

This section gives the preprocessing step, based on elementary graph theoretic methods, that both the compact drawing algorithm and the 3-bends algorithm employ. First we recall a definition from graph theory.

Definition 1. A *cycle cover* of a directed graph is a spanning subgraph that consists of directed cycles.

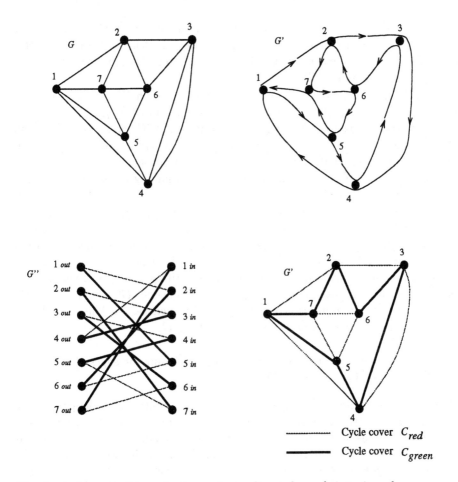

Fig. 2. *A decomposition of a 4-regular undirected graph into 2 cycle covers.*

Theorem 2. *Suppose that $G = (V, E)$ is an undirected graph of maximum degree at most 6. Then there is a directed graph $G' = (V', E')$ on the same set of vertices $V = V'$ such that*

- *each vertex of G' has indegree at most 3 and outdegree at most 3;*
- *G is a subgraph of the underlying undirected graph of G'; and*
- *the arcs of G' can be partitioned into three edge disjoint cycle covers.*

Furthermore, given $G = (V, E)$, the directed graph G' and its three cycle covers can be computed in $O(n^{3/2})$ time, where n is the number of vertices.

Proof. Pair the odd degree vertices of G and add a new edge for each pair. This can be done since the number of vertices of odd degree in any graph is even. After the addition of these new edges, each vertex has even degree at most 6. Add one self-loop (v, v) to each $v \in V$ of degree 4, and add two self-loops to each vertex of degree 2 to create a regular graph (with self-loops and multiple edges allowed) of degree 6. This graph is Eulerian, since each of its vertices has even degree. Direct its edges by following an Eulerian circuit to obtain a directed graph G' with indegree 3 and outdegree 3 at each node. Clearly these operations can be performed in $O(|V|)$ time.

Now from G', construct an undirected bipartite graph $G'' = (V_{out} \cup V_{in}, E'')$ by taking $V_{out} = \{v_{out} \mid v \in V\}$, $V_{in} = \{v_{in} \mid v \in V\}$, and $E'' = \{(u_{out}, v_{in}) \mid (u, v) \in E'\}$. Note that G'' is 3-regular and bipartite. By Hall's Theorem [13][10, p. 138] G'' contains a perfect matching; colour its edges *red* and remove them. The remaining graph is bipartite and 2-regular, so it again contains a perfect matching. Colour its edges *green* and remove them. The remaining edges form a perfect matching; colour them *blue*.

Now colour each directed arc (u, v) of G' with the colour given to (u_{out}, v_{in}) of G''. This gives each node of G' exactly one incoming and one outgoing arc of each of the three colours. Hence the arcs of G' are partitioned into three coloured subgraphs $C_{red}, C_{green}, C_{blue}$, each of which is a cycle cover for G'. Since a maximum matching in an arbitrary bipartite graph with n vertices and m edges can be computed in $O(m\sqrt{n})$ [14], and since for G'' we have $n = 2|V|$ and $m = 3|V|$, the computation of the three cycle covers can be done in $O(|V|^{3/2})$ time, including the time used to compute G' from G.

Figure 2 shows the analogous process of partitioning the edges of a 4-regular undirected graph into two cycle covers.

Preprocessing Algorithm: We call the algorithm contained in the proof of the previous theorem the *preprocessing algorithm*. To summarize, the preprocessing algorithm takes as input an undirected graph G of maximum degree 6 and computes as output a directed graph G' whose underlying undirected graph contains G; the preprocessing algorithm also computes a partition of the arcs of G' into three edge disjoint cycle covers, denoted $C_{red}, C_{green}, C_{blue}$. The algorithm runs in $O(|V|^{3/2})$ time.

Both the compact drawing algorithm and the 3-bends algorithm specify arc routes for the three cycle covers of G'. To obtain a drawing for G, the algorithms route the undirected edges of G according to the routes for the corresponding directed arcs of G'. Self-loops and arcs of G' that do not arise from edges of G are simply not drawn.

In the following sections, it is helpful to keep in mind that each node of G' has exactly one incoming and exactly one outgoing arc of each colour.

3   Compact Drawing

This section gives our compact drawing algorithm, which takes an input a graph $G = (V, E)$ of maximum degree at most 6 and produces as output a grid drawing for G having at most 7 bends per edge, maximum edge length $12\sqrt{n}$, and bounding box dimensions $(3\lceil\sqrt{n}\rceil + 2) \times 5\lceil\sqrt{n}\rceil \times 4\lceil\sqrt{n}\rceil$, where $|V| = n$.

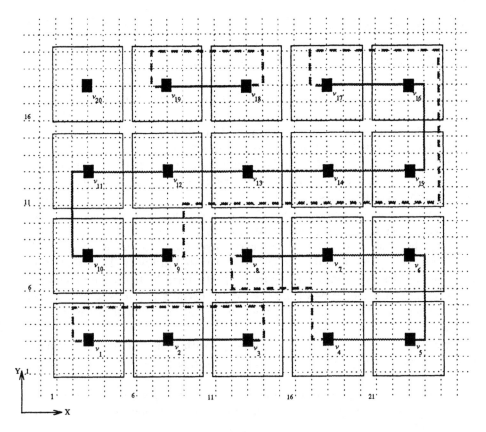

Fig. 3. *The layout of the vertices of G and the routing of edges in cycle cover* C_{red}.

Overview of the compact drawing algorithm:
The compact drawing algorithm has the following basic steps, the first of which has already been described and the rest of which are described in detail in subsequent subsections.

1. Run the preprocessing algorithm of Section 2 to construct directed graph G' and to obtain a partition of its arcs into three arc disjoint coloured cycle covers, denoted C_{red}, C_{blue}, and C_{green}.
2. Use cycle cover C_{red} to place the nodes $V' = V$ on the $Z = 0$ plane; design routes for the arcs in C_{red} that do not leave this plane and that have at most 7 bends per route.
3. Design routes for the arcs in cycle cover C_{blue} that lie on and above the $Z = 0$ plane and that have at most 7 bends per route.
4. Design routes for the arcs in cycle cover C_{green} that lie on and below the $Z = 0$ plane and that have at most 7 bends per route.
5. Draw the routes for the arcs of G' that arise from edges of G.

3.1 Processing C_{red}

Suppose cycle cover C_{red} consists of k directed cycles $c_1, c_2, \ldots, c_k$, and use these cycles to order the nodes of G' as follows. Arbitrarily choose a starting node from c_1, and order the remaining nodes of c_1 by following the cycle; then order the nodes of the remaining cycles in similar fashion, ordering the the nodes of c_i before those of c_j if $i < j$.

Next, define a square array of special grid points $p_{i,j}$ in the $Z = 0$ plane as follows. For $0 \leq i, j < \lceil \sqrt{n} \rceil$, $p_{i,j} = (5i + 3, 5j + 3, 0)$.

Rather than give explicit formulas for node placement and arc routing, we illustrate this in the context of a specific example from which the reader can easily infer the general rules.

Consider the following cycle cover C_{red}: $c_1 = \langle v_1, v_2, v_3 \rangle$, $c_2 = \langle v_4, \ldots, v_8 \rangle$, $c_3 = \langle v_9, \ldots, v_{17} \rangle$, $c_4 = \langle v_{18}, v_{19} \rangle$, and $c_5 = \langle v_{20} \rangle$.

Using the order just obtained from cycle cover C_{red}, assign the nodes of G' to grid points $p_{i,j}$ in the snake-like fashion illustrated in Fig. 3. Then route the arcs of the cycles as shown in the figure. In particular, the figure shows how to handle cycles whose nodes lie within one row of special grid points, cycles whose nodes lie in parts of two neighboring rows, and cycles whose nodes occupy more than one row.

By $Sq(i,j)$ we refer to the set of grid points within the square of side length 4 centered at special grid point $p_{i,j}$. That is, $Sq(i,j) = \{5i + k, 5j + l, 0) \mid 1 \leq k, l \leq 5\}$.

The squares themselves form a square array, so we speak of the *squares of row i* (rows are parallel to the X-axis) and the *squares of column j* (columns are parallel to the Y-axis).

Now we make two observations for future reference. These will be used in proving that no two arc routes intersect illegally.

Observation 1 *The routes for red arcs satisfy the following properties:*
i) The edge routes for arcs in C_{red} have at most 6 bends per route.
ii) The grid points having Y-coordinate one greater or less than the Y-coordinate of a special grid point $p_{i,j}$ lie in no red arc routes.
iii) The grid points contained in routes connecting nodes in the same cycle of C_{red} are entirely contained in the squares to which those nodes are assigned.

Observation 2 *Consider the positive length segments parallel to the Y-axis that are contained in routes for red arcs. Except for segments intersecting the first or last column of squares, these segments contain no grid points that differ by exactly 2 in X-coordinate from the X-coordinate of a special grid point $p_{i,j}$.*

3.2 Routing C_{blue} and C_{green}

Here we describe how to route the arcs in cycle cover C_{blue}. The arcs of C_{green} are routed similarly, but on the other side of the $Z=0$ plane.

The route of an arbitrary arc (v, w) of C_{blue} is illustrated in Fig. 4, where $v_{i,j}$ denotes the vertex placed at $p_{i,j}$.

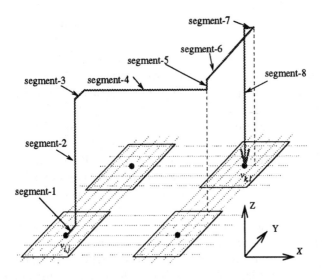

Fig. 4. *The route of arc (v, w) of π_2 with 7 bends.*

Let vertices v and w be assigned to special grid points having coordinates $(x_v, y_v, 0)$ and $(x_w, y_w, 0)$, respectively. Then the route for arc (v, w) consists of the 7 segments described in the table and shown in the figure. Here we defer until later the specification of the value z_{vw} in the table, noting for the moment that for all v, w, the value of z_{vw} will be an *odd* integer.

Segment	Start point	$\rightarrow$ finish point
1	$(x_v, y_v, 0)$	$\rightarrow (x_v, y_v + 1, 0)$
2	$(x_v, y_v + 1, 0)$	$\rightarrow (x_v, y_v + 1, z_{vw})$
3	$(x_v, y_v + 1, z_{vw})$	$\rightarrow (x_v, y_v + 2, z_{vw})$
4	$(x_v, y_v + 2, z_{vw})$	$\rightarrow (x_w + 1, y_v + 2, z_{vw})$
5	$(x_w + 1, y_v + 2, z_{vw})$	$\rightarrow (x_w + 1, y_v + 2, z_{vw} + 1)$
6	$(x_w + 1, y_v + 2, z_{vw} + 1)$	$\rightarrow (x_w + 1, y_w, z_{vw} + 1)$
7	$(x_w + 1, y_w, z_{vw} + 1)$	$\rightarrow (x_w, y_w, z_{vw} + 1)$
8	$(x_w, y_w, z_{vw} + 1)$	$\rightarrow (x_w, y_w, 0)$

3.3 Proof of Correctness

Now we prove that the routes of any two arcs do not intersect illegally.

Note that if a unit length segment contains an intersection point not located at a special grid point $p_{i,j}$, then one of its adjacent segments (of length > 1) also contains this intersection point. Thus, we need only consider the possibility of intersections of segments longer than 1, i.e, intersections among even-numbered segments.

Observe that the routing of segment-4 for every arc route takes place in plane $(*, 5j + 5, *)$, $0 \leq j < \lceil \sqrt{n} \rceil$, and that the routing of segment-6 for every arc route takes place in plane $(5i + 4, *, *)$, $0 \leq i < \lceil \sqrt{n} \rceil$. This implies that these segments cannot intersect with any segment numbered 2 or 8 from any arc route. Also, note that segment-2 obviously cannot intersect segment-8.

Observe that it is not possible for any segment-4 to intersect any segment-6. This is because the two segments are routed on parallel planes, one with even Z-coordinate, one with odd Z-coordinate. Hence the only possible intersections are between the segment-4's of two different routes or between the segment-6's of two different routes.

Finally we explain how to choose the values for z_{vw}. This is done by using the method of [9]. Consider arc (v, w). The route for this arc can intersect only with the arc routes with origin in the row of squares in which vertex v is placed and the arc routes with destination in the same column of squares in which vertex w is placed.

We construct a graph H whose vertex set is the arc set of cycle cover C_{blue}. An edge is inserted between two vertices in the graph H whenever the vertices correspond to arcs with start nodes in the same row or end nodes in the same column. H has maximum degree $2(\lceil \sqrt{n} \rceil - 1)$ and thus it has a vertex colouring by $2\lceil \sqrt{n} \rceil - 1$ colours, which can be obtained in $O(n\sqrt{n})$ time by a greedy algorithm [6, Brook's Theorem]. Say that colour $c, 1 \leq c \leq 2\lceil \sqrt{n} \rceil - 1$, has been assigned to the vertex of H that corresponds to blue arc (v, w). Then set $z_{vw} = 2c - 1$.

Now we give the main result of Section 2.

Theorem 3. *Every n-vertex maximum degree 6 graph G has a 3-dimensional, intersection-free orthogonal drawing with the following characteristics: i) at most 7 bends per edge route, ii) $(16\lceil \sqrt{n} \rceil - 7)$ maximum edge length, and iii) a bounding box of dimensions $(3\lceil \sqrt{n} \rceil + 2) \times 5\lceil \sqrt{n} \rceil \times (8\lceil \sqrt{n} \rceil - 6)$. Moreover, the drawing can be obtained in $O(n^{3/2})$ time.*

Proof. Except for their endpoints, green arc routes lie below the $Z = 0$ plane and obviously do not intersect blue arc routes illegally. Based on Observation 1 and the fact that the red arc routes lie in the $Z = 0$ plane, we conclude that they do not intersect blue or green arc routes illegally. Thus, the drawing obtained by routing edges of a graph G according to the routes for their corresponding directed arcs in G' gives a proper grid drawing.

The fact that there are at most 7 bends per edge route can be seen by inspection. The drawing fits in an axis-aligned box of dimensions $5\lceil \sqrt{n} \rceil \times 5\lceil \sqrt{n} \rceil \times (8\lceil \sqrt{n} \rceil - 6)$. The maximum possible edge route length corresponds either to a blue arc route starting from $Sq(1, \lceil \sqrt{n} \rceil)$ and going to $Sq(\lceil \sqrt{n} \rceil, 1)$ through the top of the box, or to its green counterpart directed in the opposite direction. The length of such a route is at most $(18\lceil \sqrt{n} \rceil - 9)$ units.

The maximum edge route length and the volume of the drawing can be improved by a constant factor. Based on Observation 2, each 5×5 square except the ones in the first and last column of squares can be replaced by a 3×5 rectangle. As a result of this change, the modified drawing fits in a box of dimensions $(3\lceil\sqrt{n}\rceil + 2) \times 5\lceil\sqrt{n}\rceil \times (8\lceil\sqrt{n}\rceil - 6)$ and the maximum length of a route is $(16\lceil\sqrt{n}\rceil - 7)$ units.

The time consuming parts of the compact drawing algorithm are the partition of the arcs of the 6-regular directed graph G' graph into red, green and blue cycle covers and the vertex colouring used in the routing of the blue and green arcs. Both of these operations can be completed in $O(n^{3/2})$ time.

4 The 3-Bends Algorithm

In this section we present an algorithm for constructing an orthogonal drawing of a graph $G = (V, E)$ of maximum degree at most 6 with at most 3 bends per edge. This is a substantial reduction in the number of bends from Section 3. The cost of the decrease in the number of bends is that the bounding box dimensions increase to $O(n) \times O(n) \times O(n)$, where $n = |V|$.

The 3-bends algorithm uses the preprocessing algorithm of Section 2 to obtain a 6-regular directed graph G' together with arc disjoint cycle covers C_{red}, C_{green} and C_{blue} for G'. However, it places the vertices of G (nodes of G') on the diagonal of a $3n \times 3n \times 3n$ cube. More precisely, it arbitrarily assigns numbers $1, 2, \ldots, n$ to the vertices and places vertex $a \in \{1, 2, \ldots, n\}$ at location $p_a = (3a, 3a, 3a)$.

Each pair a, b of nodes in G' determines an isothetic cube $C(a, b)$ with p_a and p_b at opposite corners. For the purpose of defining routes for possible coloured arcs of the form (a, b), we first define red, green and blue paths between p_a and p_b along the edges of the cube $C(a, b)$ as illustrated in Fig. 5. Each path of cube edges has only 2 bends.

Later, we are going to route a coloured arc (a, b) of G' close to the coloured path of cube edges of the same colour on $C(a, b)$, so that no point on the actual route for (a, b) is more than one unit away from some point on the corresponding coloured path of cube edges on $C(a, b)$. The following easy lemma shows that by doing this, we guarantee that routes for arcs that are not incident to a common node of G' do not intersect.

Lemma 4. *Suppose that $a, b, c, d \in V$ are distinct nodes of G'. Suppose that p is a point on a cube edge of $C(a, b)$ and that q is a point on a cube edge of $C(c, d)$. Then the Euclidean distance between p and q is at least 3.*

The above lemma, together with the fact that coloured paths of cube edges on the same cube get close to one another only in the vicinity of the ends of the paths, suggests that the main difficulty will be to ensure that routes do not intersect in the vicinity of their endpoints.

Given this intuition about the routing strategy, we first give an overview of the 3-bends algorithm and then give the routing details.

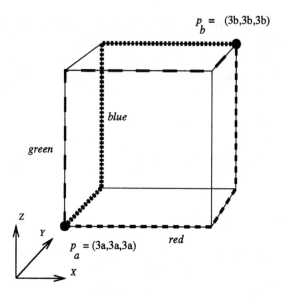

$P_b = (3b,3b,3b)$

blue

green

z

Y

$P_a = (3a,3a,3a)$

red

X

Fig. 5. *Disjoint paths*

Overview of the 3-bends algorithm:
The 3-bends algorithm has the following basic steps.

1. Use the preprocessing algorithm of Section 2 to compute the 6-regular directed graph G' and its three cycle covers C_{red}, C_{green} and C_{blue}.
2. Arbitrarily number the nodes $V' = V$ of G' 1 to $n = |V|$; for $1 \leq a \leq n$, place node a at p_a.
3. Design the routes for each coloured arc (a, b) of G' as described in detail below.
4. For each undirected edge of G, draw the route of the corresponding coloured, directed arc of G'.

To specify the arc routes in detail, it is helpful first to introduce the concept of a local minimum or maximum of a coloured cycle. Suppose, for example, that $u_1 \rightarrow u_2 \rightarrow \ldots \rightarrow u_k \rightarrow u_1$ is the red cycle through some node b of G'. Hence $b = u_i$ for some $1 \leq i \leq k$, and each u_j on the cycle is a number in the range 1 to n. The successor u_{i+1} of $b = u_i$ may be a larger or a smaller number than u_i. Hence as one moves along the red cycle, the coordinate values associated with the nodes on the cycle are sometimes increasing, sometimes decreasing. This motivates the following definition.

Definition 5. A node u_i is a *local maximum* with respect to a coloured cycle $u_1, \ldots, u_k$ if its value is greater than that of both its predecessor and its successor, i.e, if $u_{i-1} < u_i$ and $u_{i+1} < u_i$, where subscript arithmetic is modulo k.

A *local minimum* is defined analogously. An arc (u_i, u_{i+1}) is said to be *increasing* or *decreasing* if $u_i < u_{i+1}$ or $u_i > u_{i+1}$, respectively. The route for a coloured arc (u_i, u_{i+1}) will depend on whether the arc is increasing or decreasing and also, on whether u_{i+1} is a local minimum, a local maximum, or neither for that colour. Note, for example, that a node may be a local maximum with respect to one colour and a local minimum with respect to another colour.

We define four categories of arcs:

- *normal increasing arcs*: arcs (u_i, u_{i+1}) with $u_i < u_{i+1}$ and $u_{i+1} < u_{i+2}$.
- *normal decreasing arcs*: arcs (u_i, u_{i+1}) with $u_i > u_{i+1}$ and $u_{i+1} > u_{i+2}$.
- *arcs entering a local minimum*: arcs (u_i, u_{i+1}) with $u_i > u_{i+1}$ and $u_{i+1} < u_{i+2}$.
- *arcs entering a local maximum*: arcs (u_i, u_{i+1}) with $u_i < u_{i+1}$ and $u_{i+1} > u_{i+2}$.

The categories are illustrated in Fig. 6 for a cycle that passes through nodes numbered 1 to 9 with no omissions.

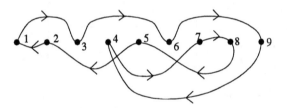

Fig. 6. *Categories*

In the cycle $1 \rightarrow 3 \rightarrow 6 \rightarrow 9 \rightarrow 4 \rightarrow 7 \rightarrow 8 \rightarrow 5 \rightarrow 2 \rightarrow 1$, arcs $(1,3)$, $(3,6)$ and $(4,7)$ are normal increasing, arcs $(8,5)$ and $(5,2)$ are normal decreasing, arcs $(9,4)$ and $(2,1)$ enter a local minimum, and arcs $(6,9)$ and $(7,8)$ enter a local maximum.

A red normal arc (u_i, u_{i+1}) (increasing or decreasing) is routed along the red path of cube edges of the cube $C(u_i, u_{i+1})$. That is, the routes for red normal arcs are:

- *normal increasing red arc*: $(3u_i, 3u_i, 3u_i) \rightarrow (3u_{i+1}, 3u_i, 3u_i) \rightarrow (3u_{i+1}, 3u_{i+1}, 3u_i)$ $\rightarrow (3u_{i+1}, 3u_{i+1}, 3u_{i+1})$ and
- *normal decreasing red arc*: $(3u_i, 3u_i, 3u_i) \rightarrow (3u_i, 3u_i, 3u_{i+1}) \rightarrow (3u_i, 3u_{i+1}, 3u_{i+1})$ $\rightarrow (3u_{i+1}, 3u_{i+1}, 3u_{i+1})$.

Note that each route for a normal red arc has two bends.

The other red arcs are routed near the red path of cube edges, but slightly offset, as described below:

- *red arcs entering a local minimum:*
  $(3u_i, 3u_i, 3u_i) \rightarrow (3u_i, 3u_i, 3u_{i+1}-1) \rightarrow (3u_i, 3u_{i+1}, 3u_{i+1}-1) \rightarrow (3u_{i+1}, 3u_{i+1}, 3u_{i+1}-1) \rightarrow (3u_{i+1}, 3u_{i+1}, 3u_{i+1})$.
  This route is illustrated in Fig. 7.
- *red arcs entering a local maximum:*
  $(3u_i, 3u_i, 3u_i) \rightarrow (3u_{i+1} + 1, 3u_i, 3u_i) \rightarrow (3u_{i+1} + 1, 3u_{i+1}, 3u_i) \rightarrow (3u_{i+1} + 1, 3u_{i+1}, 3u_{i+1}) \rightarrow (3u_{i+1}, 3u_{i+1}, 3u_{i+1})$.
  This route is illustrated in Fig. 8.

Note that each of these red arc routes has 3 bends.

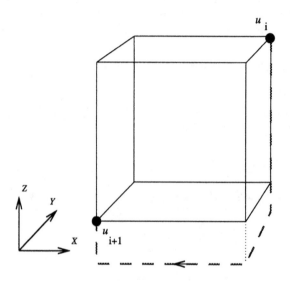

Fig. 7. *Red route into a local minimum*

The blue and green arcs are routed similarly.

Theorem 6. *Every n-vertex graph G of maximum degree at most 6 has an orthogonal grid drawing with the following characteristics: i) at most 3 bends per edge, ii) maximum edge length $9(n-1)+2$, and iii) a bounding box of dimensions $(3n - 2) \times (3n - 3) \times (3n - 2)$. Moreover, the drawing can be obtained in $O(n^{3/2})$ time.*

Proof. Observe that each segment of each arc route lies in at least one plane that is parallel to one of the coordinate planes and that contains some vertex position $p_a = (3a, 3a, 3a)$. Thus if a pair of routes intersect, they have an intersection point in such a plane. For a give generic vertex position $p_a = (3a, 3a, 3a)$, it is easy to determine the places on each of the planes $X = 3a$, $Y = 3a$ and $Z = 3a$ where route segments could possibly lie or pass through. One can exploit the similarity of the construction for routes of different colours to further simplify the task. It is straightforward to check that no illegal intersection of routes can occur.

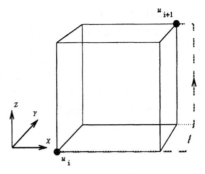

Fig. 8. *Red route into a local maximum*

The slow part of the running time is the time required to compute the cycle covers, which is $O(n^{3/2}$. The longest possible edge route would be the one connecting the extreme vertices on the diagonal. The length of this edge and the bounding box dimensions are evident.

Finally, we observe that the technique of the previous theorem can be extended in a simple manner to give a result for orthogonal grid drawing in arbitrary dimension $d \geq 3$.

Theorem 7. *Suppose G is a graph of maximum degree at most $2d$. Then there is an orthogonal grid drawing of G in dimension d with at most d bends per edge.*

5 Conclusions

We have presented two algorithms for producing grid drawings of n-vertex graphs of maximum degree 6. Both of algorithms compute an associated 6-regular directed graph G' from the input graph G together with three edge disjoint cycle covers for G'.

We note that applying this cycle cover decomposition to the algorithm of [9] eliminates the need for the dummy vertices that algorithm introduces and immediately reduces the number of bends produced from a maximum of 16 per edge route to a maximum of 8 per route. Our first algorithm further reduces this number to 7, and our second algorithm reduces it to 3, at the expense in the latter case of increased bounding box dimensions.

Our results suggest the study of trade-offs between the number of bends in routes and the dimensions of the drawing, and they contribute upper bounds for such a study.

It would be interesting to know whether a maximum of 3 bends per edge route is best possible. Note, however, that K_7, the 6-regular complete graph on 7 points, *does* have a grid drawing with at most 2 bends per edge [26].

Acknowledgment We thank David Wood for showing us that K_7 has a grid embedding with at most 2 bends per edge route; we had originally conjectured that K_7 requires at least 3 bends on some edge route.

References

1. C. Berge, *Graphs and Hypergraphs*, North Holland, Amsterdam, 1973.
2. S. Bhatt, S. Cosmadakis, "The Complexity of Minimizing Wire Lengths in VLSI Layouts", *Information Processing Letters*, Vol. 25, 1987, pp. 263–267.
3. T. Biedl, Embedding Nonplanar Graphs in the Rectangular Grid, *Rutcor Research Report 27-93*, 1993.
4. T. Biedl and G. Kant, "A Better Heuristic for Orthogonal Graph Drawings", *Proc. 2nd European Symposium on Algorithms (ESA '94)*, Lecture Notes in Computer Science, Vol. 855, Springer Verlag, 1994, pp. 24-35.
5. T. Biedl, "New Lower Bounds for Orthogonal Graph Drawings", *Graph Drawing*, Lecture Notes in Computer Science, Vol. 1027, Springer Verlag, 1995, pp. 28–39.
6. J. A. Bondy, U. S. R. Murty, *Graph Theory with Applications*, North Holland, Amsterdam, 1976.
7. G. DiBattista, P. Eades, R. Tamassia, I. Tollis, "Algorithms for Drawing Graphs: An Annotated Bibliography", *Computational Geometry: Theory and Applications*, Vol. 4, 1994, pp. 235–282.
8. D. Dolev, F. T. Leighton, H. Trickey, "Planar Embeddings of Planar Graphs", *Advances in Computing Research 2* (ed. F. P. Preparata), JAI Press Inc., Greenwich CT, USA, 1984, pp. 147–161.
9. P. Eades, C. Stirk, S. Whitesides, The Techniques of Komolgorov and Bardzin for Three Dimensional Orthogonal Graph Drawings, *TR 95-07, Dept. of Computer Science, University of Newcastle NSW, Australia*, October 1995.
10. Shimon Even, *Graph Algorithms*, Computer Science Press, 1979.
11. S. Even and G. Granot, "Rectilinear Planar Drawings with Few Bends in Each Edge", *Technical Report 797, Computer Science Department, Technion*, 1994.
12. A. Garg, R. Tamassia, On the Computational Complexity of Upward and Rectilinear Planarity Testing, TR CS-94-10, Dept. of Computer Science, Brown University, 1994.
13. P. Hall, "On Representation of Subsets", *J. London Mathematical Society*, Vol. 10, 1935, pp. 26–30.
14. J. E. Hopcroft, R. M. Karp, "An $n^{5/2}$ Algorithm for Maximum Matchings in Bipartite Graphs", *SIAM J. Comput.*, Vol. 2 (4), 1973, pp. 225–231.
15. Goos Kant, "Drawing Planar Graphs Using the lmc-Ordering", Proc. 33rd IEEE Symp. on Foundations of Computer Science, 1992, pp. 101-110.
16. A. N. Komolgorov, Ya. M. Bardzin, "About Realisation of Sets in in 3-Dimensional Space", *Problems in Cybernetics*, March 1967, pp. 261–268.
17. A. Papakostas and I. Tollis, "A Pairing Technique for Area-Efficient Orthogonal Drawings", these proceedings.
18. F. P. Preparata, "Optimal Three-Dimensional VLSI Layouts", *Mathematical Systems Theory*, Vol. 16, 1983, pp.1–8.

19. A. L. Rosenberg, "Three-Dimensional Integrated Circuitry", *Advanced Research in VLSI* (eds. Kung, Sproule, Steele), 1981, pp. 69–80.

20. A. L. Rosenberg, "Three-Dimensional VLSI: A Case Study", *Journal of the ACM*, Vol. 30 (3), 1983, pp. 397–416.

21. Markus Schäffter, "Drawing Graphs on Rectangular Grids", *Discrete Applied Math.* Vol. 63, 1995, pp. 75-89.

22. R. Tamassia, "On Embedding a Graph in the Grid with a Minimum Number of Bends", *SIAM J. Comput.*, Vol. 16 (3), 1987, pp. 421–443.

23. R. Tamassia and I. Tollis, "A Unified Approach to Visibility Representations of Planar Graphs", *Discrete and Computational Geometry*, Vol. 1, 1986, pp. 321–341.

24. R. Tamassia and I. Tollis, "Efficient Embeddings of Planar Graphs in Linear Time", *IEEE Symposium on Circuits and Systems*, 1987, pp. 495–498.

25. R. Tamassia and I. Tollis, "Planar Grid Embedding in Linear Time", *IEEE Transactions on Circuits and Systems*, Vol. 36 (9), 1989, pp. 1230–1234.

26. David Wood, "On Higher-Dimensional Orthogonal Graph Drawing", manuscript, 1996, Dept. of Computer Science, Monash U.

2-Visibility Drawings of Planar Graphs*

Ulrich Fößmeier[1]   Goos Kant[2]   Michael Kaufmann[1]

[1] Universität Tübingen, Wilhelm-Schickard-Institut, Sand 13, 72076 Tübingen, Germany
[2] Dept. of Comp. Sci., Utrecht University, Padualaan 14, 3584 CH Utrecht, Netherlands.

Abstract. In a 2-visibility drawing the vertices of a given graph are represented by rectangular boxes and the adjacency relations are expressed by horizontal and vertical lines drawn between the boxes. In this paper we want to emphasize this model as a practical alternative to other representations of graphs, and to demonstrate the quality of the produced drawings. We give several approaches, heuristics as well as provably good algorithms, to represent planar graphs within this model. To this, we present a polynomial time algorithm to compute a bend-minimum orthogonal drawing under the restriction that the number of bends at each edge is at most 1.

1 Introduction

Many algorithms for drawing graphs have been developed in the last years. Comparing them is a difficult task, because the quality of a drawing is not clearly defined, and depends highly on the application. So several models for the representation have been worked out to express the different properties of a graph [3].

One of the simplest and therefore most attractive ways of representation is to draw the edges as polygonal chains consisting of horizontal and vertical line segments. It is commonly used in the area of VLSI-design but also in data base schemes and organisational diagrams. Such drawings can be classified in various classes, where the most extreme ones are: a) *orthogonal drawings* and b) *visibility representations*. In orthogonal drawings all vertices are restricted to have a small uniform size and the edges consist of (several) horizontal and vertical segments. If there are vertices with a degree of more than four, several methods have been worked out how to solve this problem [14, 1, 11]; we adopt the model from [5]. The latter algorithm computes a drawing with the minimum number of bends preserving a given embedding. In Figure 1 we examplify the model using a graph which arises in astrophysics and was already used in the PhD. thesis of Mutzel [10]. We use this graph as a running example to distinguish the different models and approaches in this paper.

A visibility representation is a drawing where all edges are restricted to be single orthogonal line segments. No bends arise, but it is only possible to draw graphs in this way when we allow the vertices to have different sizes; it is even not possible to bound the size of the vertices. The advantages of visibility representations are: They yield very readable pictures for human spectators and the vertices have a suitable expansion in horizontal direction to write some text inside. The theory is quite developed [12, 15] for the case when the edges are restricted to be uni-directional, say

---

* This research was (partially) supported by DFG-Grant Ka812/4-1, "Graphenzeichnen und Animation" and by ESPRIT Long Term Research Project 20244 (project ALCOM IT: *Algorithms and Complexity in Information Technology*).

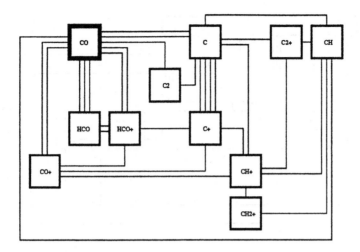

Fig. 1. An orthogonal drawing of the astro-graph

vertical. We call this model the *1-visibility* model. Several distinctions are made: The *strong* visibility model requires that the boxes can see each other if and only if the corresponding vertices are adjacent. In the *weak* visibility model they may see each other even if they are not neighbours. We will only consider the weak model.

Every planar graph can be represented in the weak model. There are efficient algorithms e.g. [7] with good bounds on the required area. 2-visibility representations are a straightforward generalization of the 1-visibility model. Here vertical *and* horizontal edges are allowed. Figure 2 shows a 2-visibility drawing for the graph of Figure 1.

Though the resulting drawings look very promising from a practical point of view, this model has not been considered very often [6, 9, 17, 18] and the known results are very preliminary. Notice that also 2-visibility representations of nonplanar graphs might be possible, though with crossing edges. In this paper we only consider planar graphs and planar representations. The purpose is to introduce this model as a practical alternative to the models used before and to demonstrate the quality of the produced drawings.

We present several heuristics and efficient algorithms to get such representations. The basic idea here is to first draw the graph orthogonally and then stretch the vertices such that they become rectangles and cover all the existing bends. Obviously we have to stretch the vertices in such a way that the final rectangles do not intersect. To make such stretchings possible we produce orthogonal drawings with special properties. Section 2 of this paper is devoted to methods for such orthogonal drawings. In Section 3 we use transform these drawings into visibility representations. Note that Tamassia & Tollis propose just the opposite way in [16]: Starting with a 1-visibility drawing they shrink the vertices and insert the corresponding edges with a number of bends to get nice orthogonal drawings.

We will present the following methods and results concerning the 2-visibility model.

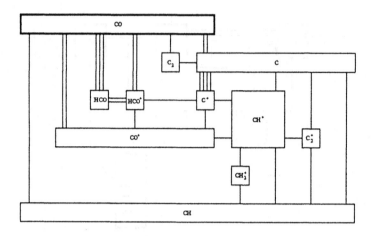

Fig. 2. A 2-visibility drawing of the astro-graph

1. In Section 2.1 we present a linear time algorithm that computes an orthogonal drawing of a planar graph with at most one bend per edge. By applying local transformations this drawing is transformed to a 1-visibility representation or a compact 2-visibility representation (Section 3.1).
2. Two polynomial time algorithms that produce 2-visibility drawings are given in Section 3.2, where the number of horizontal and vertical edges is balanced in some sence. The first algorithm is a modification of the min-cost flow approach of [13], while the latter uses the result developed in Section 2.2. There we show how to compute a minimum-bend orthogonal drawing under the restriction that each edge has at most one bend. In the resulting drawing in Section 3.2, each vertex has a uniform 'small' height.
3. In Section 4 we give an efficient algorithm based on the canonical ordering and prove upper bounds on the used area. We present a class of graphs where the 2-visibility drawing is always nearly as large as described before.
4. Finally, Section 5 contains concluding remarks and directions for further research in this practical field of graph drawings.

2 Orthogonal drawings for high degree planar graphs

In this section we present algorithms for drawing special orthogonal representations of planar graphs without any restriction on the maximum degree. We will use these results in Section 3 to achieve practical 2-visibility representations.

We take the orthogonal drawing model from [5] (notice that this model is different from the usual orthogonal drawing definition for 4-planar graphs). Because of space limitation, we refer to Figure 1 where the properties of the model can clearly be seen, instead of giving formal definitions. Most noticeably, there is at most one straight edge on each side of every vertex. The idea for achieving 2-visibility representations is that the vertices are stretched to rectangles such that the bends on the edges disappear. Note that in general this method does not work on orthogonal drawings when edges with more than one bend exist, cf. Figure 4. The stretching is possible if every edge is restricted to have at most one bend. The algorithms in our first and third approach use such orthogonal drawings as an intermediate step. The methods are probably interesting on their own.

2.1 Orthogonal drawing with at most one bend per edge

Theorem 1 *For every planar graph $G = (V, E)$ with a planar embedding there is an orthogonal drawing for G preserving the planar embedding with at most one bend on every edge and this drawing can be computed in linear time.*

Proof. At first we triangulate G, i.e., we add dummy edges to G such that every face of G is a triangle. Next we compute a *canonical ordering* for the triangulated graph G' [4]. That means that the vertices are numbered $1, \ldots, n$ such that

- the external face consists of the vertices 1, 2 and n and
- for every $i \geq 3$ there is a vertex v_i on the external face of G_i that has at least two neighbours in G_{i-1} and at least one neighbour in $G \backslash G_i$ and G_i is biconnected. The neighbours of v_i in G_{i-1} form a consecutive sequence on the outerface of the embedding of G_{i-1}. G_i is the subgraph of G' consisting of the vertices $v_1, \ldots, v_i$.

De Fraysseix, Pach & Pollack [4] show how to compute a canonical ordering of a triangulated planar graph in linear time. Our algorithm places vertices v_1 and v_2 at coordinates $(0,1)$ and $(1,0)$ and adds the rest of the vertices in the order of the canonical ordering. The vertices are placed on grid points of an integer grid such that there is only one vertex on every line and on every column. By definition of the canonical ordering, the neighbours of v_i in G_{i-1} form an interval of the external face of G_{i-1}; let v_l be the leftmost and v_r the rightmost vertex in this interval. Insert a new grid line directly below the line of v_l and a new column directly to the left of v_r, place v_i on the intersection point of the new line and the new column and draw edges with one bend per edge such that edge (v_l, v_i) is incident to v_i at its left side and all other incident edges in G_i are incident to v_i at its bottom side (see Figure 3). To ensure that this is always possible without creating crossings between edges or between an edge and a vertex we show the following invariant: The contour of the external face of G_i between v_1 and v_2 (without the edge (v_1, v_2)) is a staircase from the left to the right. It is easy to see that the new edges do not cross any old objects if the invariant holds (Figure 3) and that adding a new vertex does not destroy the invariant. This proves the theorem.

2.2 Bend-minimum drawings with at most one bend per edge

We want to apply the stretching idea to bend-minimum orthogonal drawings hoping that fewer vertices might be stretched and/or vertices are stretched by a smaller amount when we minimized the number of bends before. In this subsection we show how to produce bend-minimum orthogonal drawings under the restriction that each edge has only one bend. In Subsection 3.2 we discuss the properties of the drawing when we stretch the vertices. We also motivate this approach using the bend-minimum orthogonal drawing from Figure 1 and demonstrate in Figure 4 why it is important to restrict the number of bends per edge to be at most 1.

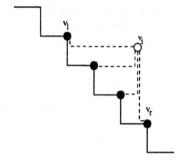

Fig. 3. A new vertex is added to G_{i-1}

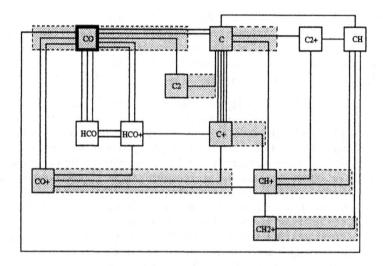

Fig. 4. From an orthogonal drawing to a visibility representation.

Theorem 2 *For every planar graph $G = (V, E)$ with a planar embedding, we can efficiently compute a bend-minimum drawing under the restrictions that the planar embedding is preserved, the area for each face is non-empty and the number of bends on each edge is at most 1.*

Proof. To restrict the drawing such that each edge has only at most one bend is easy in the case of the original model of Tamassia [13] where $0°$-angles are forbidden and the graph is 4-planar. In this case we only need to restrict the capacities of the face-to-face-arcs in the network to be 1. Then at most one bend may happen per edge since only one unit of flow may use this edge.

In the general case of higher-degree vertices we have to change the network considerably. We will motivate and describe only the changes that have to be made with respect to the approach in [5]. Note that we will again require that the faces will be represented by a polygon with a non-empty area.

The key fact again is that for each $0°$-angle there is a unique bend. We extend the number of forbidden configurations, such that e.g. two bends on the same edge corresponding to different $0°$-angles will not arise. Fortunately, the construction of the network becomes simpler than in [5]. We shortly review the original construction from [13]. The solution of such a flow problem leads to a orthogonal representation.

Tamassia defined the network N_H as follows: $N_H = (U, A, s, t, b, c)$ where $b : A \rightarrow \mathbb{R}^+$ is a nonnegative capacity function, $c : A \rightarrow \mathbb{R}$ is a cost function, U (the *nodes* of the network) $= \{s\} \cup \{t\} \cup U_V \cup U_F$, where s and t are the *source* and the *sink* of the network, U_V contains a node for every vertex of G and U_F contains a node for every face of G, A (the *arcs* of the network) contains

a) arcs from s to nodes v in U_V with cost 0 and capacity $4 - deg(v)$;

b) arcs from s to nodes f in U_F, where f represents an internal face of G with $deg(f) \leq 3$; these arcs have cost 0 and capacity $4 - deg(f)$; $deg(f)$ for a face f always denotes the number of edges in the list $H(f)$;

c) arcs from nodes f in U_F representing the external face or representing internal faces f with $deg(f) \geq 5$ to t; these arcs have cost 0 and capacity $deg(f) - 4$ if f is an internal face and capacity $deg(f) + 4$ for the external face;

d) arcs of cost 0 and capacity ∞ from nodes v in U_V to nodes f in U_F, if v is incident to an edge of $H(f)$;

e) arcs of cost 1 and capacity ∞ from a node f in U_F to a node g in U_F, whenever the faces f and g of G have at least one common edge.

Every flow unit on an arc between two faces stands for a bend on an edge between these faces. The flow on the arcs in d) defines the angles in the drawing: If $x_{v,f}$ is the flow from the node $v \in U_V$ to the node $f \in U_F$ then the angle at vertex v in face f is $(x_{v,f} + 1) \cdot 90°$. Every feasible flow of value $\Sigma_u b(s, u) = \Sigma_w b(w, t)$ with cost B leads to an orthogonal representation with exactly B bends. Thus the cost minimum solution of the flow problem corresponds to the bend minimum drawing.

Allowing $0°$-angles is easy. We extend the rules as follows:

According to the formula above such an angle corresponds to a flow of value -1 from some $v \in U_V$ to some $f \in U_F$. We interpret this as a flow of value +1 in the opposite direction, from f to v. Thus, in the network there are some additional arcs:

f) arcs of cost 0 and capacity $deg(v) - 4$ from nodes v in U_V to t, if $deg(v) \geq 5$; and

g) arcs of cost 0 and capacity 1 from a node f in U_F to a node v in U_V, whenever there is an arc of type d) from v to f.

In [5], we solved the problem that certain configurations in the network should not happen by some quite complicated modifications. Now, we also have to modify the network such that there are no two units of flow crossing one single edge. Therefore we replace the rules e) and f) by the construction shown in Figure 5.

Note that all capacities are 1 and all costs not indicated are 0. By the trick to punish the use of an arc first by costs $2c + 1$ and then to pay c costs twice back we make sure that each edge is crossed only once. A cost of 1 remains, corresponding to a single bend as before. The additional use of the nodes H_f and H_{f_e} ensures that the forbidden configurations already discussed in [5] will not occur. The introduction of nodes H_{f_e} is necessary since the flow into the face-node f across e can only go directly into the vertex-nodes adjacent to e or into the face-node f via the arc (H_{f_e}, f).

Choosing the cost parameter c sufficiently large and solving the min-cost-flow problem as usually leads to a bend-minimum orthogonal drawing of the graph.

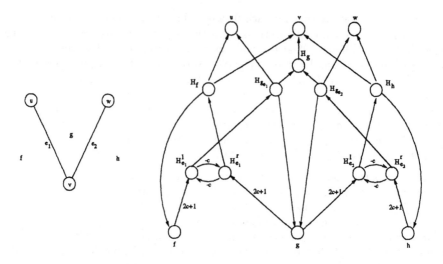

Fig. 5. Restricting the flow across each single edge

3   Three approaches to 2-visibility drawings

3.1   A first heuristic using local improvements

As a first result in this section we give an algorithm that computes a 1-visibility representation of a given planar graph preserving a given planar embedding. The idea is to start with a 1-bend orthogonal drawing as described in Section 2.1 and to delete bends by expanding vertices. In particular, the algorithm chooses a direction (w.l.o.g. horizontal), stretches the vertices horizontally such that any two neighboured edges being incident to the same vertex at its bottom side have distance at least 1 from each other; The vertex of the visibility drawing gets the y-coordinate of the vertex of the orthogonal drawing and will be extended in horizontal direction such that it covers all horizontally incident bends. The remaining edges are the vertical segments of the edges in the orthogonal drawing. It is clear that this method does not create any crossings.

This leads to a 1-visibility drawing with an area being slightly larger than the corresponding orthogonal drawing because of the first stretching of the vertices.

Our first approach to get a 'real' 2-visibility drawing (every 1-visibility drawing is a special case of a 2-visibility drawing) is to change locally the orthogonal drawing obtained by the algorithm described in Section 2.1 such that we save unnecessary bends. The resulting straight edges remain unchanged by the stretching algorithm and may run horizontally or vertically. The vertical segment of every edge that our algorithm from 2.1 creates is incident to some vertex at its bottom side; so there are two kinds of edges: Edges of type (i) are incident to the other vertex at its left side and edges of type (ii) are incident to the other vertex at its right side. So there are four ways to save a bend which are shown in Figure 6; operations (a) and (b) concern edges of type (i) and the other operations concern edges of type (ii).

A bend-saving operation can only be applied if the graph (locally) fulfills some conditions; so there are some rules to decide which operation can be realized.

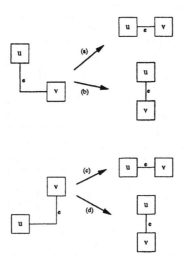

Fig. 6. How to save bends

Operation (a) can be applied if

1. e is the lowest edge being incident to vertex v at v's left side and
2. e is the rightmost edge being incident to vertex u at u's lower side and
3. there is no edge being incident to vertex u at u's right side.

Operations (b),(c) and (d) can be applied in symmetric cases. It can be seen easily that there appear no crossings if the bend-saving part of the algorithm obeys these rules. It is not hard to implement these local improvement steps such that it works in linear time overall. Applying our first algorithm to the astro-graph we get the drawing of Figure 7.

3.2 Two approaches for balanced 2-visibility drawings

In this subsection we give two more involved algorithms that lead to more balanced 2-visibility drawings. The algorithm of Section 3.1 produces vertices of uniform heights, but it clearly prefers one dimension against the other: The remainings of all the originally bending edges are drawn vertically, only edges being a result of the bend-saving step might run horizontally. Now we want to balance the two dimensions somehow. For that purpose we minimize the number of $0°$-angles between edges (a $0°$-angle arises whenever two neighboured edges are incident to a vertex at its same side). In other words: we try to use each side of the rectangle (representing the vertex) to connect edges at. Although this does not guarantee a bound for the ratio between the number of vertical edges and the number of horizontal edges, an equilibrium can be observed in practical examples.

'Non-uniform' vertices. We use a variant of the algorithms presented in Section [5] and [13], based on network flow techniques. We shortly reviewed it already in

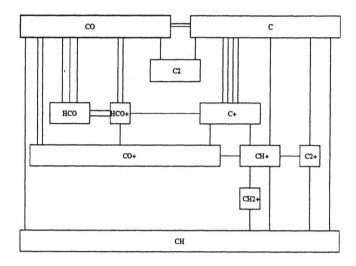

Fig. 7. Applying the local improvement algorithm to the astro-graph.

2.2. In these approaches a network is constructed having vertices and faces of the embedded planar graph as nodes. A feasible flow in this network corresponds to a drawing of the graph: A flow unit between adjacent faces characterizes a bend on an edge between these faces. A flow from a vertex v into a face f defines the angle between two edges of f having v as common vertex. In 2-visibility drawings we do not have any bends but angles of 0° are allowed arbitrarily. So we delete all arcs between two faces in the network. We add arcs from each face to each vertex being contained in the face with capacity 1 and positive cost defining the 0° angles; as a consequence a min-cost flow in this network corresponds to a no-bend drawing having a minimum number of 0° angles. Hence this yields a 2-visibility representation. Applying this algorithm to the astro graph yields the drawing of Figure 2 displayed before.

Keeping the size of the vertices 'small'. In this subsection we combine the network approach of [13] and [5], refined in Section 2.2 with the stretching idea already used in Section 3.1. We showed how to produce a bend-minimum orthogonal drawing where each edge has only one bend. Now we can easily stretch the vertices such that they cover all bends, even using only one direction (see Figure 4). Moreover, we can choose the stretching direction under some criteria like 'the used area ' or 'balance of horizontal and vertical edges' or 'sizes of the rectangles of the vertices'. Suppose we only stretch the vertices in horizontal direction. Then, since in our orthogonal drawing the vertices have a squarish shape we get a drawing where all vertices have the same 'small' height only depending on the degree of the graph. This avoids high and skinny rectangles which are possible in the second approach and enables a reasonable vertex labeling.

However, notice that the width of the rectangles can increase arbitrarily; if the user insists also on a small width which only depends on the size of the labeling, we propose the following technique: shrink the width of the vertices, such that only a few adjacent vertices are not visible anymore. Inserting the edges now with some bends is an easy task and mostly leaves the size of several vertices unchanged. Obviously, this contradicts the model of 2-visibility, nevertheless it might be a practical approach for taking the size of the labeling into account.

4 Upper and lower bounds on the area

In the following, we present an alternative linear time algorithm and analyse its behaviour with respect to the used area. For counting the area, we determine the corner coordinates of the rectangles to be integers and the rectangles to have at least a size of 1×1. The edges are placed at half-integer coordinates. So, the area in Figure 9 (b) is 6×7.

Let $G = (V, E)$ be the embedded planar graph.
If G is not triangulated, add dummy edges to it to make it so.
Compute the canonical ordering of G, denoted by $v_1, \ldots, v_n$.
Place the vertices v_1 and v_2 as boxes of size 2×1 and 1×1 in an L-shape.
for $i := 3$ **to** n **do**
    Let $v_{\alpha_1}, \ldots, v_{\alpha_k}$ be the neighbours of v_i in G_{i-1} from left to right.
    Let α_t be the maximum index, with $1 \leq t \leq k$.
    Place v_i above and/or to the right of its neighbours in the drawing of G_{i-1}
        s.t. the edge (v_{α_1}, v_i) will be horizontal and (v_{α_k}, v_i) will be vertical.
    Stretch all rectangles to the right or to the top such that
        they can see v_i if they are adjacent to v_i and such that the
        stair-case invariant is maintained.

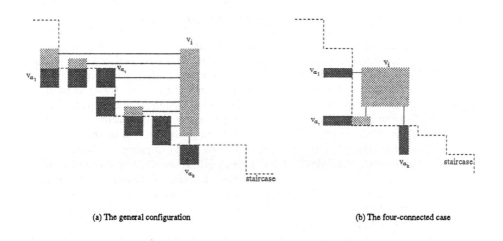

(a) The general configuration   (b) The four-connected case

Fig. 8. Two cases: Before and after insertion of vertex v_i.

From the stretching approach, described in Section 2.1, we have to come to an exact computation of every rectangle in the representation. For this computation a simple adaption of the 'Shift'-method of Chrobak & Payne yields the desired result [2]. In Figure 9 an example is given.

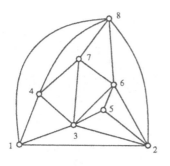

(a) A canonical numbered triangulated planar graph

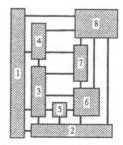

(b) The corresponding 2–visibility representation

Fig. 9. Example of the algorithm.

Observe that the layout will be stretched when inserting vertex v_i by the number of incoming edges to $v_i - 1$. The -1 comes from the fact that we do not need to stretch the highest indexed vertex v_{α_t}. Summing this up over all vertices ($\sum_i indegree(v_i) - 1 \leq 3n - 6 - n$) gives a bound of $2n$ on the sum of the height and the width of the layout. We state this result in the following

Theorem 3 *For every planar graph $G = (V, E)$ with $|V| = n$, there is a 2-visibility drawing for G that uses an area $x \times y$ where $x + y \leq 2 \cdot n$.*

The bound of $x + y \leq 2 \cdot n$ is tight, since it is not difficult to construct a triangulated planar graph, for which this algorithm indeed requires this area. Notice also that the required area heavily depends on the chosen canonical ordering. Again, it is not difficult to construct a planar graph, requiring an area of size $n \times n$ for one ordering, and an area of size $n \times 3$ for another canonical ordering. Using some more refinements of the canonical ordering, it is not hard to improve the theorem above by the fact $x \leq n$ and $y \leq n$, but it is certainly not trivial to improve the $2 \cdot n$ bound.

However, for 4-connected planar graphs we can easily improve this area bound. For this purpose, we use an observation made in [8]. When we insert vertex v_i with the incoming edges from $v_{\alpha_1}, ..., v_{\alpha_k}$ there is only one local minimum in the sequence of these vertices. Let v_{α_t} be the vertex with minimal index in the sequence. Note that in the current representation v_{α_k} lies at the point where a vertical segment of the staircase hits a horizontal one.

Here it is sufficient to stretch only this single vertex v_{α_t} by one unit such that it can see v_i, and to update the staircase-structure (see Figure 8). This means that inserting vertex v_i stretches the layout by only one unit. Moreover, similar as in the 3-connected case, we have the choice whether we stretch in horizontal or vertical direction. Hence we can balance width and height of the layout such that our result is the following:

Theorem 4 *For every 4-connected planar graph $G = (V, E)$ with $|V| = n$, there is a 2-visibility drawing for G that uses an area $x \times y$ where $x = y \leq n/2$.*

The following lower bounds shows that the upper bounds achieved by the linear time algorithms above are getting close to the lower bounds.

Theorem 5 *There is a planar graph $G = (V, E)$ with $|V| = n$, such that all 2-visibility drawings for G have an area $x \times y$ where $x + y \geq 5/3 \cdot n$, $x \geq 2/3 \cdot n$ and $y \geq 2/3 \cdot n$.*

Proof. Let G be the graph of $n/3$ nested triangles, which are internally triangulated. Note that each triangle needs at least 2 horizontal and 2 vertical lines for the realization. Since the triangles are nested, the width and the height of the layout is at least $2n/3$.

We can get an even better bound on the area if we determine the triangulation in the way as shown in the next figure. Extensive case analysis shows that for the realization of the triangulation an extra horizontal or vertical line is necessary, such that height + width is increased by 5 units when we add a new triangle.

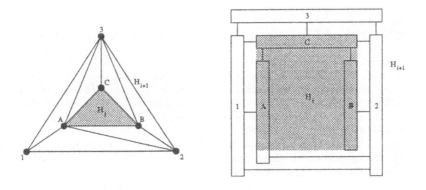

Fig. 10. The lower bound example drawn in height + width $\leq 5/3 \cdot n$

Applying the same approach for a set of nested rectangles, we get the following:

Theorem 6 *There is a 4-connected planar graph $G = (V, E)$ with $|V| = n$, such that all 2-visibility drawings for G have an area $x \times y$ where $x \geq n/2$ and $y \geq n/2$.*

5 Conclusion

In this paper we presented several practical approaches for producing 2-visibility drawings of planar graphs. The purpose of this paper is to emphasize this model and to demonstrate the quality of the resulting drawings. Moreover, we also presented theoretical upper bounds with respect to the required area. Our opinion is that that for practical purposes 'weak' visibility drawings have no disadvantages compared to 'strong' visibility drawings, and will lead to improved practical drawings.

On the other hand, there is certainly a lack of theory. Up to now there are hardly any properties and theorems derived for 2-visibility as it happened for 1-visibility. The reason is that the 2-visibility concept is much stronger as it is used here. We only used it for planar drawings, but even some thickness-two graphs (sparse non-planar graphs) can be drawn in this model, of course yielding crossings of horizontal and vertical edges. Unfortunately, unclear is the exact characterization of the class of graphs, admitting a 2-visibility representation with crossing edges. Hence, more theoretical work is required on the area bounds, as well as a precise characterization of the balance between horizontal and vertical edges. Also theory with respect to the maximum and total edge length is a field for further research.

On the practical side, it is interesting to test the behaviour of the presented algorithms for several different subclasses of planar graphs. Especially the influence of the edge balance on the size of the rectangles for the vertices and the area yields very interesting questions.

References

1. Biedl T. and G. Kant, A better heurisitic for orthogonal graph drawings, *Proc. 2nd Ann. European Symposium on Algorithms (ESA '94)*, LNCS 855, Springer-Verlag, pp. 24-35, 1994.
2. Chrobak, M., and T.H. Payne, A Linear Time Algorithm for Drawing Planar Graphs on the Grid, *Tech. Rep. UCR-CS-90-2*, Dept. of Math. and Comp. Science, University of California at Riverside, 1990.
3. Di Battista G., P. Eades, R. Tamassia and I.G. Tollis, Algorithms for automatic graph drawing: an annotated bibliography, *Computational Geometry: Theory and Practice 4*, pp. 235–282, 1994.
4. Fraysseix, H. de, J. Pach and R. Pollack, How to draw a planar graph on a grid, *Combinatorica 10*, pp. 41–51, 1990.
5. Fößmeier, U., and M. Kaufmann, Drawing high degree graphs with low bend numbers, *Proc. 4th Symposium on Graph Drawing (GD'95)*, LNCS 1027, Springer-Verlag, pp. 254-266, 1995.
6. Hutchinson, J.P., T. Shermer and A. Vince, On representation of some thickness-two graphs, *Proc. 4th Symposium on Graph Drawing (GD'95)*, LNCS 1027, Springer-Verlag, pp. 324-332, 1996.
7. Kant, G., A more compact visibility representation, *Proc. 19th Intern. Workshop on Graph-Theoretic Concepts in Comp. Science (WG'93)*, LNCS 790, Springer-Verlag, pp. 411–424, 1994.
8. Kant, G., and X. He, Two algorithms for finding rectangular duals of planar graphs, *Proc. 19th Intern. Workshop on Graph-Theoretic Concepts in Comp. Science (WG'93)*, LNCS 790, Springer-Verlag, pp. 396–410, 1994.
9. Kirkpatrick, D.G., and S.K. Wismath, Weighted visibility graphs of bars and related flow problems, *Proc. 1st Workshop Algorithms Data Structures (WADS'89)*, LNCS 382, Springer-Verlag, pp. 325-334, 1989.
10. Mutzel, P., *The Maximum Planar Subgraph Problem*, Doctoral Dissertation, Köln 1994.
11. Papakostas A. and I. Tollis, Improved algorithms and bounds for orthogonal drawings, *Proc. DIMACS Workshop on Graph Drawing (GD'94)*, LNCS 894, Springer-Verlag, pp. 40–51, 1994.
12. Rosenstiehl, P., and R.E. Tarjan, Rectilinear planar layouts and bipolar orientations of planar graphs, *Discrete Comput. Geom. 1*, pp. 343-353, 1986.

168

13. Tamassia, R., On embedding a graph in the grid with the minimum number of bends, *SIAM Journal of Computing* 16, pp. 421-444, 1987.
14. Tamassia, R., G. Di Battista and C. Batini, Automatic graph drawing and readability of diagrams, *IEEE Trans. on Systems, Man and Cybernetics* 18, pp. 61–79, 1988.
15. Tamassia R. and I. Tollis, A unified approach to visibility representations of planar graphs, *Discrete and Computational Geometry* 1, pp. 321–341, 1986.
16. Tamassia, R., and I.G. Tollis, Efficient embedding of planar graphs in linear time, in: *Proc. IEEE Int. Symp. on Circuits and Systems*, Philadelphia, pp. 495–498, 1987.
17. Thomassen, C., Rectilinear drawings of graphs, *J. Graph Theory* 12, pp. 335–341, 1988.
18. Wismath, S.K., Characterizing bar line-of-sight graphs, *Proc. 1st Annual ACM Symp. on Computational Geometry*, pp. 147–152, 1985.

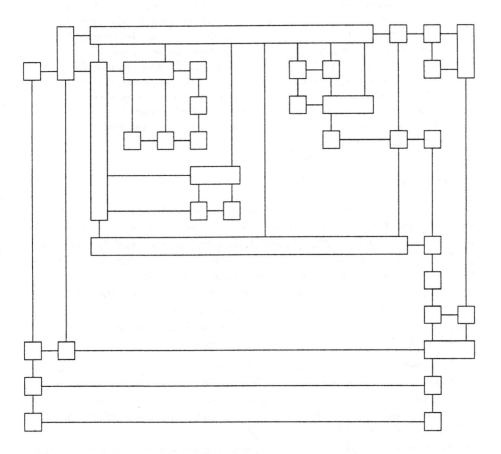

Fig. 11. Applying the first algorithm of Section 3.2 to the A-graph of the competition in GD'95 (two edges are omitted due to planarity)

Upper Bounds on the Number of Hidden Nodes in Sugiyama's Algorithm

Arne Frick[**]

Tom Sawyer Software, 804 Hearst Avenue, Berkeley CA 94710

Abstract. This paper analyzes the exact and asymptotic worst-case complexity of the *simplification* phase of SUGIYAMA's algorithm [12] for drawing arbitrary directed graphs.

The complexity of this phase is determined by the number of hidden nodes inserted. The best previously known upper bound for this number is $O(\max\{|V|^3, |E|^2\})$. This paper establishes a relation between both partial results and gives upper bounds for many classes of graphs. This is achieved by constructing a worst-case example for every *legal configuration* $C = (h, n, m)$ of the input hierarchy for the simplification phase. These results provide further insight into the worst-case runtime and space complexity of SUGIYAMA's algorithm. Possible applications include their use as feasibility criteria, based on simply derived quantitative information on the graph.

1 Introduction

SUGIYAMA's algorithm [12] is a well-known technique for drawing arbitrary directed graphs $G = (V, E)$. It is being widely used in current graph-drawing systems, such as daVinci [3], dag and dot [4], GraphEd [6], the Graph Layout Toolkit [8], Edge [9], and vcg [10]. Despite its importance and wide-spread use, little is known about the time and space complexity of several parts of the algorithm.

This paper improves this situation by analyzing the exact and asymptotic worst-case complexity of an important intermediate phase of SUGIYAMA's algorithm. There are several reasons to devote research into the complexity of the algorithm itself and into this particular phase of it.

1. It is desirable in graph drawing systems to have good estimates on the additional memory required to store information on the hidden nodes.
2. Accurate and efficiently computable *a priori* estimates on the number of hidden nodes can be used as a criterion to determine the feasibility of using the algorithm at all for a given graph.
3. The runtime complexity of the simplification phase can make the difference between usability and uselessness of SUGIYAMA's algorithm as a whole, as it is an intermediate phase that modifies the original graph, thereby influencing the complexity of the following phases.

---

[*] This work was performed while the author was employed at Universität Karlsruhe, Institut für Programmstrukturen und Datenorganisation.
[**] EMail: africk@tomsawyer.com

The remainder of this paper is organized as follows. Section 2 gives basic definitions and introduces our notation. A brief overview and discussion of SUGIYAMA's algorithm follows in Sect. 3. Section 4 discusses related work. In Sect. 5, the worst-case analysis on the maximal number of hidden nodes is given, including the main theorem. A summary and suggested directions for further research conclude the paper.

2 Fundamentals

Before the problem we consider can be stated precisely and solved subsequently, we introduce several basic definitions from graph theory and the notation used in this paper. As an overall assumption, we do not consider graphs with *multi-edges*, i.e. edges $e = (u,v), e' = (u',v') \in E$ with $u = u'$ and $v = v'$. The following two definitions recall standard notation from graph theory.

Definition 1. A *topological ordering* of a directed acyclic graph (dag) $G = (V, E)$ is a numbering

$$\lambda : v \mapsto \lambda(v) \in \mathbb{N}$$

with $\lambda(u) < \lambda(v)$ for all $e = (u,v) \in E$.

Definition 2. The length $\delta(G)$ of the longest path in an acyclic digraph G is called the *diameter* of G.

Next, the notions of a layering and, stricter, a hierarchy, are defined.

Definition 3. An *h-layering* of a directed graph (digraph) $G = (V, E)$ is a partition $V = \bigcup_{i=1\ldots h} V_i$ of the vertices of G into *layers* $V_i, i = 1, \ldots, h$ with

$$\forall (u, v) \in E, 1 \leq i, j \leq n : \quad u \in V_i, v \in V_j \Rightarrow i < j,$$

where n is the *height* of the layering. The *rank* $r(v)$ of a vertex v is defined as the index of its layer. For each edge $e = (u, v) \in E$, the difference $s(u, v) = r(v) - r(u)$ is called the *span* of the edge. Layerings with $\forall e : \quad s(e) > 0$ are called a *hierarchy*. A hierarchy is called *simple* if $\forall e \in E : \quad s(u, v) = 1$.

Hierarchies exist only for acyclic digraphs. By definition, any simple hierarchy has only edges between adjacent layers, inducing a partition of the edge set E of G:

$$E = \bigcup E_i, \quad E_i \subseteq V_i \times V_{i+1}.$$

Given a hierarchy $\mathcal{H}$ of a directed acyclic graph G, a topological ordering of G can be constructed by enumerating the vertices according to their rank r, starting with the vertices of rank 1. By placing every vertex on a different layer, every acyclic digraph with $|V| = n$ admits has an n-hierarchy.

The following Lemma combines definitions 2 and 3.

Lemma 4. *Let $\delta(G)$ be the diameter of a dag $G = (V, E)$. There is a $\delta(G) + 1$-hierarchy of G, but no $\delta(G)$-hierarchy.*

This observation leads to the following definition of a compact hierarchy.

Definition 5. A $\delta(G) + 1$-hierarchy of G is called *compact*.

For the remainder of this paper, we shall assume that drawings of hierarchies are ordered horizontally, i.e. the edges run from top to bottom, i.e. the layer belonging to rank 1 is on top of the drawing. As a direct consequence of definition 3, we have

Lemma 6. *Every acyclic digraph $G = (V, E)$ admits a compact hierarchy with configuration $C(G) = (\delta(G) + 1, n, m)$.*

Configurations $C = (h, n, m)$ induced by graphs G and an associated compact hierarchy $\mathcal{H}$ are called *legal*.

3 The Sugiyama algorithm

This section reviews SUGIYAMA's algorithm and prepares the ground to state the precise problem being solved in this paper. Fig. 1 shows a high-level description of the algorithm. The algorithm uses the aesthetic criteria

- minimization of backward edges
- minimization of the maximal edge length
- minimization of the number of edge crossings
- approximately even layer sizes

Program 1 The five phases of SUGIYAMA's algorithm.

```
-- Input:   a directed graph  G = (V, E)
-- Output:  mappings ρ, σ assigning positions to each node
--              and curves to each edge
(1) cycle breaking
(2) computation of a compact hierarchy
(3) simplification of the hierarchy
(4) reduction of crossings between adjacent layers
(5) fine-tuning the vertex positions
```

The reason for proceeding in phases instead of optimizing for all of the criteria at once can be seen as follows. It is well-known that each of these criteria is already $\mathcal{NP}$-hard to optimize for by itself. In addition, a good solution for one of the criteria may conflict with the one or more of the remaining criteria. Therefore, an algorithm cannot be expected to compute the globally optimal solution within a reasonable amount of time. The key idea of [12] is to split the global optimization into phases, and to employ heuristics in each phase to optimize for a single (or several mutually compatible) criterion. The optimization order is carefully chosen to preserve the quality of partial solutions found in prior phases as much as possible.

In phase (1), existing cycles in the input graph are broken by reversing the edges leading to cycles. The problem here is to reverse the minimal possible number of edges in order to maintain the graph structure as well as possible.

Phase (2) computes a hierarchy. Aesthetic considerations based on the criteria listed above suggest that the embedding should be compact to achieve an approximate width/height balance. Also, the total edge span $S(E) = \sum_{e \in E} s(e)$, as phase (3), to be discussed next, introduces $s(e) - 1$ new nodes into the graph. It is desirable for complexity and aesthetic reasons to keep this number low. We shall only mention some of the latter here, since the former are going to be a major concern in the remainder of this paper.

1. If edges are represented as polylines in the final drawing, then minimizing the number of hidden nodes in general also minimizes the number of bends in the polyline, another well-established aesthetic criterion for drawings of graphs.
2. If, alternatively, edges are represented as splines, then the positions of the hidden nodes can be used as interpolation points to draw the original long-span edge.
3. As a side-effect in both cases, long edges cannot cross vertices in the drawing, and the number of edge-crossings for long edges can also be reduced by phase (4).

In [4], a technique to compute a hierarchy with minimal number of hidden nodes is described. Their algorithm, however, has no *a priori* estimate on the size of this number.

As already mentioned, phase (3) introduces new nodes. Each edge e with span $s(e) > 1$ is replaced by a path of length $s(e)$ according to Prog. 2. The interior vertices of the path are called *hidden, invisible* or *dummy* nodes, for obvious reasons.

Note that hierarchy simplification does not improve the layout quality, but is an intermediate computation that adapts the output of phase (2) to the input of phase (4), which is based on the precondition that no edges of span > 1 exist. Consider the situation arising if the crossing-reduction phase for layers i and $i + 1$ could not take into account edges passing from layer $i' < i$ to layer $i'' > i$. Such edges might cross vertices and edges in layer i, going completely unnoticed and dealt with by phase (4).

Consequently, the crossing-reduction phase takes the hidden nodes into account, and its runtime complexity depends on $|V''|$ instead of $|V|$. An upper bound for the runtime complexity of phase (4) is shown as follows. Every known algorithm for the reduction of the number of edge crossings requires the computation of the *number* of crossings. As the reduction phase only considers adjacent

---

Program 2 An algorithm to simplify a hierarchy.

```
V'':=V'; E'':=E';
forall e = (u, v) ∈ G'' = (V'', E'') do
  if s(e) > 1 then
    E'' := E'' \ {e};
    d₀:=u;
    for i := 1 to s(e) do
      V'':=V'' ∪ dᵢ;
      r(dᵢ):=r(u) + i;
      E'':=E'' ∪ {(dᵢ₋₁, dᵢ)};
    od;
    E'':=E'' ∪ {dₛ₍ₑ₎, v)};
  fi;
od;
```

---

layers, it suffices to have an upper bound for an arbitrary layer i. Let V_i and V_{i+1} be the sets of vertices of the original graph on layers i and $i + 1$, respectively. Both layers are connected by a subset $E_i \subseteq E$ of edges in the original graph. If $|V_i| = |V_{i+1}| = O(1)$, the number of crossings of edges of the original graph must also be constant, which can obviously be computed in constant time.

Consider now the case that phase (3) introduces a total of $O(|V|^2)$ hidden nodes, which is quite possible, as we shall see below (cf. Sect. 4). Assume that $O(|V|)$ of these new nodes are placed on layers i and $i + 1$ in the above scenario, respectively. Using the sweep-line algorithm described in [11], the number of crossings of a bipartite graph with $|V_1|$ and $|V_2|$ can be computed in time $O(|V_1| + |V_2| + |E| + c)$, where c is the number of actual crossings. In the scenario we have developed above, this may result in quadratic runtime complexity, depending on the number of crossings in the augmented graph, thereby providing evidence of the significant impact that the number of hidden nodes introduced in phase (3) may have on the runtime complexity of phase (4).

Phase (5) involves fine-tuning the computed vertex positions and is of no further interest here.

4 Related Work

A literature survey mainly based on a current version of [1] revealed very few results on upper bounds on the number of hidden nodes. The authors of [2] state that the number of vertices may grow quadratically. This number turns out to be too optimistic, as a consideration of configuration $C = (n, n, n(n - 1)/2)$ shows. This configuration uniquely describes the complete directed graph of size n, that has a path of length $n - 1$ [5]. There exists a legal hierarchy for C with as many as $\Theta(|V|^3)$ hidden nodes, which can be shown by a simple counting argument. Another estimate, given in [7], is based on the configuration $(h, 2h - 1, 2h - 2)$.

She derives an upper bound of $O(|E|^2)$. In summary, the prior state of the art, as known to the author, can be summarized as follows.

Lemma 7. *The number of hidden nodes in an h-hierarchy of a graph $G = (V, E)$ is at most $O(\max(|V|^3, |E|^2))$.*

Note that Lemma 7 does not relate the bounds between n and m, as it only describes the one-dimensional boundary of d, depending on either n or m.

5 Analysis of the simplification phase

In this section, we analyze the worst-case runtime and space complexity of simplifying an arbitrary h-hierarchy $G' = (V, E')$, $E' \subseteq E$ of a directed acyclic graph $G = (V, E)$. This problem is equivalent to maximizing the edge span sum over all hierarchies of G and may serve as a lower bound for the worst-case complexity for the simplification phase, since simplification requires at least the replacement of all long-span edges by hidden nodes. The analysis depends on the fact that the input hierarchy G' is compact, which is guaranteed in this case by phase 2 of Prog. 1.

Program 2 showed an algorithm for simplifying an arbitrary hierarchy that replaces every edge $e = (u, v)$ of span $s(e) > 1$ by a path consisting of hidden nodes on layers $\lambda \in \{r(u) + 1, \ldots, r(v) - 1\}$, respectively. The algorithm is asymptotically optimal, if the number of set insertion and deletion operations is used as a complexity measure, since every algorithm that effectively performs a simplification of a given hierarchy must perform at least as many set operations as Prog. 2. The remainder of this section analyzes the number of hidden nodes introduced by Prog. 2. We consider the number of hidden nodes of a hierarchy as a function d of their configuration $C = (h, n, m)$.

The number h_e of hidden nodes generated by Prog. 2 for edge e in a compact hierarchy of graph G depends only on e. This allows us to consider all edges independently, which allows for a greedy incremental strategy to construct worst-case graphs for a given legal configuration $C = (h, n, m)$, which in turn helps to prove our main result:

Theorem 8. *An h-hierarchy of an acyclic digraph $G = (V, E)$ with $n = |V|, m = |E|$ and diameter $h - 1$ can have at most*

$$d(h, n, m) \leq \sum_{i=1}^{k-1} (n - h + i) \cdot (h - i - 1) + \tag{1}$$

$$\min\left\{ n - h + k, m - \sum_{j=1}^{k-1} (n - h + j) \right\} \cdot (h - k - 1) \tag{2}$$

hidden nodes, where $h \leq n, m \geq n - 1$ and

$$k = \min\left\{ h - 2, \lfloor \sqrt{(n-h)^2 + n + 2m - 3h + 9/4} - n + h + 1/2 \rfloor \right\}.$$

The proof is based on an arbitrary, but fixed choice of h and requires several further observations. In order to span an h-hierarchy, h vertices are required, which are connected by $h - 1$ edges. These form the so-called h-*spine* (cf. Fig. 1) of the hierarchy.

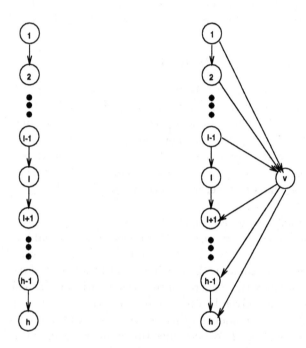

Fig. 1. The h-spine. **Fig. 2.** Adding a vertex at level l.

The spine can extended into a worst-case graph by successively adding the remaining vertices and edges, making locally optimal choices. This approach is valid by the above independence observation. An additional vertex v in layer $l, 1 \leq l < h$ can be connected with the $l - 1$ spine vertices of lower rank, and with the $h - l$ spine vertices with higher rank (cf. Fig. 2). In total, vertex v can be connected to other vertices by at most $(h - l) + (l - 1) = h - 1$ edges. The total edge span is maximized by successively adding edges from v to the farthest remaining spine vertex, until no edges remain.

For symmetry reasons, we can safely assume that

$$l - 1 \leq h - l \Leftrightarrow l \leq \lfloor h/2 \rfloor.$$

Using this assumption, the first $h - 2(l - 1) - 1 = h - 2l + 1$ edges connect to the vertices on levels i for $i = h, h - 1, h - 2, \ldots, h - 2l + 1$. The remaining $2l - 2$ edges pair-wise have the same span i for $i = 1, 2, \ldots, l - 1$.

Lemma 9. *The total edge span $S_{h,l}$ induced by connecting an additional vertex on level l to an h-spine is minimal for*

$$l = \lfloor \frac{h+1}{2} \rfloor.$$

Proof. For $h \leq \lfloor h/2 \rfloor$,

$$
\begin{aligned}
S_{h,l} &= \sum_{i=2l}^{h}(i-l) + 2\sum_{i=1}^{l-1} i \\
&= \sum_{i=l}^{h-l} i + 2\sum_{i=1}^{l-1} i \\
&= \sum_{i=1}^{h-l} i + \sum_{i=1}^{l-1} i \\
&= \frac{(h-l)(h-l+1)}{2} + \frac{(l-1)l}{2} \\
&= (h^2 - 2l + 2l^2 + h - 2l)/2.
\end{aligned}
$$

which is a quadratic function in l that has a integer minimum at $\lfloor h/2 \rfloor$. By symmetry, the same is true for $h \geq \lfloor h/2 \rfloor$.

Therefore, the maximum value is assumed at $l = 1$ and $l = h$. Together with the independence observation, which proves ▯

Corollary 10. *For arbitrary graphs G and h-hierarchies with $n > h$, the number of hidden nodes introduced is maximized iff all $n - h$ non-spine nodes v are assigned rank $r(v) = 1$ or $r(v) = h$.*

From now on, we are only concerned with hierarchies of the type characterized in Corollary 10 (see also Fig. 3). We shall map non-spine vertices to the top layer and say that such hierarchies have the *right* type. The behavior of d depending on $m = |E|$ is analyzed next. Note that $h - 1$ edges are already part of the spine, and that edges connecting spine nodes have to be considered appropriately.

Lemma 11. *Legal configurations induce hierarchies of the right type with*

$$
\begin{aligned}
& n - h + 1 \text{ edges with span } h - 2 \\
& n - h + 2 \text{ edges with span } h - 3 \\
& \qquad\qquad \vdots \\
& n - h + (k-1) \text{ edges with span } h - k \\
\min\{n - h + k, m - &\textstyle\sum_{j=1}^{k-1}(n-h+j)\} \text{ edges with span } h - (k+1),
\end{aligned}
\tag{3}
$$

where k is implicitly defined by the inequality

$$\sum_{i=1}^{k-1}(n-h+i) \leq m - h + 1 < \sum_{i=1}^{k}(n-h+i).$$

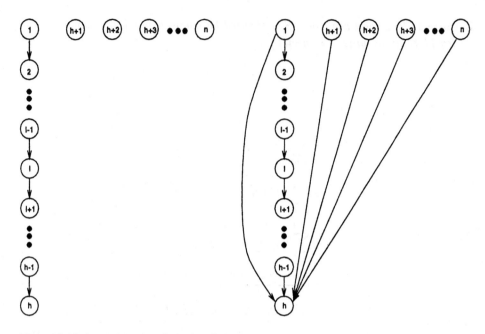

Fig. 3. Added remaining $n - h$ vertices in layer 1.

Fig. 4. Added edges with span $h - 2$.

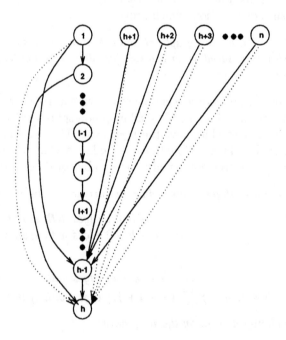

Fig. 5. Added edges with span $h - 3$.

Proof. In hierarchies of the right type, there are $n-h+1$ vertices in layer 1 that can be connected to the vertex in layer h by edges of span $h-2$ (cf. Fig. 4). For span $h-3$, in addition to the $n-h+1$ vertices in layer 1 that can be connected to the single vertex in layer $h-1$, there is another edge from 2 to h (cf. Fig. 5). This procedure is iterated through the layers of the spine until there are no more edges. In step j, we can connect the $n-h+1$ vertices in layer 1 with vertex $h-j+1$, thus creating an edge with span $h-(j+1)$. Furthermore, there exist exactly $j-1$ possibilities to place edges of span $h-(j+1)$ on the spine, totalling the claimed number.

The final index k can be determined exactly, depending only on the (legal) configuration $C = (h, n, m)$. Let $a := n-h$, $b := m-h+1$. Then

$$
\begin{aligned}
&\sum_{i=1}^{k-1}(n-h+i) &&\leq m-h+1 &&< \sum_{i=1}^{k}(n-h+i)\\
\Leftrightarrow\ &\sum_{i=1}^{k-1}(a+i) &&\leq b &&< \sum_{i=1}^{k}(a+i)\\
\Leftrightarrow\ &(k-1)a+k(k-1)/2 \leq b &&&&< ka+k(k+1)/2\\
\Leftrightarrow\ &2a(k-1)+k^2-k &&\leq 2b &&< 2ka+k^2+k\\
\Leftrightarrow\ &k^2+2k(a-1/2)-2a \leq 2b &&&&< k^2+2k(a+1/2)\\
\Leftrightarrow\ &k^2+2k(a-1/2)-2a \leq 2b+(a-1/2)^2 &&&&< k^2+2k(a+1/2)\\
&\quad +(a-1/2)^2 &&&&\quad +(a-1/2)^2\\
\Leftrightarrow\ &(k+(a-1/2))^2 &&\leq 2(b+a)+(a-1/2)^2 &&< k^2+2k(a+1/2)+\\
&&&&&\quad a^2-a+1/4+2a\\
&&&&&= (k+(a+1/2))^2\\
\Leftarrow\ &k+a-1/2 &&\leq \sqrt{2(a+b)+(a-1/2)^2} &&< k+(a+1/2)\\
\Leftrightarrow\ &k &&\leq \sqrt{2(a+b)+(a-1/2)^2} &&< k+1\\
&&&\quad -a+1/2
\end{aligned}
$$

As k is an index and therefore by definition an integer, we get for $m > n$:

$$
\begin{aligned}
k &= \lfloor \sqrt{2(a+b)+(a-1/2)^2}-a+1/2 \rfloor \qquad (4)\\
&= \lfloor 1/2\sqrt{4(n-h)^2+4n-12h+9+8m}-n+h+1/2 \rfloor.
\end{aligned}
$$

Theorem 8 is now a corollary of Lemma 11 and Equation 5. For a given configuration $C = (h, n, m)$, Theorem 8 allows the exact computation of k and d. As an application, the asymptotic upper bounds cited in Sect. 4 are derived exactly. For $C = (t, 2t-1, 2t-2)$, the substitution $m = 2t-2$ results in

$$
\begin{aligned}
d(t, 2t-1, 2t-2) &= d((m+2)/2, m+1, m)\\
a &= m/2\\
b &= m/2\\
k &= \lfloor \sqrt{2(m/2+m/2)+(m/2-1/2)^2}-m/2+1/2 \rfloor\\
&= \lfloor 1/2(\sqrt{m^2+7m+1}-m+1) \rfloor
\end{aligned}
$$

Upper and lower bound estimates for k are given by

$$k \leq \lfloor 1/2(\sqrt{m^2 + 8m + 16} - m + 1) \rfloor$$
$$= \lfloor 3/2 \rfloor = 1$$

$$k \geq \lfloor 1/2(\sqrt{m^2 + 6m + 9} - m + 1) \rfloor$$
$$= \lfloor 2/2 \rfloor = 1$$

In this case, $k = 1$ for all t, resulting in

$$d((m+2)/2, m+1, m) = \min\{m/2, m+1 - (m+2)/2 + 1\} \cdot ((m+2)/2 - 2)$$
$$= \min\{m/2, m/2 + 1\} \cdot (m/2 - 1)$$
$$= m/2(m/2 - 1) = m^2/4 - m/2.$$

The complete directed graph with n vertices must have the globally maximal number d for fixed $|V| = n$, since every other graph with $|V| = n$ is a subgraph of the complete graph. In [5] it is shown that every complete directed graph of size $|V| = n$ has a path of length $h - 1$. Therefore, the worst-case configuration $C = (n, n, n(n-1)/2)$. We know that

$$k = \lfloor 1/2\sqrt{4n(n-1) - 8n + 9} + 1/2 \rfloor$$
$$= \lfloor 1/2\sqrt{4n^2 - 12n + 9} + 1/2 \rfloor$$
$$= \lfloor \sqrt{n^2 - 3n + 9/4} + 1/2 \rfloor$$
$$= \lfloor (n - 3/2) + 1/2 \rfloor$$
$$= n - 1$$

This implies that

$$d(n, n, n(n-1)/2) = \sum_{i=1}^{n-1} i \cdot (n - i - 1)$$
$$= \sum_{i=1}^{n-2} i \cdot (n - i - 1)$$
$$= (n - 1) \sum_{i=1}^{n-2} i - \sum_{i=1}^{n-2} i^2$$
$$= (n - 1)^2(n - 2)/2 + (n - 2)(n - 1)(2n - 3)/6$$
$$= \frac{(3n^3 - 12n^2 + 15n - 6) - (2n^3 - n^2 + 13n - 6)}{6}$$
$$= \frac{1}{6}n^3 - \frac{1}{2}n^2 + \frac{1}{3}n$$

Due to the complexity of the equations for the behavior of the functions d and k, it is not possible in general to describe them in a closed form. Instead, table 1 summarizes the asymptotic behavior of k and d for configurations $C = (h, n, m)$. The table should be read as follows: $c_1, c_2 > 0$, and the choice of c_1 and c_2 must ensure that $h \leq n$ and $m \leq n(n-1)/2$. The case $h = c_1 n, m = c_2 n$ gives trivial results and is omitted from the table. Instead, we shall consider this practically important case separately and in more detail. The values were computed using the Computer Algebra system MAPLE [14]. We used the following bounds for k and d to work the estimates:

$$k \leq 1/2\sqrt{4(n-h)^2 + 4n - 12h + 9 + 8m} - n + h + 1/2 =: k_{\text{high}} \qquad (5)$$

$$k \geq 1/2\sqrt{4(n-h)^2 + 4n - 12h + 9 + 8m} - n + h - 1/2 =: k_{\text{low}} \qquad (6)$$

$$d \leq \sum_{i=1}^{k_{\text{high}}} (n - h + i)(h - i - 1) =: d_{\text{high}} \qquad (7)$$

$$d \geq \sum_{i=1}^{k_{\text{low}}-1} (n - h + i)(h - i - 1) + 1 =: d_{\text{low}} \qquad (8)$$

The special case $h = n$ (not contained in Table 1) results in an upper bound that is limited by

$$d(n, n, m) \geq nm - n^2 + \frac{9}{8}n + 1 + (\frac{1}{12}n - \frac{1}{3}m - \frac{5}{24})\sqrt{8m - 8n + 9}$$
$$= nm + o(nm)$$
$$d(n, n, m) \leq nm - n^2 - 4m + \frac{49}{8}n - \frac{13}{2} + (\frac{13}{12}n - \frac{1}{3}m - \frac{53}{24})\sqrt{8m - 8n + 9}$$
$$= nm + o(nm)$$

The coefficient 1 of the leading term nm is sharp for $m = o(n^2)$. The maximal error of this estimate can be obtained by computing the maximum of the difference between upper and lower bounds

$$d_{\text{high}} - d_{\text{low}} = 5n + n\sqrt{8m - 8n + 9} - \frac{15}{2} - 2\sqrt{8m - 8n + 9} - 4m.$$

This difference is a function that grows quadratically in n, which can be seen from the partial derivative (cf. Fig. 6) of

$$\partial(d_{\text{high}} - d_{\text{low}})/\partial m = 0 \Leftrightarrow m = 1/8n^2 + 1/4n.$$

h	m	$f(n)$	$\lim_{n\to\infty} k(h,n,m)/f(n)$	$g(n)$	$\lim_{n\to\infty} d(h,n,m)/g(n)$
c_1	c_2	1	1	n	$[1-c_1, c_1-2]$
c_1	$c_2\sqrt{n}$	1	1	n	$[1-c_1, c_1-2]$
c_1	$c_2\log n$	1	1	n	$[1-c_1, c_1-2]$
c_1	$c_2 n/\log n$	1	1	n	$[1-c_1, c_1-2]$
c_1	$c_2 n$	1	c_2	n	$(+2c_1 + 2c_1c_2 - 4 - 5c_2 - c_2^2)/2$
c_1	$c_2 n^2$	n	$\sqrt{2c_2+1}-1$	n^3	$\left(3c_2 + 1 - (2c_2+1)^{3/2}\right)/3$
$c_1\sqrt{n}$	$c_2\sqrt{n}$	1	1	$n^{3/2}$	$(0, c_1]$
$c_1\sqrt{n}$	$c_2\log n$	1	1	$n^{3/2}$	$(0, c_1]$
$c_1\sqrt{n}$	$c_2 n/\log n$	1	1	$n^{3/2}$	$(0, c_1]$
$c_1\sqrt{n}$	$c_2 n$	1	c_2	n^2	$[c_1c_2 - c_1, c_1c_2 + c_1]$
$c_1\sqrt{n}$	$c_2 n\sqrt{n}$	$\sqrt{n}$	c_2	n^2	$c_1c_2 - c_2^2$
$c_1\sqrt{n}$	$c_2 n^2$	n	$\sqrt{2c_2+1}$	n^3	$\left(3c_2 + 1 - (2c_2+1)^{3/2}\right)/3$
$c_1\log n$	$c_2\log n$	1	1	$n\log n$	$(0, c_1]$
$c_1\log n$	$c_2 n/\log n$	1	1	$n\log n$	$(0, c_1]$
$c_1\log n$	$c_2 n$	1	c_2	$n\log n$	$[c_1c_2 - 1, c_1c_2 + 1]$
$c_1\log n$	$c_2 n^2$	n	$\sqrt{2c_2+1}-1$	n^3	$\left(3c_2 + 1 - (2c_2+1)^{3/2}\right)/3$
$c_1 n/\log n$	$c_2 n/\log n$	$n\log n$	$1/4$	n^3	$1/24$
$c_1 n/\log n$	$c_2 n$	$n\log n$	$1/4$	n^3	$1/24$
$c_1 n/\log n$	$c_2 n^2$	n	$\sqrt{2c_2+1}-1$	n^3	$\left(3c_2 + 1 - (2c_2+1)^{3/2}\right)/3$
$c_1 n$	$c_2 n\sqrt{n}$	$\sqrt{n}$	$c_2/\|1-c_2\|$	$n^{5/2}$	$c_2/\|1-c_1\|$
$c_1 n$	$c_2 n^2$	n	$\sqrt{2c_2+1+c_1^2-2c_1}$ $+c_1 - 1$	n^3	$\left(3c_1^2 + 1 - c_1^3 - 3c_1\right.$ $\left. - \left((1-c_1)^2 + 2c_2\right)^{3/2}\right)$

Table 1. Asymptotic behavior of $k(h,n,m)$ and $d(h,n,m)$.

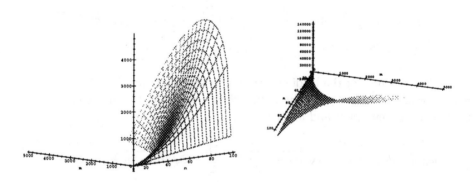

Fig. 6. Growth of the difference $d_h - d_l$ depending on n, m.

6 Conclusion

We have presented a detailed analysis of the simplification phase of SUGIYAMA's algorithm [12]. Previous work has been mainly concerned with gross estimates.

The complexity of the simplification phase may influence the runtime behavior of subsequent phases dramatically. Consequently, possible applications of our results include improved time estimates on the space and time complexity of the simplification phase and, even more importantly, the crossing-reduction phase of SUGIYAMA's algorithm.

Future work should focus on studying which parameter combinations actually arise in practice. Combinatorial results on the expected diameter of a directed graph may lead to estimates on the expected behavior of d, which should be complemented by experimental results.

We would like to conclude with an interesting open problem: Give a precise characterization of the parameter domain for legal configurations. It is easy to see that a configuration $C = (h, n, m)$ can exist only for $1 \leq h \leq n$ and $h-1 \leq m$, but the contrary is not true due to limitations imposed by the hierarchy condition. For example, there can be no legal $(h, h + 1, h(h + 1)/2)$ configuration, because that would require a hierarchy with a layer containing two vertices. The fact that there are $h + 1$ vertices and $h(h + 1)/2$ edges identifies the graph uniquely as the complete digraph with $h + 1$ vertices. The contradiction follows from the fact that there cannot be an edge between the vertices sharing the same layer.

References

1. G. di Battista, P. Eades, R. Tamassia, and I. G. Tollis. Algorithms for drawing graphs: an annotated bibliography. Report, June 1993.
2. P. Eades and K. Sugiyama. How to draw a directed graph. *Journal of Information Processing*, 14(4):424–437, 1990.
3. M. Fröhlich and M. Werner. Demonstration of the interactive graph-visualization system davinci. In Tamassia and Tollis [13].
4. E. R. Gansner, E. Koutsofios, S. C. North, and K.-P. Vo. A technique for drawing directed graphs. *IEEE Transactions on Software Engineering*, 19(3):214–230, March 1993.
5. F. Harary. *Graph Theory*. Series in Mathematics. Addison Wesley Publishing Company, 1969.
6. M. Himsolt. Graphed: A graphical platform for the implementation of graph algorithms. In Tamassia and Tollis [13].
7. I. Lemke. Entwicklung und implementierung eines visualisierungswerkzeuges für anwendungen im Übersetzerbau. Diplomarbeit, Universität des Saarlandes, FB 14 Informatik, 1994.
8. B. Madden, P. Madden, S. Powers, and M. Himsolt. Portable graph layout and editing (system demonstration). In Franz Brandenburg, editor, *Proceedings of Graph Drawing'95*, volume 1027 of *Lecture Notes in Computer Science*, pages 385–395. Springer Verlag, 1996.
9. F. Newbery-Paulisch and W. F. Tichy. Edge: An extendible graph editor. *Software – Practic and Experience*, 20(S1):S1/63–S1/88, June 1990.

10. G. Sander. Graph layout through the VCG tool. In Tamassia and Tollis [13], pages 194–205.

11. G. Sander. *Visualisierungstechniken fuer den Compilerbau*. PhD thesis, Univ. des Saarlandes, FB 14 Informatik, Saarbrücken, 1996.

12. K. Sugiyama, S. Tagawa, and M. Toda. Methods for visual understanding of hierarchical system structures. *IEEE Transactions on Systems, Man and Cybernetics*, SMC-11(2):109–125, February 1981.

13. R. Tamassia and I. Tollis, editors. *Proceedings of Graph Drawing'94*, volume 894 of *Lecture Notes in Computer Science*. DIMACS Workshop on Graph Drawing, Springer Verlag, 1995.

14. Waterloo Maple Software. *Maple V Release 3*, 1994.

Integration of Declarative Approaches
(System Demonstration)

Arne Frick[1] *, Can Keskin, Volker Vogelmann[2]

[1] Tom Sawyer Software, 804 Hearst Avenue, Berkeley CA 94710, EMail:
africk@tomsawyer.com
[2] Universität Karlsruhe, Institut für Telematik, Postfach 6980, D-76128 Karlsruhe, Germany
EMail: vogelmann@teco.uni-karlsruhe.de

Abstract. This demonstration shows the GOLD system, an extensible software
architecture integrating several declarative layout strategies, including spring-
embedders, local constraints and genetic algorithms. The underlying paradigm
is to consider graph layout problems as geometric constraint satisfaction problems [4].

In addition to satisfying global aesthetics criteria, the system allows for the interactive specification of local criteria per vertex (edge).

1 Introduction

In general, spring embedder algorithms produce "good" drawings of undirected graphs
according to the following global criteria

- edge lengths should be approximately equal,
- adjacent vertices should be closer together than non-adjacent ones, and
- inherent symmetries should be displayed.

In many cases, however, there exist aesthetically more pleasing drawings, which
may rely on other aesthetic criteria unavailable to spring embedder algorithms. Often,
such criteria can be expressed declaratively. Declarative graph drawing strategies include *simulated annealing* [3], *constraint-satisfaction* [10], *Graph Grammars* [1], and
genetic algorithms [11].

In [7], a proposal was made towards the integration of declarative and algorithmic
approaches. Their goal was to combine the strengths, while overcoming their respective difficulties. Algorithmic approaches are fast, but difficult to modify. Declarative
approaches, on the other hand, are usually easy to adapt to changing user requirements,
but inherently slow, as they employ general problem-solving frameworks.

2 Genetic algorithms for Graph Drawing

In this section, we briefly review genetic algorithms and show how to apply this paradigm
to the field of graph drawing.

---

* This research was performed while the author was working at Universität Karlsruhe, Institut
für Programmstrukturen und Datenorganisation.

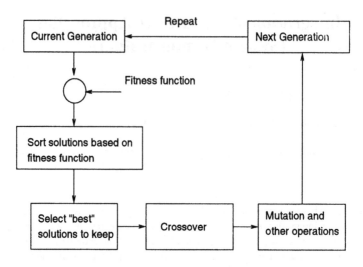

Fig. 1. The basic steps in a genetic algorithm.

2.1 Overview

Genetic algorithms (GAs) [5] belong to a family of optimization techniques known as *evolutionary algorithms (EAs)* that are based on principles of natural evolution. Maintaining a *population* of potential solutions, they perform a *selection* operation based on the *fitness* of individuals, modified by *recombination* and *mutation* operators to respect genetic diversity.

A GA can be interpreted as a modified random search. As with *simulated annealing (SA)*, it is a probabilistic method to solve hard optimization problems. These algorithms do not guarantee the optimum value, but the error probability can be made arbitrarily small.

The application of GA's to optimization problems requires a suitable encoding of candidate solutions. Traditional GA's represent possible solutions by binary bit strings (*chromosomes*), while newer work is also investigating other forms of representation as well. The initial population for a GA search is usually selected at random.

The basic structure of a GA is shown in 1. A part of the population is selected based on the principle of *survival of the fittest* by an *objective* or *fitness* function. The population size affects both the performance and the efficiency of the GA. A large population discourages premature convergence to suboptimal solutions. On the other hand, a large population requires more evaluations per generation, resulting in slower convergence. Crossover rate and mutation rate also affect the convergence to suboptimal solutions. The *generation gap* controls the percentage of the population being replaced during each generation according to a *selection strategy*.

GAs are known to be slow in practice, which is not really a surprise, since they are used to solve hard optimization problems. In many cases, the efficiency of a GA can be enhanced by combining it with other heuristics. For example, the initial population can be seeded with the results of a problem-specific heuristic.

2.2 Constraint satisfaction using genetic algorithms

Despite their common problems achieving satisfactory runtime efficiency, genetic algorithms have traditionally been used for constraint solving problems in cases where optimal or near-optimal solutions are required.

Genetic algorithms being used for constraint-solving problems have to deal with the problem of candidate solutions violating the constraints [12]. The most popular strategy is to generate potential solutions without considering the constraints at first. In a second phase, candidate solutions in violation of constraints are penalized by reducing their fitness. Other strategies have drawbacks exist, but have drawbacks such as being computationally expensive or posing problems to express the constraints.

2.3 Graph Drawing with genetic algorithms

This section shows how to use genetic algorithms to solve graph drawing problems. Usually, the formulation of a graph drawing problem as a GA involves the solution of a numeric optimization problem.

Previous research in this area includes [6, 8, 9] and [11]. In [6], a parallel GA for network-diagram layout is presented, in which perceptual organization is preferred over aesthetic layout. For a directed graph with 12 nodes, it took about 2 minutes to create a 2-D layout on 4096-processor machine. In [9] a GA is used for interactive two dimensional directed graph layout. The user can modify constraints like "two specified nodes have the same x-coordinate". Although the parameters (i.e. crossover rate, and mutation rate) were published, no data on the computational complexity of their approach is available. In view of the remarks make above regarding the tendency of GA's to be computationally slow, it is probably safe to assume that the runtime complexity does not stand out in particular.

To apply the GA paradigm to graph drawing, a set of candidate graphs is maintained as the population. Global aesthetics and local criteria are expressed as geometric constraints. In order to apply the penalty method these constraints become part of the fitness function (also called "energy constraints" in [14]). Each candidate graph is then evaluated and assigned a fitness value. The problem of minimizing the fitness function is equivalent to finding a solution for the constraint set.

The use of constraints provides a mechanism for quantifying the quality of a given layout, thus allowing objective comparison of the quality of two layouts.

3 The GOLD Architecture

The aim of GOLD is to combine the high speed of the spring-embedder paradigm with the ability of other declarative approaches to consider arbitrary geometric constraints in addition to the few global criteria employed by spring-embedders. To this end, we have developed a flexible, modular architecture comprised of three components (cf. Fig. 2):

- A *graphical user interface* (GUI) module provides for control mechanisms to steer the layout process, as well as a mechanism to visually specify local constraints on the graph.

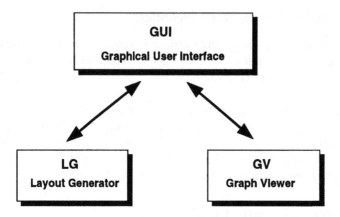

Fig. 2. The GOLD architecture.

- Several *layout generator* (LG) modules. Each LG module realizes a graph drawing strategy. In principle, arbitrary strategies conforming to a common protocol could be plugged in here. According to our goal stated above, we have implemented a random layout module to get uniform input graph layouts, a GEM-3D spring embedder module according to [2], and a genetic algorithm module. For the purpose of research an incremental constraint solver is also integrated, which solves local constraints with heuristic techniques.
- A *graph visualization* (GV) module is based on GeomView [13], an interactive 3-D geometry viewer with the usual rotation, translation and zoom mechanisms for interactive exploration.

The system creates 3-D drawings. All modules are realized as separate processes and are exchangeable. The key features of GOLD are

- extensibility. Layout strategies and viewer modules can be added or exchanged easily.
- the ability to specify constraints textually and visually. Vertices, edges and groups of these can be selected and assigned constraints.
- an extensible set of predefined local constraints.
- perspective 3-D views with real-time interactive graph exploration, e.g. navigation, zoom, camera flights.

In the GOLD system, global aesthetic constraints such as the desired inter-node distance and edge length are pre-defined. In addition, an extensible set of local constraints includes the following:

- Edges should enclose a specified angle.
- Nodes should lie in a plane.
- Nodes should lie on a line.
- Desired edge length for a particular edge.
- Equal length of a set of edges.

is provided. There also exist user controls for quality, speed, and constraint priority (i.e. aesthetic tradeoffs).

4 Applications

Layout creation in GOLD is usually performed in two distinct phases. In the first phase, only global constraints are active. Second, both global and local constraints are active. Then, local constraints are considered to have a higher priority than global constraints, as they are specified by the user, and we would like to take these constraints as a strong hint as to what the user would like to see.

This two-phase approach allows factoring out the first phase and use a fast spring-embedder algorithm. Compared to the use of genetic algorithms or spring-embedders by themselves, the integration of both combines their respective strengths, i.e. it leads to better results than spring embedders, while being faster than traditional genetic algorithms. Fig. 3 demonstrates this using two examples. For example, Fig. 3a shows a 3-D drawing of a wheel, which a user would like to see as a planar drawing. Using only a 3-D spring embedder, it is virtually impossible to achieve a plane drawing. Adding a few simple constraints, however, achieves the desired result.

The remainder of this section describes two experiments to measure the speedup and quality improvement achieved by the integrated strategy and the spring-embedder approach, respectively.

The first experiment measures the convergence speed and quality are measured for the graphs in Fig. 3. Each graph was drawn using two different initial seedings for the GA. The result for the Wheel is shown in Fig. 4. The figure indicates a significant speedup and quality improvement for an initial seeding based on the output of the GEM-3D spring embedder algorithm, compared to a random initial seeding. The corresponding figure for the Chair graph, which cannot be shown here for lack of space, displays the same behavior.

In the second experiment, the quality of four different graph layouts was measured both visually and by their respective fitness values. The constraints used in these examples were as follows. The Chair example (see Fig. 5) was solved by constraining its legs to have equal distance, and the nodes of its back to fall within the same plane. The Tree (see Fig. 7) was solved by defining a preferred direction for edges. The House (see Fig. 6) was solved by a set of constraints on angles, plane and lines. The hardest graph was the Wheel (see Fig. 8). To make it look like a planar wheel in 3-D, constraints were defined that restrict its spokes and the outer edges to have equal lengths, respectively.

The performance results are given in Table 1, which displays the input parameters and the resulting fitness, the latter being defined as the logarithm of the sum of all penalty terms, i.e. lower fitness values indicate better results. The data suggest that the Wheel is the hardest problem in this experiment: Although it took much longer, its fitness is worse than the other's.

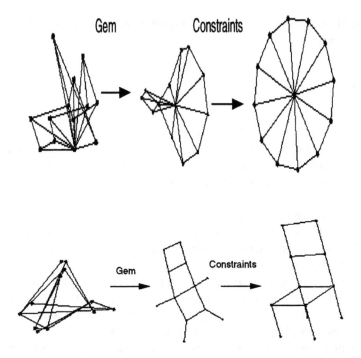

Fig. 3. Two results of the integrated GA approach with a spring-embedder used to seed the GA. The quality improvement is clear.

5 Summary

The use of spring-embedder algorithms allows to quickly generate initial drawings. Measurements indicate large speedups for the integrated approach, as compared to randomly-seeded genetic algorithms. Also, we have found evidence of quality improvements of an integrated approach vs. an approach that was based on combining spring-embedders with local constraint propagation techniques. For example, the 61-vertex Wheel in Fig. 8 was no problem at all for the integrated approach, while 13-vertex Wheel was found to be a hard instance of the combination of spring-embedders with local constraints. Overall, the GOLD architecture has proved to be a valuable tool to explore the effects of combining several layout strategies.

6 Acknowledgements

We would like to thank Frank Schwellinger for his initial implementation of our genetic algorithm in his M.Sc. thesis.

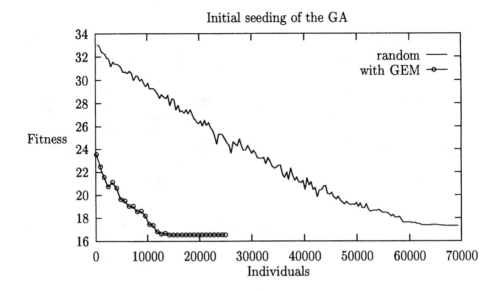

Fig. 4. Speedup and performance improvement displayed by a spring-embedder seeded GA, versus one seeded randomly. The measurement was performed on a 61-vertex **Wheel**.

Parameter	Chair			House		Tree		Wheel	
Population Size	15	20	50	20	50	20	20	5	5
Mutation Rate	0.0	0.001	0.001	0.001	0.001	0.002	0.001	0.002	0.001
Crossover Rate	0.5	0.0	0.6	0.5	0.0	0.5	0.6	0.7	0.6
Generation Gap	1.0	1.0	1.0	1.0	1.0	1.0	1.0	1.0	1.0
Selection Strategy	P	P	P	P	P	E	P	P	P
Fitness	10.822	10.637	10.588	17.477	17.437	14.606	15.563	23.285	23.356
Time[s]	1	2	5	1	3	3	2	58	55

Table 1. Performance of the integrated layout method based on genetic algorithms on the example graphs.

References

1. F. J. Brandenburg. Designing graph drawings by layout graph grammars. In Roberto Tamassia and Ioannis Tollis, editors, *Proceedings of Graph Drawing'94*, volume 894 of *Lecture Notes in Computer Science*, pages 416–427. DIMACS Workshop on Graph Drawing, Springer Verlag, 1995.
2. I. Bruß and A. Frick. Fast interactive 3-D graph visualization. In Franz Brandenburg, editor, *Proceedings of Graph Drawing'95*, volume 1027 of *Lecture Notes in Computer Science*, pages 99–110. Springer Verlag, 1996.

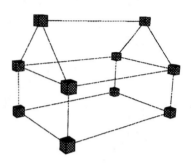

Fig. 5. Chair

Fig. 6. House

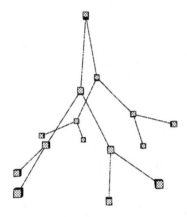

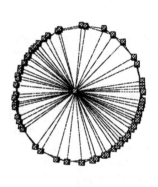

Fig. 7. Tree

Fig. 8. Wheel

3. I. F. Cruz and J. P. Twarog. 3-D graph drawing with simulated annealing. In F. J. Brandenburg, editor, *Graph Drawing*, volume 1027 of *Lecture Notes in Computer Science*, pages 162–165. Springer-Verlag, 1995.

4. E. Dengler, M. Friedell, and J. Marks. Constraint-driven diagram layout. In *Proceedings of the 1993 IEEE Workshop on Visual Languages*, pages 330–335, 1993.

5. J. Holland. *Adaptation in Natural and Artificial Systems*. University of Michigan Press, 1975.

6. C. Kosak, J. Marks, and S. Shieber. A parallel genetic algorithm for network-diagram layout. In *Proc. 4th Int. Conf. on Genetic Algorithms (ICGA91)*, 1991.

7. T. Lin and P. Eades. Integration of declarative and algorithmic approaches for layout creation. In R. Tamassia and I. G. Tollis, editors, *Graph Drawing*, volume 894 of *Lecture Notes in Computer Science*, pages 376–387. DIMACS, Springer-Verlag, October 1994. ISBN 3-540-58950-3.

8. E. Mäkinen and M. Sieranta. Genetic algorithms for drawing bipartite graphs. Technical report, Department of Computer Science, University of Tampere, 1994.

9. T. Masui. Graphic object layout with interactive genetic algorithms. In *Proceedings of the 1992 IEEE Workshop on Visual Languages*, pages 74–87, Seattle, Washington, 1992.

10. S. Matsuoka, S. Takahashi, T. Kamada, and A. Yonezawa. A general framework for bidirectional translation between abstract and pictorial data. *TOIS*, 10(4):408–437, 1992.

11. Z. Michalewicz. *Genetic algorithms + data structures = evolution programs*. Springer-Verlag, 1992.

12. Z. Michalewicz and C. Janikow. Handling constraints in genetic algorithms. In R. Belew and L. Booker, editors, *Genetic Algorithms*, pages 151–157, 1991.

13. Phillips, Levy, and Munzner. Geomview: An interactive geometry viewer. *Notices of the American Mathematical Society*, 40, 1993.

14. A. Witkin, K. Fleischer, and A. Barr. Energy constraints on parameterized models. In Maureen C. Stone, editor, *Computer Graphics (SIGGRAPH '87 Proceedings)*, volume 21, pages 225–232, July 1987.

GIOTTO3D: A System for Visualizing Hierarchical Structures in 3D*

Ashim Garg and Roberto Tamassia

Department of Computer Science
Brown University
Providence, RI 02912–1910, USA
{ag,rt}@cs.brown.edu

Abstract. Hierarchical structures represented by directed acyclic graphs are widely used in visualization applications (e.g., class inheritance diagrams and scheduling diagrams). 3D information visualization has received increasing attention in the last few years, motivated by the advances in hardware and software technology for 3D computer graphics. We present GIOTTO3D, a system for visualizing hierarchical structures in 3D. GIOTTO3D uses a new technique combining 2D drawing methods with a lifting transformation that exploits the third dimension to visualize hierarchical relations among the vertices. GIOTTO3D also employs several graphical aids such as user-defined coloring, showing/hiding sub-hierarchies, "footprints", and representation of edges as "Bezier tubes" to improve the effectiveness of its visualizations.

1 Introduction

The effective visualization of hierarchical structures is an important problem in the area of advanced visual interfaces. Hierarchical structures occur in a wide variety of information visualization applications, including WWW-navigation, business graphics, multimedia documents, software engineering, algorithm animation, planning, database design, and visual languages.

A hierarchical structure can be formally modeled by a *directed acyclic graph* (*DAG*), i.e., a graph with directed edges and no directed cycles. Hence, throughout this paper, we will use the two terms interchangeably. Also we will use the terms visualization, drawing, layout and representation to denote the same concept. A hierarchical structure is usually visualized with a *hierarchical drawing* where each edge is drawn as a curve monotonically non-increasing in the y-direction for 2D drawings and in the z-direction for 3D drawings.

Three dimensions offer many advantages over two dimensions for visualizing hierarchical structures. The extra dimension gives greater flexibility for placing the vertices and edges of the graph. The presence of navigational operations such as rotation, translation, and zooming not only helps the user in understanding the structure, but also results in a more effective use of screen space because not every object has to be shown in front on the screen (as is the case with 2D-visualization). Empirical studies have also shown that 3D visualizations are generally easier to understand than 2D visualizations [21, 26]. Many

---

*Research supported in part by the National Science Foundation under grant CCR-9423847, and by the U.S. Army Research Office under grant DAAH04-96-1-0013.

virtual-reality packages also accept descriptions of 3D layouts produced by other graphics packages.

Research in graph drawing has generally concentrated on constructing two-dimensional visualizations because of the two-dimensional geometry of display surfaces such as paper and computer screen. Recent advances in hardware and software technology for computer graphics open up the possibility of displaying three-dimensional (3D) visualizations on a variety of low-cost workstations, and several researchers (and film makers**) have begun to explore the possibility of displaying graphs and networks using this new technology. Indeed, it is expected that in the next few years, 3D visualization will become widespread in Web documents, thanks to the VRML language for modeling three-dimensional scene graphs. Recent work on WWW navigation [18], software visualization [19], information visualization [20], and algorithm animation [1] has also advocated the use of 3D visual representations.

Previous research on 3D graph drawing (both hierarchical and non-hierarchical) has focused on the development of visualization systems (see, e.g., [2, 5, 17]). Recent theoretical work has been reported in [3, 4, 15].

We now review previous work on visualizing hierarchical structures in 3D. In a *cone tree* [20], each subtree is associated with a cone such that the vertex at the root of the subtree is placed at the apex of the cone and its children are circularly arranged around the base of the cone. SemNet [10] is a system for displaying knowledge bases. GraphVizualizer3D [27] emphasizes manual layouts. COMAIDE [9] uses a force-directed method. The systems GMB [16] and PLUM [19] extend to 3D the layering approach [23, 13] which was conceived for constructing 2D hierarchical drawings. The drawings constructed by these two systems have characteristics similar to their 2D counterparts.

2 GIOTTO3D

GIOTTO3D is a system for visualizing hierarchical structures in 3D. Namely, given a DAG G as input, it constructs a 3D hierarchical drawing of G.

In Section 2.1, we describe the 3D drawing algorithm used by GIOTTO3D and discus graphical aspects. In Section 2.2, we give examples of drawings produced by and applications of GIOTTO3D.

2.1 Drawing Algorithm

GIOTTO3D uses a novel combination of 2D undirected layout methods, which guarantee the satisfaction of several aesthetic criteria, and a lifting transformation that uses the third dimension to visualize the hierarchical relationships between the vertices.

There is considerable work on 2D drawings of undirected graphs [6], and many algorithms have been developed for constructing drawings satisfying one or more aesthetic criteria.

** An important plot element in the movie *Jurassic Park* involves a 3D virtual-reality traversal of a tree representing a Unix file system.

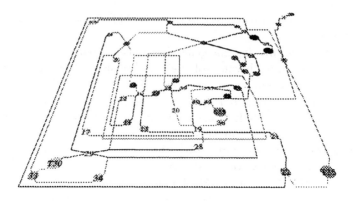

Fig. 1. 2D polyline drawing constructed by Phase *Draw-Flat* of GIOTTO3D.

GIOTTO [25] is a successful general-purpose drawing algorithm based on the *planarization* approach and a bend-minimization method. Namely, given a non-planar graph as input, it first transforms the graph into a planar graph by replacing each crossing with a fictitious vertex, and then constructs a 2D orthogonal drawing with the bend-minimization method of [24]. GIOTTO produces high-quality layouts and has been widely used in software visualization systems. In [7, 8], the performance of GIOTTO on more than 11,000 graphs derived from "real-life" software engineering and database applications was experimentally evaluated. This study show that GIOTTO performs very well with respect to several quality measures such as crossings, bends, total edge-length and area.

The main idea behind the approach of GIOTTO3D is to leverage the aesthetic quality of the 2D layouts produced by GIOTTO. Given a DAG G as input, GIOTTO3D constructs a 3D drawing of G in three phases:

1. *Draw-Flat:* Construct a 2D polyline drawing of G in the XY-plane using a variation of GIOTTO.
2. *Lift:* Assign z-coordinates to the vertices and to the bends of the edges such that their placement reflects the hierarchy.
3. *Reshape:* Draw vertices as spheres and edges as Bezier tubes [22], and create a footprint of the 3D drawing.

We now describe each step in more detail.

Draw-Flat. In this phase, GIOTTO3D constructs a 2D polyline drawing of G in the XY-plane using a variation GIOTTO. Figure 1 shows an example 2D drawing constructed in Step *Draw-Flat* by GIOTTO3D.

GIOTTO works in two steps:. In the planarization step, G is converted into a planar graph G' by replacing crossings with fictitious vertices. In the orthogonalization step, an orthogonal drawing D of G' is constructed using the bend-minimization algorithm of [24]. In the drawing D produced by GIOTTO, the vertices are drawn as rectangles with possibly different sizes. Thus, we have added a postprocessing step that shrinks all the vertices to the same size. Additional segments are added to the edges to maintain connectivity, whenever needed. The output of this phase is a 2D polyline drawing of G.

Lift. In this phase, we assign z-coordinates to the vertices, and to the bends of the edges such that each edge is drawn as a polygonal chain nonincreasing in the z-direction. We use the following approach for reducing the total length of the edges, which is common to many layering based methods (see, e.g., [13]). We call a *source* of G a vertex without incoming edges. The z-coordinates of the vertices are given by an optimal solution to the following integer linear program:

$\min(\sum_{v \to w} z(v) - z(w))$
subject to: $z(v) \geq z(w) + 1$ for each edge $v \to w$ of G, and
$z(u) = 0$ for each source vertex of G.

Although solving an integer linear program is NP-hard in general, we have implemented a fast heuristic based on topological numbering for finding a good solution.

Reshape GIOTTO3D draws vertices as spheres. We chose this representation because of the good visual properties of spheres, such as reflectivity, roundness, smoothness, and symmetry. Also, after the postprocessing step of the *Draw-Flat* phase, the end-segments of each edge are directed towards the centers of their incident vertices. Thus, by drawing vertices in 3D as spheres, it is likely that their incident edges will enter along a direction normal to the surface of the sphere. This feature adds to the visual appeal of the drawing. The vertices can be assigned colors by an application. This allows the users to encode application-specific information with colors.

Each edge is represented by a *Bezier tube* [22] whose control points are given by the bend-points of the edge. A Bezier tube is a Bezier surface [11] that is shaped like a hollow tube. Bezier tubes were introduced in [22], and were shown to have good visual properties, such as reflectivity, roundness, and smoothness. We have also experimented with drawing edges as Bezier curves and as a set of connected rigid pipes. Bezier tubes give by far the best results. Other than [22], we do not know of any other visualization system that uses Bezier tubes. Edges are given randomly generated colors over a wide range. This reduces the possibility that two edges have the same color, and helps in distinguishing them if they are drawn close to each other.

Previous approaches have recommended displaying idealized shadows of the objects of a 3D structure to assist in understanding it [20]. GIOTTO3D follows a different approach. Instead of displaying idealized shadows, GIOTTO3D displays the *footprints* of the graphical objects. A footprint of an object is its projection on the XY-plane(see Fig. 2(a)). The footprint of a displayed structure is similar to its 2D-drawing constructed by Step *Draw-Flat*.

The footprint displays the hierarchy-independent connectivity information of the structure, and has few crossings. The use of the same coloring scheme for both 3D visualization and footprint helps in maintaining the correspondence between them.

GIOTTO3D provides the usual 3D navigational operations: rotation, translation, and zooming. Fog and appropriate lighting can be used to enhance depth perception. A user can also select a subgraph and perform various operations on it, including showing (hiding) the selected subgraph, or showing (hiding) the descendants of selected vertices. Users can also choose to either show an outline of the hidden objects or not show them at all.

2.2 Example Layouts and Applications

Hierarchical structures are common in real-life applications. In this section, we illustrate the use of GIOTTO3D in three applications. Sample GIOTTO3D visualizations are also shown (with color images) in [14].

Education. Figure 2 shows two different views (from different camera angles and at different zooms) of a DAG depicting the evolution-history of UNIX, taken from [13].

Economic Planning. Figure 3(a) shows a drawing of a subgraph of the dynamic model of world interactions given in [12], which shows the interdependency of five important economic variables — population, capital investments, natural resources, fraction of capital devoted to agriculture, and pollution.

Data Structures. Figure 3(b) shows a complete binary tree with levels. While a complete binary tree can be effectively visualized in 2D, a 3D representation uses screen space more efficiently than a 2D representation [20].

References

1. M. Brown and M. Najork. Algorithm animation using 3D interactive graphics. In *ACM Symp. on User Interface Software and Tech.*, pages 93–100, 1993.
2. I. Bruß and A. Frick. Fast interactive 3-d graph visualization. *Graph Drawing (Proc. GD '95). LNCS*, 1027:99–110, Springer-Verlag. 1996.
3. M. Chrobak, M. T. Goodrich, and R. Tamassia. Convex drawings of graphs in two and three dimensions. In *Proc. 12th Annu. ACM Sympos. Comput. Geom.*, pages 319–328, 1996.
4. R. F. Cohen, P. Eades, T. Lin, and F. Ruskey. Three-dimensional graph drawing. In R. Tamassia and I. G. Tollis, editors, *Graph Drawing (Proc. GD '94)*, volume 894 of *Lecture Notes in Computer Science*, pages 1–11. Springer-Verlag, 1995.
5. I. F. Cruz and J. P. Twarog. 3d graph drawing with simulated annealing. *Graph Drawing (Proc. GD '95). LNCS*, 1027:162–165, Springer-Verlag. 1996.
6. G. Di Battista, P. Eades, R. Tamassia, and I. G. Tollis. Algorithms for drawing graphs: an annotated bibliography. *Comput. Geom. Theory Appl.*, 4:235–282, 1994.

7. G. Di Battista, A. Garg, G. Liotta, R. Tamassia, E. Tassinari, and F. Vargiu. An experimental comparison of three graph drawing algorithms. In *Proc. 11th Annu. ACM Sympos. Comput. Geom.*, pages 306–315, 1995.

8. G. Di Battista, A. Garg, G. Liotta, R. Tamassia, E. Tassinari, and F. Vargiu. An experimental comparison of four graph drawing algorithms. *Comput. Geom. Theory Appl.*, 1996. to appear.

9. D. Dodson. Comaide: Information visualization using cooperative 3d diagram layout. *Graph Drawing (Proc. GD '95). LNCS*, 1027:190–201, Springer-Verlag. 1996.

10. K. M. Fairchild, S. E. Poltrock, and G. W. Furnas. Semnet: Three-dimensional graphic representation of large knowledge bas es. In *Cognitive Sc. and its Applns for Human-Computer Interaction,* pages 201–233. Lawrence Erlbaum Assoc., 1988.

11. J. D. Foley, A. van Dam, S. K. Feiner, and J. F. Hughes. *Computer Graphics: Principles and Practice.* Addison-Wesley, Reading, MA, 1990.

12. J. W. Forrester. *World Dynamics.* Wright-Allen, Cambridge, Mass., 1971.

13. E. R. Gansner, E. Koutsofios, S. C. North, and K. P. Vo. A technique for drawing directed graphs. *IEEE Trans. Softw. Eng.*, 19:214–230, 1993.

14. A. Garg and R. Tamassia. Effective visualization of hierarchical structures in 3D. Manuscript, Dept. of Computer Sci., Brown University, 1996. Available at *http://www.cs.brown.edu/people/rt/fadiva/giotto3d.html.*

15. A. Garg, R. Tamassia, and P. Vocca. Drawing with colors. In *Proc. 4th Annu. European Sympos. Algorithms (ESA '96)*, 1996.

16. D. Jablonowsky and V. A. Guarna. GMB: A tool for manipulating and animating graph data structures. *Softw. – Pract. Exp.*, 19(3):283–301, 1989.

17. B. Monien, F. Ramme, and H. Salmen. A parallel simulated annealing algorithm for generating 3D layouts of undirected graphs. In F. J. Brandenburg, editor, *Graph Drawing (Proc. GD '95). LNCS*, 1027:396–408, Springer-Verlag. 1996.

18. S. Mukherjea. Visualizing the information space of hypermedia systems. Technical report, Graphics and Visualization Center, GeorgiaTech.

19. S. P. Reiss. 3-D visualization of program information. *Graph Drawing (Proc. GD '94). LNCS*, 894:12–24, Springer-Verlag. 1995.

20. G. G. Robertson, J. D. Mackinlay, and S. K. Card. Cone trees: Animated 3D visualizations of hierarchical information. In *Proc. CHI'91*, pages 189–193.

21. R. Sollenberger and M. P. The effects of stereoscopic and rotational displays in a three-dimensional path-tracing task. *Human Factors.* 35(3), 483–500.

22. L. Spratt and A. Ambler. Using 3D tubes to solve the intersecting line representation problem. In *Proc. IEEE Symp. on Visual Lang., 1994*, pages 254–261.

23. K. Sugiyama, S. Tagawa, and M. Toda. Methods for visual understanding of hierarchical systems. *IEEE Trans. Syst. Man Cybern.*, SMC-11(2):109–125, 1981.

24. R. Tamassia. On embedding a graph in the grid with the minimum number of bends. *SIAM J. Comput.*, 16(3):421–444, 1987.

25. R. Tamassia, G. Di Battista, and C. Batini. Automatic graph drawing and readability of diagrams. *IEEE Trans. Syst. Man Cybern.*, SMC-18(1):61–79, 1988.

26. C. Ware and G. Franck. Evaluating stereo and motion cues for visualizing information nets in three dimensions. *ACM Transactions on Graphics.* to appear. *http://www.omg.unb.ca/hci/projects/hci-gv3D.html.*

27. H. D. Ware, C. and G. Franck. Visualizing object oriented software in three dimensions. In *CASCON '93*, pages 621–620, 1993. *http://www.omg.unb.ca/hci/projects/hci-gv3D.html.*

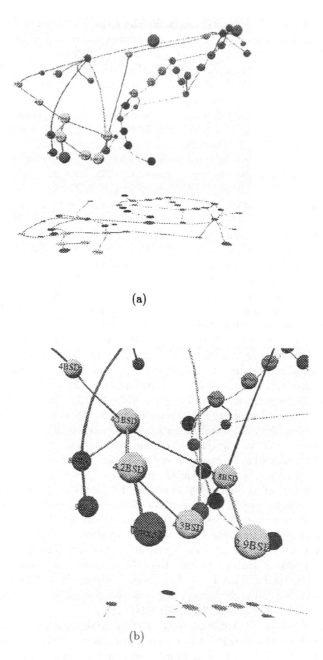

(a)

(b)

Fig. 2. Evolution History of UNIX

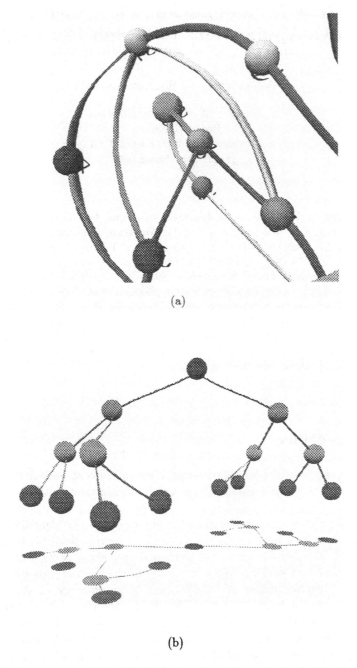

(a)

(b)

Fig. 3. (a) A Dynamic Model of World Interactions (from the book *World Dynamics* by Jay W. Forrester); (b) A Complete Binary Tree

A New Minimum Cost Flow Algorithm with Applications to Graph Drawing[*]

Ashim Garg and Roberto Tamassia

Department of Computer Science
Brown University
Providence, RI 02912–1910, USA
{ag,rt}@cs.brown.edu

Abstract. Let N be a single-source single-sink flow network with n nodes, m arcs, and positive arc costs. We present a pseudo-polynomial algorithm that computes a maximum flow of minimum cost for N in time $O(\chi^{3/4} m \sqrt{\log n})$, where χ is the cost of the flow. This improves upon previously known methods for networks where the minimum cost of the flow is small. We also show an application of our flow algorithm to a well-known graph drawing problem. Namely, we show how to compute a planar orthogonal drawing with the minimum number of bends for an n-vertex embedded planar graph in time $O(n^{7/4} \sqrt{\log n})$. This is the first subquadratic algorithm for bend minimization. The previous best bound for this problem was $O(n^2 \log n)$ [19].

1 Introduction

Minimum cost flow is a fundamental problem in network optimization, and a large body of literature exists on theoretical and practical methods for solving it [1].

While sophisticated polynomial and strongly-polynomial algorithms for minimum cost flow have been recently devised [1], their complexity is $\Omega(nm \log n)$, where n and m denote the number of nodes and arcs, respectively, of the flow network, and they may perform worse than some of the simpler pseudo-polynomial algorithms when the magnitude and/or cost of the flow are small. For example, let the magnitude of the optimal flow be ϕ. One can achieve running time $O(\phi \, m \log n)$ with a simple minimum cost flow algorithm based on successive flow augmentations along a minimum cost path determined with Dijkstra's method.

This paper is organized as follows: In Section 2, we provide background material on network flow algorithms and present some preliminary results. In Section 3, we present our new pseudo-polynomial algorithm that computes a minimum cost flow for a flow network with positive arc costs in time $O(\chi^{3/4} m \sqrt{\log n})$,

---

[*] Research supported in part by the National Science Foundation under grant CCR–9423847.

where χ is the cost of the flow. This improves upon all previously known methods for networks with positive arc costs such that $\chi = o(n^{4/3}\sqrt{\log n})$ and $\chi = o(\phi^{4/3}\sqrt{\log n})$.

In Section 4, we give an application of our minimum cost flow algorithm to an important graph drawing problem. Namely, we show how to compute a planar orthogonal drawing with the minimum number of bends for an n-vertex embedded planar graph in time $O(n^{7/4}\sqrt{\log n})$. This is the first subquadratic algorithm for bend minimization. The previous best bound for this problem was $O(n^2 \log n)$ [19]. Improving the time complexity of bend minimization was mentioned as one of the major open problems of graph drawing in a standard bibliographic survey of the field [3].

2 Preliminaries

In this section, we review some basic concepts and definitions related to network flows. We also give a brief description of Algorithms *AugPath*, *BlockFlow* and *PrimDua*, which embody three well-known techniques for computing flows in flow networks. The notation followed is generally the one of [15].

2.1 Definitions

A *flow network* N is a directed graph such that N has two disjoint non-empty sets of distinguished nodes, called its *sources* and *sinks*, and each arc e of N has a cost $c(e)$ and a capacity $u(e)$ associated with it, where $c(e)$ is an integer and $u(e)$ is a positive integer. Let m and n be the number of arcs and nodes, respectively, of N. We assume that N is connected so that $m \geq n - 1$. N is a flow network with *positive arc costs* if the cost of each arc is at least 1. N is a *single-source single-sink* flow network if it has only one source and only one sink.

Let $inarc(v)$ and $outarc(v)$ be the set of incoming and outgoing arcs of a node v of N. A *flow* f in N is an assignment of a non-negative integer label $f(e)$ to each arc e of N such that:

- for each arc e, $f(e) \leq u(e)$;
- for each node v, where v is not a source or a sink,

$$\sum_{e \in inarc(v)} f(e) = \sum_{e \in outarc(v)} f(e) \text{ (flow conservation)};$$

- for each source s,

$$\sum_{e \in inarc(s)} f(e) \leq \sum_{e \in outarc(s)} f(e); \text{ and}$$

- for each sink t,

$$\sum_{e \in inarc(t)} f(e) \geq \sum_{e \in outarc(t)} f(e).$$

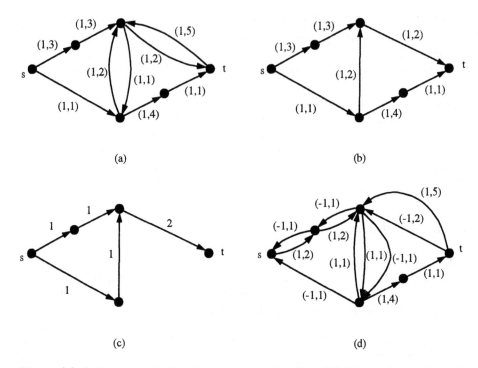

Fig. 1. (a) A flow network N with source s and sink t. (b) The shallowest layered network $L(N, \mathbf{0})$ of N with respect to zero flow; L is also the admissible flow network of N with respect to zero flow because each arc of N has cost 1. (c) A Blocking flow f of L. (d) Residual network $R(N, f)$ of N with respect to f. In parts (a), (b), and (d) we have labeled each arc e by the pair $(c(e), u(e))$. In part (c) we have labeled each arc by the amount of flow in it, and have shown only the arcs with non-zero flow.

We say that $f(e)$ is the flow in arc e *due to* flow f. For notational convenience, we do not distinguish between a flow in an arc and its magnitude. An arc e is *saturated* by flow f if $f(e) = u(e)$. Let S be the set of the sources of N. The *magnitude* of flow f, denoted by $|f|$, is defined as $|f| = \sum_{s \in S}(\sum_{e \in outarc(s)} f(e) - \sum_{e \in inarc(s)} f(e))$, i.e., the "net" flow going out of the sources of N.

The *cost* of a flow f, denoted by $c(f)$, is defined as the sum of the costs of the flows due to f in the arcs of N. A *maximum* flow of N is a flow with maximum magnitude among all the flows of N. A *minimum cost* flow of N is a maximum flow with minimum cost among all the maximum flows of N. A *zero* flow, denoted by $\mathbf{0}$, is a flow with zero magnitude. A *non-zero* flow is a flow with non-zero magnitude. For simplicity of discussion, we assume in the rest of this section that N is a flow network with a single source s and a single sink t.

The *cost* of a directed path of N is equal to the sum of the costs of its arcs. A *shortest* directed path from a node u to a node v is a directed path with the least number of arcs among all the directed paths from u to v.

The *residual* flow network $R(N, f)$ of N with respect to a flow f (see Figure 1(c–d)) is the flow network such that for every arc $e(u, v)$ of N, it consists of arcs $e' = e(u, v)$ and $e'' = (v, u)$ such that $c(e') = -c(e'') = c(e)$, $u(e') = u(e) - f(e)$, and $u(e'') = f(e)$ (notice that from the definition of a flow network, $R(N, f)$ can only have arcs with non-zero capacities, hence if e' or e'' has zero capacity, then we delete it from $R(N, f)$). A directed path of $R(N, f)$ from s to t is an *augmenting path* of N with respect to flow f. A flow g in $R(N, f)$ corresponds to a flow f^* in N as follows: for every arc e of N, $f^*(e) = f(e) + g(e') - g(e'')$; we say that flow g is the new flow *pushed* in N (and also in $R(N, f)$). Notice that because $u(e'') = f(e)$, we have that $g(e'') \leq u(e'') \leq f(e)$, and hence, $f^*(e) \geq 0$. It can also be shown easily that f^* satisfies the capacity and demand-supply constraints, and the flow conservation property. Therefore, from a high-level perspective, the concepts of residual flow networks and augmenting paths give us a convenient way of introducing more flow in the original flow network by reducing flow in certain arcs and increasing it in others.

A *layered* flow network $L(N, f, d)$ of N with respect to a flow f is the maximal subgraph of the residual flow network $R(N, f)$ such that $L(N, f, d)$ contains both s and t, all the directed paths of $L(N, f, d)$ from s to t have the same length d, and each arc of $L(N, f, d)$ is in a directed path of $L(N, f, d)$ from s to t. The *depth* of $L(N, f, d)$ is equal to d. Figure 1(b) shows a layered network with depth 3. $L(N, f, d)$ is the *shallowest* layered network of N with respect to flow f if all the layered networks of N with respect to f have depth at least d (see Figure 1(b)). A *blocking flow* g of $L(N, f, d)$ is one that saturates at least one arc of every directed path from s to t, i.e., every directed path from s to t has an arc e such that $g(e) = u(e)$(see Figure 1(c)). Notice that a blocking flow of a flow network may not be a maximum flow of the network. For example, the layered flow network of Figure 1(b) admits a maximum flow of magnitude 3 and also admits a blocking flow of magnitude 2 (which is shown in Figure 1(c)). The concept of layered flow networks allows us to introduce more flow in a flow network by pushing it through the augmenting paths with smallest lengths.

Let p be a directed path of $R(N, f)$ from s to t with least cost among all the directed paths of $R(N, f)$ from s to t. Let $c(N, f)$ be the cost of p. The *admissible* flow network $A(N, f)$ of N with respect to a flow f is the maximal subgraph of the residual flow network $R(N, f)$ such that $A(N, f)$ contains both s and t, all the directed paths of $A(N, f)$ from s to t have the same cost, equal to $c(N, f)$, and each arc of $A(N, f)$ is in a directed path of $A(N, f)$ from s to t(see Figure 1(b)). Hence the concept of an admissible flow network is similar to that of a layered flow network except that we consider the costs of directed paths instead of their lengths. For the flow network of Figure 1(a), the admissible and layered flow networks are the same (shown in Figure 1(b)), but for a general flow network, they may be different. The concept of admissible flow networks allows us to introduce more flow in a flow network by pushing it through the augmenting paths with least costs.

Lemma 1. *Let $R(A(N, f), g)$ be the residual flow network of $A(N, f)$ with respect to a flow g. All the directed paths of $R(A(N, f), g)$ from s to t have the same cost, namely, $c(N, f)$.*

Proof. Our proof is based on the proof of Lemma 8.4 of [21](chapter 8, pg. 110).

Let p be a directed path of $R(A(N, f), g)$ from s to t, and v be a node of p. Let $c(v, p, g)$ denote the cost of the subpath from s to v of p. It follows from the definition of $A(N, f)$ that in $A(N, f)$, all the paths from s to a node w have the same cost, which we denote by $c(w)$, and that $c(t) = c(N, f)$. It is also easy to see that if (u, w) is an arc of $A(N, f)$, then $c(w) - c(u) = c(u, w)$ (recall that $c(u, w)$ denotes the cost of the arc (u, w)).

We claim that if p is a directed path of $R(A(N, f), g)$ from s to t and v is a node of p, then $c(v, p, g) = c(v)$, and if (u, v) is an arc of p, then $c(u, v) = c(v) - c(u)$. This will give immediately that all the directed paths of $R(A(N, f), g)$ from s to t have the same cost, which is equal to $c(t) = c(N, f)$. We prove our claim using induction over the magnitude of flow g.

If $|g| = 0$, our claim is trivially true. Let h be a flow in N with magnitude $|g| - 1$ such that pushing a flow with unit magnitude through a directed path p' of $R(A(N, f), h)$ from s to t gives us flow g in $A(N, f)$. From the inductive hypothesis, it follows that if (u, v) is an arc of p', then $c(u, v) = c(v) - c(u)$. The only arcs of $R(A(N, f), g)$ that are not arcs of $R(A(N, f), h)$ also are all of the form (y, x) where (x, y) is an arc of p', and $c(y, x) = -c(x, y) = c(x) - c(y)$.

Let p be a directed path of $R(A(N, f), g)$ from s to t. Let v be a node of p. Let (u, v) be an arc of p. We first show that $c(u, v) = c(v) - c(u)$, and then using this show that $c(v, p, g) = c(v)$.

If (u, v) was an arc of $R(A(N, f), h)$ also then from the inductive hypothesis, $c(u, v) = c(v) - c(u)$. However, if (u, v) was not an arc of $R(A(N, f), h)$ then also from the above discussion it follows that $c(u, v) = c(v) - c(u)$.

A simple argument based on induction over the length of the subpath from s to v of p shows that $c(v, p, g) = c(v)$. ☐

2.2 Three Basic Flow Algorithms

Research in the area of network flows has a rich tradition (see [1] for an extensive survey). A number of algorithms have been proposed for finding maximum flows and minimum cost flows in flow networks. In this section, we review three well-known basic flow algorithms, which we call *AugPath*, *BlockFlow* and *PrimDua*, respectively.

Algorithm *AugPath* (see Figure 2) is the classic method developed by Ford and Fulkerson [10] that finds a maximum flow in a single-source single-sink flow network by successively pushing flow through augmenting paths.

Lemma 2 [10]. *Let N be a single-source single-sink flow network with n nodes and m arcs. Algorithm* AugPath *computes a maximum flow for N in time $O(\phi \cdot m)$, where ϕ is the maximum flow magnitude.*

Figure 3 shows Algorithm *BlockFlow*, which computes a maximum flow in a single-source single-sink flow network N. Each phase of Algorithm *BlockFlow* computes a blocking flow in the shallowest layered network of N with respect to the flow already computed. See [15] for further details about this algorithm.

Algorithm *Augpath(N)*: /*N is a flow network with a single source s and a single sink t*/
begin
    $R_1 \leftarrow N$; $F_1 \leftarrow \mathbf{0}$; $i \leftarrow 1$;
    while N has an augmenting path with respect to flow F_i
    begin (Phase i)
        Using R_i, find an augmenting path p_i of N with respect to flow F_i;
        Push a non-zero flow f_i from s to t through p_i
        Let F_{i+1} be the total flow in N after pushing flow f_i in R_i;
        Let R_{i+1} be the residual flow network of N with respect to flow F_{i+1};
        $i \leftarrow i + 1$;
    end
end

Fig. 2. Algorithm *AugPath*.

Algorithm *BlockFlow(N)*: /*N is a single-source single-sink flow network */
begin
    $L_1 \leftarrow N$; $i \leftarrow 1$;
    while L_i admits a non-zero blocking flow
    begin (Phase i)
        Find a blocking flow f_i of L_i;
        Let F_{i+1} be the total flow in N after pushing flow f_i in L_i;
        Let L_{i+1} be the shallowest layered network of N with respect to flow F_{i+1};
        $i \leftarrow i + 1$;
    end
end

Fig. 3. Algorithm *BlockFlow*.

Lemma 3 [15] (chapter IV, Section 9). *The following properties hold for each phase i of Algorithm* BlockFlow:

1. *The magnitude of flow in N is increased by at least one, i.e., $|F_i| \geq |F_{i-1}|+1$.*
2. *The depth of L_i is strictly greater than the depth of L_{i-1}.*
3. *Phase i can be executed in time $O(m \log n)$, where n and m denote the number of nodes and arcs, respectively, of N.*

Algorithm *PrimDua*, shown in Figure 4, reduces the problem of computing a minimum cost flow in a single-source single-sink flow network N to the problem of computing maximum flows in a sequence of intermediate flow-networks. Each of these maximum flows can be computed using any algorithm for computing maximum flows. Algorithm *PrimDua* (short for primal-dual) was developed first by Ford and Fulkerson [9, 10]. This algorithm computes f in a sequence of stages, where each stage computes a maximum flow in the admissible network of N with respect to the flow already pushed in it. See [1, 9, 10] for details.

Algorithm *PrimDua(N)*: /*N is a single-source single-sink flow network*/
begin
    Let A_1 be the admissible network of N with respect to a zero-flow;
    $i \leftarrow 1$;
    while A_i admits a non-zero maximum flow
    begin (Stage i)
        Find a maximum flow f_i of A_i;
        Let F_{i+1} be the total flow in N after pushing flow f_i in A_i;
        Let A_{i+1} be the admissible network of N with respect to flow F_{i+1};
        $i \leftarrow i + 1$;
    end
end

Fig. 4. Algorithm *PrimDua*.

Notice that from the definition of an admissible flow network, all the directed paths of A_i from s to t have the same cost, which we denote by c_i (if A_i has no directed path from s to t, then c_i is ∞). Lemma 4 follows directly from the discussion on the primal-dual algorithm in [1].

Lemma 4. *For each stage i of Algorithm* PrimDua, $c_{i+1} \geq c_i + 1$.

The following corollary is immediate.

Corollary 5. *If N is a flow network with positive arc costs, then for each stage i of Algorithm* PrimDua, $c_i \geq i$.

3 A New Minimum Cost Flow Algorithm

Let N be a single-source single-sink flow network with n nodes, m arcs, and positive arc costs. Let s and t be the source and sink, respectively, of N. Theorem 10, which is the main result of this section, shows that a minimum cost flow in N can be computed in $O(\chi^{3/4} m \sqrt{\log n})$ time, where χ is the cost of the flow. This computation is done using Algorithm *AugBlock* shown in Figure 5. Algorithm *AugBlock* is a variation of Algorithm *PrimDua* described in Section 2. The main feature of Algorithm *AugBlock* is that, in each stage i, it computes f_i by running algorithms *AugPath* and *BlockFlow* in parallel. Flow f_i is the flow computed by the algorithm terminating first.

Let χ be the cost of the minimum cost flow f computed by Algorithm *Aug-Block*. Let ϕ_i and χ_i be the magnitude and cost, respectively, of flow f_i. Let $S = \{f_i | 1 \leq \phi_i \leq \sqrt{\chi} \log n\}$. Let $B = \{f_i | \phi_i > \sqrt{\chi} \log n\}$. We call the stages of S as the *light* stages and the stages of B as the *heavy* stages. Let $\phi_S = \sum_{f_i \in S} \phi_i$, and $\phi_B = \sum_{f_i \in B} \phi_i$. Let $\chi_S = \sum_{f_i \in S} \chi_i$, and $\chi_B = \sum_{f_i \in B} \chi_i$. Therefore, ϕ_S and χ_S (ϕ_B and χ_B) denote the total magnitude and total cost, respectively, of the flow pushed in the light (heavy) stages.

Algorithm *AugBlock(N)*: /*N is a single-source single-sink flow network with positive arc costs*/

begin

    Let A_1 be the admissible network of N with respect to a zero-flow;

    $i \leftarrow 1$;

    while A_i admits a non-zero maximum flow

    begin (Stage i)

        To find a maximum flow of A_i, run *AugPath*(N_i) and *BlockFlow*(N_i) in parallel. Go to the next step when one of them terminates after computing a maximum flow f_i;

        Let F_{i+1} be the total flow in N after pushing flow f_i in A_i;

        Let A_{i+1} be the admissible network of N with respect to flow F_{i+1};

        $i \leftarrow i + 1$;

    end

end

Fig. 5. Algorithm *AugBlock*.

Lemma 6. *The total flow pushed in the light stages of Algorithm* AugBlock *is* $O(\chi^{3/4}\sqrt{\log n})$, *i.e.*, $\phi_S = O(\chi^{3/4}\sqrt{\log n})$.

Proof. Suppose set S consists of flows $f_{i_1}, f_{i_2}, \ldots, f_{i_l}$, where $1 \le i_1 < i_2 \ldots < i_l$. From the definition of ϕ_S, $\phi_S = \sum_{1 \le j \le l} \phi_{i_j}$, and each ϕ_{i_j} is at most $\sqrt{\chi}\log n$. From the definition of flow network A_{i_j}, all the directed paths of A_{i_j} from s to t have the same cost c_{i_j}. Therefore, $\chi_{i_j} = c_{i_j} \cdot \phi_{i_j}$. Since N is a flow network with positive arc costs, from Corollary 5, it follows that $\chi_{i_j} = c_{i_j} \cdot \phi_{i_j} \ge i_j\phi_{i_j} \ge j\phi_{i_j}$. Therefore, $\chi \ge \chi_S = \sum_{1 \le j \le l} \chi_{i_j} \ge \sum_{1 \le j \le l} j\phi_{i_j}$. In other words, $\chi \ge \phi_{i_1} + 2\phi_{i_2} + \cdots + l\phi_{i_l}$. Since ϕ_{i_j} is at most $\sqrt{\chi}\log n$, it follows that for a given χ, ϕ_S is maximum when the values of l and the ϕ_{i_j}'s are such that $\phi_{i_1} = \phi_{i_2} = \cdots = \phi_{i_l} = \sqrt{\chi}\log n$ and $\chi = \sum_{1 \le j \le l} j\sqrt{\chi}\log n = (l(l+1)/2)\sqrt{\chi}\log n$. From some mathematical manipulation, it follows that when ϕ_S is maximum, l is such that $l^2 + l = 2\sqrt{\chi}/\log n$, and hence, if we denote by l^*, the value of l for which ϕ_S is maximum, we have that $l^* = O(\chi^{1/4}/\sqrt{\log n})$. Therefore, $\phi_S \le \sum_{1 \le j \le l^*} \sqrt{\chi}\log n = l^*\sqrt{\chi}\log n = O(\chi^{1/4}/\sqrt{\log n})\sqrt{\chi}\log n = O(\chi^{3/4}\sqrt{\log n})$.
□

Lemma 7. *Let* $k(\alpha, \beta)$, *where* $0 < \alpha < 1$ *and* $\beta > 0$, *be the number of stages of Algorithm* AugBlock *for which* $\phi_i > \chi^\alpha\beta$. *Then we have*

$$k(\alpha, \beta) < \sqrt{2/\beta}\chi^{(1-\alpha)/2}$$

Proof. Let $f_{i_1}, f_{i_2}, \ldots, f_{i_l}$, where $1 \le i_1 < i_2 \ldots < i_l$ and $l = k(\alpha, \beta)$, be the set of all the flows for which $\phi_{i_j} > \chi^\alpha\beta$. From the definition of flow network A_{i_j}, all the directed paths of A_{i_j} from s to t have the same cost c_{i_j}. Therefore, $\chi_{i_j} = c_{i_j} \cdot \phi_{i_j}$. Since N is a flow network with positive arc costs, from Corollary 5, it follows that $\chi_{i_j} = c_{i_j} \cdot \phi_{i_j} \ge i_j\phi_{i_j} \ge j\phi_{i_j}$. We have that,

$\chi \geq \sum_{1 \leq j \leq l} \chi_{i_j} = \sum_{1 \leq j \leq l} j\phi_{i_j} > \sum_{1 \leq j \leq l} j\chi^{\alpha}\beta = (l(l+1)/2)\chi^{\alpha}\beta$. From some mathematical manipulation, we get that $(2/\beta)\chi^{1-\alpha} > l^2 + l \geq l^2$. Hence, $k(\alpha, \beta) = l < \sqrt{2/\beta}\chi^{(1-\alpha)/2}$. □

Since $|B| = k(1/2, \log n)$, by Lemma 7 we have,

Corollary 8. *The total number of heavy stages of Algorithm* AugBlock *is less than* $\sqrt{2/\log n}\chi^{1/4}$, *i.e.,* $|B| < \sqrt{2/\log n}\chi^{1/4}$.

Lemma 9 yields an upper bound on the time used by Algorithm *BlockFlow* for computing the maximum flow f_i of A_i.

Lemma 9. *In Stage* i *of Algorithm* AugBlock, *Algorithm* BlockFlow *computes the maximum flow* f_i *of* A_i *in time* $O(\sqrt{\chi}m_i \log n_i)$, *where* n_i *and* m_i *are the number of nodes and arcs, respectively, of* A_i.

Proof. Let r be the number of phases used by Algorithm *BlockFlow* to compute f_i. We now show that r is at most $3\sqrt{\chi}$. From Lemma 3 (Property 3) it will then follow that Algorithm *BlockFlow* computes f_i in time $O(\sqrt{\chi}m_i \log n_i)$.

We consider two cases:

Case 1. $\phi_i < \sqrt{\chi}$: From Lemma 3 (Property 1), each phase of Algorithm *BlockFlow* increases flow in A_i by at least 1. Therefore, $r \leq \phi_i < \sqrt{\chi}$.

Case 2. $\phi_i \geq \sqrt{\chi}$: Let l be the phase that increases the flow in A_i to at least $\phi_i - \sqrt{\chi}$. From Lemma 3 (Property 1), each phase of Algorithm *BlockFlow* increases the flow in A_i by at least 1. Therefore, the number of phases after phase l is at most $\sqrt{\chi}$. Hence, if we show that $l \leq 2\sqrt{\chi}$, we are done. From Lemma 3 (Property 2), each phase of Algorithm *BlockFlow* increases the depth of the shallowest layered network used for pushing more flow by at least 1. Therefore, if we can show that the depth d of the shallowest layered network L_l of A_i used for phase l is at most $2\sqrt{\chi}$, it will follow that $l \leq 2\sqrt{\chi}$. Now we show that $d < 2\sqrt{\chi}$.

Our proof follows this approach: We first compute the cost of the new flow pushed in A_i in the phases $l, l+1, \ldots, r$. Using the costs of this new flow, we compute the cost $c(F_{i+1})$ of the total flow in N at the end of stage i. Then we show that because $c(F_{i+1})$ is at most χ, d is less than $2\sqrt{\chi}$. We now give the details of the proof.

Let g be the total flow pushed in A_i in phases $1, 2, \ldots, l-1$. Let $R(A_i, g)$ be the residual flow network of A_i with respect to flow g. Let h be the total flow pushed in $R(A_i, g)$ in phases $l, l+1, \ldots, r$. Let μ be the magnitude of flow h. From the definition of phase l, we have that $\mu \geq \sqrt{\chi}$. Flow h can be decomposed into μ flows $h_1, h_2, \ldots, h_\mu$ where each h_j has unit magnitude. From Lemma 1 and Corollary 5, it follows that the cost of each h_j is greater than 0. Also, since the depth of L_l is d, we have that the length of each directed path of $R(A_i, g)$ from s to t is at least d, and therefore, each h_j uses at least d arcs of $R(A_i, g)$.

Let W_j and Z_j be the sets consisting of the arcs of h_j with positive and negative costs, respectively. Let $c(W_j) = \sum_{e \in W_j} c(e)$, i.e., the total cost of the arcs of h_j with positive costs. Similarly, let $c(Z_j) = \sum_{e \in Z_j} c(e)$. We now show that $c(W_j) > d/2$. We have shown earlier that h_j uses at least d arcs of $R(A_i, g)$. Therefore, $|W_j| + |Z_j| \geq d$. Since each arc of N and therefore, each arc of $R(A_i, g)$ has non-zero cost, we have that $c(W_j) \geq |W_j|$ and $|c(Z_j)| = -c(Z_j) \geq |Z_j|$. Hence, $c(W_j) + |c(Z_j)| \geq |W_j| + |Z_j| \geq d$. We have shown earlier that the cost of h_j is greater than 0. Since the cost of h_j is equal to $c(W_j) + c(Z_j)$, we have that $c(W_j) + c(Z_j) > 0$. It follows that $c(W_j) > -c(Z_j) = |c(Z_j)|$. Therefore, $2c(W_j) > c(W_j) + |c(Z_j)| \geq d$. Hence, $c(W_j) > d/2$.

Recall that F_i denotes the total flow pushed in N by Algorithm *AugBlock* (since the beginning of its execution) at the end of Stage $i - 1$ (See Figure 5). Let K be the set of the arcs of $R(A_i, g)$ with negative costs. Also recall that h is the total flow pushed in $R(A_i, g)$ in phases $l, l + 1, \ldots, r$. We have that

$$
\begin{aligned}
c(F_{i+1}) &= c(F_i) + c(g) + \sum_{1 \leq j \leq \mu} c(h_j) \\
&\geq c(F_i) + c(g) + \sum_{1 \leq j \leq \mu} (c(W_j) + c(Z_j)) \\
&\geq \sum_{1 \leq j \leq \mu} c(W_j) + c(F_i) + c(g) + \sum_{1 \leq j \leq \mu} c(Z_j) \\
&\geq \sum_{1 \leq j \leq \mu} c(W_j) + c(F_i) + c(g) + \sum_{1 \leq j \leq \mu} \sum_{e \in Z_j} c(e) \\
&= \sum_{1 \leq j \leq \mu} c(W_j) + c(F_i) + c(g) + \sum_{e \in K} h(e) c(e) \qquad (1)
\end{aligned}
$$

Let $e = (y, x)$ be an arc of $R(A_i, g)$ with negative cost. Arc e corresponds to an arc $e^* = (x, y)$ of N such that $u(e) = F_i(e^*) + g(e^*)$ and $c(e) = -c(e^*)$. The cost of the flow in e^* due to flows F_i and g is $F_i(e^*)c(e^*) + g(e^*)c(e^*) = (F_i(e^*) + g(e^*))c(e^*) = -u(e)c(e) \geq -h(e)c(e)$. Summing both the sides of the inequality over all the arcs of $R(A_i, g)$ with negative costs, it follows that $c(F_i) + c(g) \geq -\sum_{e \in K} h(e)c(e)$, and hence, $c(F_i) + c(g) + \sum_{e \in K} h(e)c(e) \geq 0$. Informally speaking, this means that the (negative) costs of the flows due to flow h in the arcs of $R(A_i, g)$ with negative costs is "accounted for" by the (positive) cost of the flows due to F_i and g in the arcs of N.

We have shown earlier that $c(W_j) > d/2$. Hence, $\sum_{1 \leq j \leq \mu} c(W_j) > \mu d/2$. We have also shown that $\mu \geq \sqrt{\chi}$. Therefore, $\sum_{1 \leq j \leq \mu} c(W_j) > \sqrt{\chi} d/2$.

From Eq. 1 it follows now that $c(F_{i+1}) > \sqrt{\chi} d/2$. Since $\chi \geq c(F_{i+1})$, we have that $\chi \geq c(F_{i+1}) > \sqrt{\chi} d/2$, and therefore, $d < 2\sqrt{\chi}$. □

The following theorem summarizes the main result of this section.

Theorem 10. *Let N be a single-source single-sink flow network with n nodes, m arcs, and positive arc costs. Algorithm AugBlock computes a minimum cost flow f of N in time $O(\chi^{3/4} m \sqrt{\log n})$, where χ is the cost of f.*

Proof. The proof is based on the following idea: We can bound the running-time of a stage of Algorithm *AugBlock* by the running-time (for that stage) of either of Algorithm *AugPath* and Algorithm *BlockFlow*. We can do this because the flow computed in a stage is the one given by the algorithm that terminates first. Suppose we bound the running time for a light stage by the running-time of Algorithm *AugPath*, and for a heavy stage by the running-time of Algorithm *BlockFlow*. From Lemma 6, the total flow computed in the light stages is $O(\chi^{3/4}\sqrt{\log n})$. Therefore, from Lemma 2, it follows that the total running-time of light stages is bounded by $O((\chi^{3/4}\sqrt{\log n})m)$. As for the heavy stages, from Corollary 8, we have that the number of heavy stages is at most $\sqrt{2/\log n}\chi^{1/4}$. Since from Lemma 9, Algorithm *BlockFlow* takes time $O(\sqrt{\chi}m\log n)$ for computing a flow in a stage of Algorithm *AugBlock*, the total running time of the heavy stages is bounded by $\sqrt{2/\log n}\chi^{1/4} \cdot O(\sqrt{\chi}m\log n) = O(\chi^{3/4}m\sqrt{\log n})$. This gives the total time-complexity of $O(\chi^{3/4}m\sqrt{\log n})$ for Algorithm *AugBlock*.

We now give the details of the proof.

Let r be the total number of stages used by *Algorithm AugBlock* to compute the optimal flow f. Consider stage i, where $1 \leq i \leq r$. Stage i increases the amount of flow in N by at least 1. Therefore, $\phi_i \geq 1$. Let n_i and m_i be the number of nodes and arcs, respectively, of A_i. Clearly, $n_i \leq n$ and $m_i \leq 2m$ (the factor of 2 appears because there may be some backward arcs in A_i). Let T_i, T_{S_i}, and T_{B_i} be the time taken by algorithms *AugBlock*, *AugPath* and *BlockFlow*, respectively, for computing f_i in A_i. Flow f_i is the one computed by the algorithm that terminates first. Therefore, $T_i = \min(T_{S_i}, T_{B_i})$. Hence, $T_i = T_{S_i}$ if $f_i \in S$, and $T_i = T_{B_i}$ if $f_i \in B$ are valid upper bounds for T_i.

Let T be the total time taken by *Algorithm AugBlock* to compute f. We have:

$$T = \sum_{1 \leq i \leq r} T_i$$

$$= \sum_{f_i \in S} T_{S_i} + \sum_{f_i \in B} T_{B_i}$$

From Lemma 2, $T_{S_i} = O(\phi_i \cdot m_i)$, and from Lemma 9, $T_{B_i} = O(\sqrt{\chi}m_i \log n_i)$. Therefore,

$$T = \sum_{f_i \in S} O(\phi_i \cdot m_i) + \sum_{f_i \in B} O(\sqrt{\chi}m_i \log n_i)$$

$$= O(m) \sum_{f_i \in S} \phi_i + (\sum_{f_i \in B} 1) \cdot O(\sqrt{\chi}m \log n)$$

$$= \phi_S \cdot O(m) + |B| \cdot O(\sqrt{\chi}m \log n)$$

From Lemma 6, we have that $\phi_S = O(\chi^{3/4}\sqrt{\log n})$, and from Corollary 8, we have that $|B| < \sqrt{2/\log n}\chi^{1/4}$. Therefore, $T = O(\chi^{3/4}\sqrt{\log n} \cdot m) + O(\sqrt{2/\log n}\chi^{1/4} \cdot \sqrt{\chi}m \log n) = O(\chi^{3/4}m\sqrt{\log n})$. □

4 Faster Bend Minimization

Orthogonal drawings of graphs, where the edges are drawn as polygonal chains with alternating horizontal and vertical segments, are widely used in visualization applications, and they have been extensively studied (see, e.g., [2, 4, 6, 8, 11, 12, 13, 14, 16, 18, 20, 22]).

An important quality measure for orthogonal drawings is the total number of bends along the edges. Bend minimization is the core of a practical drawing technique [5] for general graphs, called Giotto, which performs a preliminary planarization followed by bend-minimization. Giotto has has been widely used in software and data visualization systems [5, 7]. A sample drawing is shown in Figure 6(d)). The extensive experiments conducted by Di Battista et al. [4] on general-purpose orthogonal drawing algorithms, which use 11,582 graphs derived from "real-life" software engineering and database applications, show that Giotto outperforms all other known orthogonal drawing algorithms in quality measures such as area, number of bends, and aspect-ratio. However, the bottleneck in the running time of Giotto is the execution of the bend minimization step.

Let G be an embedded planar graph with maximum degree 4. As shown in [19], a drawing of G with the minimum number of bends can be computed by an algorithm consisting of the following two main phases:

1. computation of an *orthogonal shape* for G, where only the bends and the angles of the orthogonal drawing are defined;
2. assignment of integer lengths to the segments of the orthogonal shape.

Phase 1 uses a transformation into a network flow problem (Figure 6(a–c)), where each unit of flow is associated with a right angle in the orthogonal drawing. Hence, angles are viewed as a commodity that is produced by the vertices, transported across faces by the edges through their bends, and eventually consumed by the faces. It is easier to to describe this flow problem on a network with lower bounds on flows in arcs (in addition to arc-capacities) and supplies/demands on the sources and sinks. From the embedded graph G we construct such a flow network N as follows. The nodes of network N are the vertices and faces of G. Let $\deg(f)$ denote the number of edges of the circuit bounding face f. Each vertex v supplies $\sigma(v) = 4$ units of flow, and each face f consumes $\tau(f)$ units of flow, where

$$\tau(f) = \begin{cases} 2\deg(f) - 4 & \text{if } f \text{ is an internal face} \\ 2\deg(f) + 4 & \text{if } f \text{ is the external face} \end{cases}$$

By Euler's formula, $\sum_v \sigma(v) = \sum_f \tau(f)$, i.e., the total supply is equal to the total consumption.

Network N has two types of arcs:

– arcs of the type (v, f), where f is a face incident on vertex v; the flow in (v, f) represents the angle at vertex v in face f, and has lower bound 1, capacity 4, and cost 0;

– arcs of the type (f, g), where face f shares an edge e with face g; the flow in (f, g) represents the number of bends along edge e with the right angle inside face f, and has lower bound 0, capacity $+\infty$, and cost 1.

The conservation of flow at the vertices expresses the fact that the sum of the angles around a vertex is equal to 2π. The conservation of flow at the faces expresses that fact that the sum of the angles at the vertices and bends of an internal face is equal to $\pi(p-2)$, where p is the number of such angles. For the external face, the above sum is equal to $\pi(p+2)$.

It can be shown that every flow ϕ in network N corresponds to an admissible orthogonal shape for graph G, whose number of bends is equal to the cost of flow ϕ. Hence, an orthogonal shape for G with the minimum number of bends can be computed from a minimum cost flow in G.

Phase 2 uses a simple compaction strategy derived from VLSI layout, where the lengths of the horizontal and vertical segments are computed independently after a preliminary refinement of the orthogonal shape that decomposes each face into into rectangles.

The best previous time bound for bend minimization is $O(n^2 \log n)$ [19], which is achieved with standard flow-augmentation techniques. Our new minimum cost flow method yields a faster bend minimization algorithm:

Theorem 11. *Let G be an embedded planar graph with n vertices and maximum vertex degree 4. An orthogonal drawing of G with the minimum number of bends can be computed in $O(n^{7/4}\sqrt{\log n})$ time.*

Sketch of Proof: Modify the flow network N associated with G to a get a new single-source single-sink flow network N' with positive arc costs by

– assigning unit cost to all the arcs of the type (v, f),
– adding two new nodes s and t, designating them the source and sink, respectively,
– for every vertex v, adding an arc (s, v) with capacity $\sigma(v)$ and cost 1, and
– for every face f, adding an arc (f, t) with capacity $\tau(f)$ and cost 1.

It is easy to see a flow f is of minimum cost in N if and only if it is of minimum cost in N'. Network N' has $O(n)$ nodes and arcs. Also, since the minimum number of bends for G is $O(n)$ (see, e.g., [20]), the minimum cost of the flow in N' is $O(n)$. Hence, we can use Algorithm $AugBlock$ to compute a minimum cost flow for N' in time $O(n^{7/4}\sqrt{\log n})$ (see Theorem 10).

□

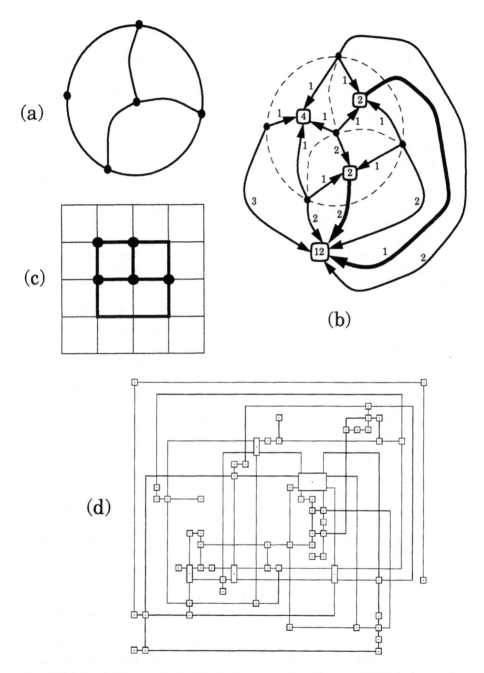

Fig. 6. (a) Embedded graph G. (b) Minimum cost flow in network N associated with G: the flow is shown next to each arc; arcs with zero flow are omitted; arcs with unit cost are drawn with thick lines; a face f is represented by a box labeled with $\tau(f)$. (c) Planar orthogonal grid drawing of G with minimum number of bends. (d) Orthogonal grid drawing of a nonplanar graph produced by Giotto.

References

1. R.K. Ahuja, T.L. Magnanti, and J.B. Orlin. *Network Flows: Theory, Algorithms, and Applications.* Prentice Hall, Englewood Cliffs, NJ, 1993.

2. T. Biedl and G. Kant. A better heuristic for orthogonal graph drawings. In *Proc. 2nd Annu. European Sympos. Algorithms (ESA '94)*, volume 855 of *Lecture Notes in Computer Science*, pages 24–35. Springer-Verlag, 1994.

3. G. Di Battista, P. Eades, R. Tamassia, and I. G. Tollis. Algorithms for drawing graphs: an annotated bibliography. *Comput. Geom. Theory Appl.*, 4:235–282, 1994.

4. G. Di Battista, A. Garg, G. Liotta, R. Tamassia, E. Tassinari, and F. Vargiu. An experimental comparison of three graph drawing algorithms. In *Proc. 11th Annu. ACM Sympos. Comput. Geom.*, pages 306–315, 1995.

5. G. Di Battista, A. Giammarco, G. Santucci, and R. Tamassia. The architecture of Diagram Server. In *Proc. IEEE Workshop on Visual Languages (VL'90)*, pages 60–65, 1990.

6. G. Di Battista, G. Liotta, and F. Vargiu. Spirality of orthogonal representations and optimal drawings of series-parallel graphs and 3-planar graphs. In *Proc. Workshop Algorithms Data Struct.*, volume 709 of *Lecture Notes in Computer Science*, pages 151–162. Springer-Verlag, 1993.

7. G. Di Battista, G. Liotta, and F. Vargiu. Diagram Server. *J. Visual Lang. Comput.*, 6(3):275–298, 1995. (special issue on Graph Visualization, edited by I. F. Cruz and P. Eades).

8. S. Even and G. Granot. Grid layouts of block diagrams — bounding the number of bends in each connection. In R. Tamassia and I. G. Tollis, editors, *Graph Drawing (Proc. GD '94)*, volume 894 of *Lecture Notes in Computer Science*, pages 64–75. Springer-Verlag, 1995.

9. L.R. Ford and D.R. Fulkerson. A primal-dual algorithm for the capacitated hitchcock problem. *Naval Research Logistics Quarterly*, 4:47–54, 1957.

10. L.R. Ford and D.R. Fulkerson. *Flows in Networks.* Princeton University Press, Princeton, NJ, 1962.

11. U. Fößmeier and M. Kaufmann. On bend-minimum orthogonal upward drawing of directed planar graphs. In R. Tamassia and I. G. Tollis, editors, *Graph Drawing (Proc. GD '94)*, volume 894 of *Lecture Notes in Computer Science*, pages 52–63. Springer-Verlag, 1995.

12. A. Garg and R. Tamassia. On the computational complexity of upward and rectilinear planarity testing. In R. Tamassia and I. G. Tollis, editors, *Graph Drawing (Proc. GD '94)*, volume 894 of *Lecture Notes in Computer Science*, pages 286–297. Springer-Verlag, 1995.

13. G. Kant. Drawing planar graphs using the canonical ordering. *Algorithmica*, 16:4–32, 1996. (special issue on Graph Drawing, edited by G. Di Battista and R. Tamassia).

14. Y. Liu, P. Marchioro, R. Petreschi, and B. Simeone. Theoretical results on at most 1-bend embeddability of graphs. Technical report, Dipartimento di Statistica, Univ. di Roma "La Sapienza", 1990.

15. K. Mehlhorn. *Graph Algorithms and NP-Completeness*, volume 2 of *Data Structures and Algorithms*. Springer-Verlag, Heidelberg, West Germany, 1984.

16. A. Papakostas and I. G. Tollis. Improved algorithms and bounds for orthogonal drawings. In R. Tamassia and I. G. Tollis, editors, *Graph Drawing (Proc. GD '94)*, volume 894 of *Lecture Notes in Computer Science*, pages 40–51. Springer-Verlag, 1995.

17. D.D. Sleator. *An O(nm log n) Algorithm for Maximum Network Flow*. PhD thesis, Dept. Comput. Sci., Stanford Univ., Palo Alto, California, 1980.
18. J. A. Storer. On minimal node-cost planar embeddings. *Networks*, 14:181–212, 1984.
19. R. Tamassia. On embedding a graph in the grid with the minimum number of bends. *SIAM J. Comput.*, 16(3):421–444, 1987.
20. R. Tamassia and I. G. Tollis. Planar grid embedding in linear time. *IEEE Trans. Circuits Syst.*, CAS-36(9):1230–1234, 1989.
21. R. E. Tarjan. *Data Structures and Network Algorithms*, volume 44 of *CBMS-NSF Regional Conference Series in Applied Mathematics*. Society for Industrial Applied Mathematics, 1983.
22. L. Valiant. Universality considerations in VLSI circuits. *IEEE Trans. Comput.*, C-30(2):135–140, 1981.

Constrained Graph Layout

Weiqing He and Kim Marriott

Computer Science Department,
Monash University,
Clayton 3168, Australia

{whe,marriott}@cs.monash.edu.au

Abstract. Most current graph layout technology does not lend itself to interactive applications such as animation or advanced user interfaces. We introduce the constrained graph layout model which is better suited for interactive applications. In this model, input to the layout module includes suggested positions for nodes and constraints over the node positions in the graph to be layed out. We describe three implementations of layout modules which are based on the constrained graph layout model. The first two implementations are for undirected graph layout and the third is for tree layout. The implementations use active set techniques to solve the layout. Our empirical evaluation shows that they are quite fast and give reasonable layout.

1 Introduction

Most research on graph layout has concentrated on how to layout a graph in some fixed style in isolation from the rest of the application. However, this model of graph layout, while very simple, does not lend itself to interactive applications such as animation or advanced user interfaces. This is for two main reasons. The first reason is that, in many interactive applications, the graph is repeatedly modified (by either the user or the application program) and redisplayed. When the graph is redisplayed, the new layout should preserve the *mental map* of the user [11, 32, 28], that is the new layout should not move an existing node unless the current position leads to poor layout. The second reason is that almost all existing graph layout algorithms are quite restrictive in how graphs can be laid out since they encapsulate fixed layout aesthetics. The application program cannot place constraints on the layout which take into account the underlying semantics of the object represented by the graph.

To overcome these problems, we introduce a general model of *constrained graph layout* which is better suited for interactive applications involving graph layout, see Fig. 1. In constrained graph layout, the graph layout module takes: the graph, a set of constraints over the x and y positions of the nodes, and a partial assignment of suggested values for the node coordinates. The graph layout module is responsible for finding an assignment to variables representing the node coordinates which is *feasible*, that is satisfies the constraints, gives a good layout, and assigns values to the variables which are as close as possible to the suggested values. The constraints enable the layout module to take additional semantic information about the graph into account, and the suggested values allow the layout module to try and preserve the current layout of the graph.

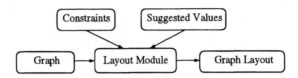

Fig. 1. The Constrained Graph Layout Model

Different graph layout modules allow different classes of constraints and may also embody different layout algorithms and aesthetic criteria. For different applications one must use the appropriate graph layout module. Our major technical contribution is the implementation and design of three different constrained graph layout modules. All three modules allow arbitrary linear arithmetic constraints. The first two modules are for undirected connected graphs, while the third module is for tree layout. The first module is based on Kamada's aesthetic cost function [22, 23]. The second module is a modification of this cost function which removes the non-polynomial terms. Unfortunately, Kamada's algorithm for graph layout is not suitable for constrained graph layout since it cannot handle interaction between variables. Instead we use active set methods developed to solve constrained optimization problems with an arbitrary non-linear objective function. The tree-layout module relies on methods developed to solve quadratic programming.

We also present an empirical evaluation of the three modules. Our evaluation shows that our second layout module for undirected connected graphs is not significantly slower than Kamada's original algorithm for unconstrained graph layout — thus there is very little performance penalty in using this more flexible model. Second, we show that the second module gives faster and more robust layout, but sometimes the layout is inferior to that given by the model based on Kamada's original cost aesthetic. Third, we show that the tree layout module is significantly faster than the other two modules. Thus, if you are only laying out trees you should use this specialized module.

Until recently there has been relatively little work on graph layout in the presence of both constraints and suggested values. The work most closely related is that of Kamps and Klein [24]. They also look at layout in the presence of constraints. The class of constraints they allow, called geometric constraints, is less expressive than the arbitrary linear arithmetic constraints considered here. For example, they cannot specify that a node should be placed at the mid-point of two nodes. Furthermore, they do not allow suggested values for nodes. Other closely related work is described in Luders *et al* [27] who use a two-phase combinatorial optimization algorithm for graph layout. They also allow constraints, which are inequalities between node positions. Again, their constraints are not as expressive as those considered here. Dengler *et al* [9] also look at constraint-driven layout. Their motivation is quite different. In their approach constraints are generated automatically and encode good graphic design rules. They do not require all of the constraints to be satisfied, rather the layout algorithm just tries to satisfy as many of these constraints as possible.

There is, of course, a considerable body of literature to do with constraint based generation of diagrams, in which layout is hardwired into the rules of generation. See for example, Brandenburg [4], Helm and Marriott [19, 18], Weitzman and Wittenburg [38] and Cruz *et al* [7]. There is also a considerable body of literature on constraint based diagram manipulation. See for instance, Garnet [29], QOCA [20] and ThingLab [2]. More general related work is described in the surveys [12, 1]. In particular, force-directed graph drawing algorithms for unconstrained graph layout can be found in [8, 23, 36, 15, 34, 10, 14], and [5] contains an experimental comparison between [36, 14, 8, 23, 15]. Algorithms for incremental unconstrained graph layout are given in [25, 30], while [26] discusses the approach of integration of declarative and algorithmatic graph layout.

2 Constrained Graph Layout

Existing research on graph layout has tended to focus on how to layout a graph statically using a fixed pre-defined style. Unfortunately, this model of graph layout, while very simple, does not lend itself to interactive applications since when a graph is modified and redisplayed the new layout may not preserve the mental map of the user [11, 32] and the application program cannot add constraints on the layout which take into account the underlying semantics of the object represented by the graph.

For these reasons we have introduced the *constrained graph layout* model which overcomes these problems and so is better suited for interactive applications. In constrained graph layout, the graph layout module takes three parameters. The model is shown in Fig. 1. The first is a *graph*, $G = \langle V, E \rangle$, where $V = \{1, \ldots, n\}$ is the set of nodes in the graph where each node is represented by a (unique) integer and E is a set of edges $\langle i, j \rangle \in E$ for $i, j \in V$. The second parameter is a set of *constraints* over the x and y position of the nodes, where x_i and y_i are variables denoting the x and y coordinate of node i in the layout. Note that the constraints may also refer to other variables than the node coordinates. The third parameter is an assignment, ψ, of *suggested values* for the variables representing the node coordinates. For example, $\psi(x_i)$ is the desired value for the x coordinate of node i. Each assignment to a variable, v, has an associated weight, $w(v)$, indicating the importance of the suggested value. The larger the weight, the more the value is desired. If the weight is zero, i.e. $w(v) = 0$, then the suggested value is ignored. In the constrained graph layout model, the constraints enable the layout module to take additional semantic information about the graph into account, and the suggested values allow the layout module to try and preserve the current layout of the graph.

The graph layout module is responsible for finding an assignment to variables representing the node coordinates which satisfies the constraints, gives a good layout, and assigns the values to the variables which are as close as possible to the suggested values. More exactly, the graph layout module embodies an algorithm to solve the optimization problem:

minimize $\phi(\mathbf{v}) + dist(\psi(v), v)$
with respect to $C(\mathbf{v})$.

where $\mathbf{v} = (x_1, x_2, \ldots, x_n, y_1, y_2, \ldots, y_n)$ and $\phi(\mathbf{v})$ captures the graph layout aesthetic as an objective function over the nodes' x coordinates and y coordinates, $C(\mathbf{v})$ is a set of constraints and $dist$, which takes into account the weight, is some metric over the variable values.

Different graph layout modules support different classes of constraints. For example, one layout module might accept arithmetic equalities while another might allow linear arithmetic equality and inequality constraints. Of course there is a trade off between the speed of layout and the expressiveness of the allowed constraints. Different graph layout modules may also embody different layout algorithms and aesthetic criteria. For example, a layout module might be specialized for tree layout in which the criteria is to minimize the size of the tree or else the module might be for general directed graph layout, in which the criteria for layout are to minimize the number of edge crossings and to represent isomorphic sub-graphs identically. The final way in which layout modules may differ, is in the choice of metric by which solutions are compared. For instance, the metric might be the Euclidean norm or it might be the function which gives 0 if the values are identical and 1 otherwise.

For different applications one must use the appropriate graph layout module. That is, one should choose a module which allows expressive enough constraints, yet is also efficient enough. In the next two sections we describe three different graph layout modules which we have implemented.

3 Implementation for Undirected Graphs

In this section we detail two constrained graph layout models for general undirected connected graphs. Both models handle arbitrary arithmetic linear equality and inequality constraints and use the square of the Euclidean distance as the metric over solution values with the contribution of each variable multiplied by the associated weight. The basic idea is to use a spring model energy function as the aesthetic cost function ϕ, however the models differ in the choice of ϕ.

The first model, *Model A*, uses the aesthetic cost function suggested by Kamada [23]. This is

$$\phi_A = \sum_{i=1}^{n-1} \sum_{j=i+1}^{n} \frac{1}{2} k_{ij} (|p_i - p_j| - l_{ij})^2 \qquad (1)$$

where: k_{ij} is spring factor between node p_i and node p_j; and
l_{ij} is a desirable length between node p_i and node p_j.

We can rewrite this cost function to the following:

$$\phi_A = \frac{1}{2} \sum_{i=1}^{n-1} \sum_{j=i+1}^{n} k_{ij} \{ (x_i - x_j)^2 + (y_i - y_j)^2 - 2 \cdot l_{ij} \cdot \sqrt{(x_i - x_j)^2 + (y_i - y_j)^2} \} \qquad (2)$$

The model places a "spring" between each pair of nodes which tries to position the nodes so that the distance between them is the desired length. The cost function is a measure of the energy in the springs.

Our second model, *Model B*, is a polynomial approximation of *Model A*.

$$\phi_B = \sum_{i=1}^{n-1} \sum_{j=i+1}^{n} k_{ij}^2 ((x_i - x_j)^2 + (y_i - y_j)^2 - l_{ij}^2)^2 \qquad (3)$$

In this model the cost function is the sum of the squared differences between the desired distance between nodes and the actual distance. The definitions of k_{ij} and l_{ij} are the same as those in (2).

The primary disadvantage of *Model B* over *Model A* is the weaker repulsive force between nodes which arises from the lack of $\sqrt{(x_i - x_j)^2 + (y_i - y_j)^2}$ like terms in the aesthetic cost function of *Model B*. This means that *Model B* may sometimes produce a layout with some coincident nodes. In the case of *Model A*, however, since ϕ_A has $\sqrt{(x_i - x_j)^2 + (y_i - y_j)^2}$ like terms, nodes can never be assigned the same location since

$$\frac{\partial}{\partial x_i} \left(\sqrt{(x_i - x_j)^2 + (y_i - y_j)^2} \right) = \frac{(x_i - x_j)}{\sqrt{(x_i - x_j)^2 + (y_i - y_j)^2}},$$

which has no definition when $x_i = x_j$ and $y_i = y_j$. In other words, ϕ_A will have some points where its partial derivatives do not exist, and so any local minimum search method will never go to these points.

On the other hand, lack of smoothness of the partial derivatives of ϕ_A means that the computation of the minimum of ϕ_A may be sensitive to the initial configuration and initial feasible solution. This is because the optimization method may not be capable of leaping over a point where the partial derivative does not exist from the current feasible solution to a solution closer to the local minimum. Since ϕ_B has no points whose partial derivatives do not exist, we would expect numerical optimization techniques to be more stable when solving *Model B*. This is of particular importance in the case of constrained optimization because the presence of constraints makes it more difficult to find an initial feasible solution which is close to the global minimum.

The other main advantage of *Model B* over *Model A* is that we would expect to be able to compute a minimum of ϕ_B faster than we can compute a minimum of ϕ_A. This is because one of the main time costs in computing a minimum in almost any numerical method is the need to calculate partial derivatives and computing partial derivatives of ϕ_A will take longer because of the $\sqrt{(x_i - x_j)^2 + (y_i - y_j)^2}$ terms. In particular, for popular optimization techniques such as *gradient descent method* and *conjugate gradient descent method* [13], the increment in line search can be determined symbolically if the aesthetic cost function has polynomial form while only numeric methods can be used if the aesthetic cost function includes $\sqrt{(x_i - x_j)^2 + (y_i - y_j)^2}$ terms.

The most important question is how we can efficiently solve the resulting optimization problems when using *Model A* or *Model B* for constrained graph layout. In both cases we must solve a constrained optimization problem of the following form:

$$\begin{aligned} minimize \quad & \phi^*(\mathbf{v}) \\ subject\ to \quad & \mathbf{C} \end{aligned} \tag{4}$$

where: $\mathbf{v} = (x_1, x_2, \ldots, x_n, y_1, y_2, \ldots, y_n)$;

$\mathbf{C}$ is a set of linear equality and inequality constraints, and

$\phi^*(\mathbf{v})$ is $\phi_A(\mathbf{v}) + \sum_{v \in \mathbf{v}} w(v) \cdot (\psi(v) - v)^2$ in the case of *Model A*, and
$\phi^*(\mathbf{v})$ is $\phi_B(\mathbf{v}) + \sum_{v \in \mathbf{v}} w(v) \cdot (\psi(v) - v)^2$ in the case of *Model B*.

Kamada [22] gives a simple and efficient algorithm to minimize ϕ_A in the case that there are no constraints. Kamada claims that the $2n$-dimensional Newton-Raphson method cannot be directly applied because $\sum_1^n \frac{\partial \phi_A}{\partial x_i} = 0$, and $\sum_1^n \frac{\partial \phi_A}{\partial y_i} = 0$, which means the $2n$ partial derivatives are not independent of one another. Instead his algorithm repeatedly recomputes the position of each node, one at a time, by solving two linear equations, involving calculation of derivatives only for that one node, to obtain x- and y-increments of the node to be moved while the other nodes are temporarily frozen. The algorithm terminates when a local minimum is reached. Unfortunately, it seems impossible to use Kamada's algorithm to solve problems of the form of (4) because the constraints introduce interaction between variables. This means it is not possible to recompute the position of a node independently from the current position of the other nodes. Therefore we need some other method for solving optimization problems of the form of Equation (4).

Our implementation for *Model A* and *Model B* is instead based on the *Active Set Method* [13]. This is an iterative technique developed by the operations research community to solve constrained optimization problems with inequality constraints. It is reasonably robust and quite fast. The key idea behind the algorithm is to solve a sequence of constrained optimization problems O_1, ..., O_n, which only have equality constraints. This set of equality constraints, $\mathcal{A}$, is

(a) Compute l_{ij}, k_{ij} for $1 \leq i \neq j \leq n$;

(b) Compute initial feasible solution $\mathbf{v}^{(1)}$ and active set $\mathcal{A} := \mathcal{A}(\mathbf{v}^{(1)})$;
$\phi_{old}^* = \phi^*(\mathbf{v}^{(1)})$;

(c) Compute $\mathbf{d}$ by solving (5);
    if $(\mathbf{d} \geq \varepsilon)$ **goto** (e);

(d) Let $\lambda_q^{(k)}$ solve $min \ \lambda_i^{(k)}$, for all $\mathbf{a}_i \in \mathcal{A}$ and c_i is an inequality constraint;
    if $(\lambda_q^{(k)} \geq 0)$
        terminate with $\mathbf{v}^{(k)}$ as the solution;
    else{
        remove $\mathbf{a}_q$ from $\mathcal{A}$;
        goto (c); }

(e) Compute $\overline{\alpha}^{(k)} := min \ \frac{b_i - \mathbf{v}^{(k)} \cdot \mathbf{a}_i}{\mathbf{d} \cdot \mathbf{a}_i}$ for $\mathbf{a}_i$ in $\overline{\mathcal{A}}$ and $\mathbf{d} \cdot \mathbf{a}_i < 0$
    Choose α as an increment along the line search direction $\mathbf{d}$;
    if ($\overline{\alpha}^{(k)}$ *exist*)
        $h := min(\overline{\alpha}^{(k)}, \alpha)$;
    else
        $h := \alpha$;

(f) $\mathbf{v}^{(k+1)} := \mathbf{v}^{(k)} + h \cdot \mathbf{d}$;
    while $(\phi^*(\mathbf{v}^{(k+1)}) \geq \phi_{old}^*)$ {
        REDUCTION(h);
        $\mathbf{v}^{(k+1)} = \mathbf{v}^{(k)} + h \cdot \mathbf{d}$; }

(g) **if**($h = \overline{\alpha}$) {
    let p be a constraint index which holds $\overline{\alpha}$ in (e);
    add $\mathbf{a}_p$ to $\mathcal{A}$;}

(h) $\phi_{old}^* = \phi^*(\mathbf{v}^{(k+1)})$;
    $k = k + 1$;
    goto (c);

Fig. 2. Constrained Graph Layout Algorithm

called the *active set*. It consists of the original equality constraints plus those inequality constraints which are required to be equalities. The other inequalities are ignored.

Essentially, each optimization problem O_i is solved using a *gradient descent method* which iteratively computes a solution which is feasible with respect to the equality constraints in O_i and which is in a search direction $\mathbf{d}$ reducing the objective function. However, it may be that the new solution while feasible with respect to the active set of O_i, holds $\mathbf{v} \cdot \mathbf{a} = b$ for an inequality constraint $\mathbf{v} \cdot \mathbf{a} \geq b$ of the original problem which is not in the active set. In this case the corresponding equality $\mathbf{v} \cdot \mathbf{a} = b$ is added to the active set, giving rise to a new optimization problem O_i. Constraints may also be taken out of the active set, when a better search direction can be found "away" from the constraint in the direction satisfying the original inequality.

The precise algorithm is given in Fig. 2. Assume that $\mathbf{C} = \{c_1, c_2, \ldots, c_m\}$

and that $\mathbf{a} = (\mathbf{a}_1, \mathbf{a}_2, \ldots, \mathbf{a}_m)$ and $\mathbf{b} = (b_1, b_2, \ldots, b_m)$ where $\mathbf{a}_i$ and b_i are coefficient column vector and right-hand side constant of c_i. At each step $\mathcal{A} = \{\mathbf{a}_i : c_i \text{ is active}\}$ is a column vector set representing the active set.

The algorithm in Fig. 2 proceeds as follows. In step(b), an initial feasible solution is found, and the initial active set $\mathcal{A}$ is set to $\mathcal{A}(\mathbf{v}^{(1)})$, which is the set of $\mathbf{a}_i$ such that $\mathbf{v}^{(1)} \cdot \mathbf{a}_i = b_i$ holds. Step(c) to step(h) are then performed iteratively. In each iteration $\mathbf{v}^{(k)}$ is a feasible point and step(c) attempts to solve the optimization problem

$$\begin{aligned} minimize \qquad & \phi^*(\mathbf{v}^{(k)} + \mathbf{d}) \\ subject\ to \qquad & \mathbf{d} \cdot \mathbf{a}_i = 0, \mathbf{a}_i \in \mathcal{A}. \end{aligned} \tag{5}$$

The solution to this problem is the search direction $\mathbf{d}$. If $\mathbf{d}$ is small enough, that means $\mathbf{v}^{(k)}$ is an acceptable solution, and the Lagrange multipliers $\boldsymbol{\lambda}^{(k)}$ of the problem are examined to determine if a local minimum has been reached.[1] Denote $\lambda_q^{(k)} = \min \lambda_i^{(k)}$ for all $\mathbf{a}_i \in \mathcal{A}$ and c_i is an inequality constraint. The algorithm terminates with $\mathbf{v}^{(k)}$ as the solution if $\lambda_q^{(k)}$ is non-negative because if all λ_i are non-negative there exist no better solution near $\mathbf{v}^{(k)}$, then $\mathbf{v}^{(k)}$ is a local minimum. Otherwise c_q should be removed from the active set, i.e. remove a_q from $\mathcal{A}$, then go to step(c). If $\mathbf{d}$ in step(c) is not small enough, step(e) and step(f) are used to work out a new feasible solution $\mathbf{v}^{(k+1)} = \mathbf{v}^{(k)} + h \cdot \mathbf{d}$. In step(e), to make sure that h will keep $\mathbf{v}^{(k+1)}$ feasible for (4), h is chosen to be the minimum of α and $\overline{\alpha}$ if $\overline{\alpha}$ can be found. Otherwise h is set to α if $\overline{\alpha}$ does not exist. The **while** loop in step(f) is designed to guarantee that $\phi^*(\mathbf{v}^{(k+1)}) < \phi^*(\mathbf{v}^{(k)})$. This is how the new feasible solution is produced. If $\overline{\alpha}$ has finally been chosen as an increment h in step(e) and step (f), then the c_i that holds $\overline{\alpha}$ is added to the active set, and before starting a new iteration from step(c), step(h) recomputes $\phi^*(\mathbf{v}^{(k+1)})$ and k. For further details about the *active set method*, please see [13].

After variable elimination, (5) can be reduced to an unconstrained optimization problem, and then any appropriate unconstrained optimization technique can be applied in step(c) to solve it. We first use gradient descent method and then the conjugate gradient descent method, which is more efficient when a point is close to a minimum.

The most important consideration when using the active set method is how to compute the initial feasible solution in step(b) of Fig. 2. If the initial solution is close to the global optimum, then convergence of the active set algorithm will be fast. Conversely, if the initial solution is too far from the global minimum, convergence may be slow, and the algorithm may return a local minimum rather than the global minimum.

---

[1] Essentially, the minimum of a function $f(x_1, x_2, \ldots, x_n)$ subject to equality constraints

$$e_j(x_1, x_2, \ldots, x_n) = 0 \text{ for } j = 1, 2, \ldots, s,$$

is to be found among the turning points of the *Lagrangian form*

$$\Phi(\mathbf{x}, \lambda) = f(\mathbf{x}) + \sum_{j=1}^{s} \lambda_j \cdot e_j(\mathbf{x})$$

where $\lambda = (\lambda_1, \ldots, \lambda_s)$ are known as Lagrange multipliers [3].

Kamada's algorithm for unconstrained minimization of ϕ_A also requires a good initial solution. In this case, however, because there are no constraints any assignment to the variables is feasible and so it is easier to find a good "guess" for the initial solution. If there are n nodes, Kamada's algorithm uses the assignment, θ_{init}, which assigns values to the nodes x and y coordinates so that the nodes are placed on the vertices of the regular n-polygon circumscribed by the largest circle which can fit in the display rectangle [23].

Unfortunately, in our case we cannot directly use θ_{init} as the initial solution as it may not be feasible. Also, we would like to use the suggested values for variables wherever possible. This suggests that we use as the initial solution the assignment, ψ_{init}, defined by

$$\psi_{init}(v) = \begin{cases} \psi(v) & \text{if } w(v) \neq 0, \\ \theta_{init}(v) & \text{otherwise} \end{cases}$$

where ψ is the suggested value for the node placements. However, we cannot use ψ_{init} as the initial solution because it may not be feasible. Instead we compute the closest solution to ψ_{init} which satisfies the constraints. In other words, our initial solution, ψ'_{init}, is the solution to the following constrained optimization problem:

$$\begin{aligned} minimize \quad & \sum_{v \in \mathbf{v}} (v - \psi_{init}(v))^2 \\ subject\ to \quad & \mathbf{C}. \end{aligned} \tag{6}$$

Problem (6) is an example of a *Quadratic Programming* problem. Such problems are well-behaved since they are convex, so that the local minimum is the global minimum. There are many fast algorithms for solving quadratic programming problems. For example, interior point methods with guaranteed polynomial worst-case behavior could be used. We chose to use a specialized active set method which has been found to be fast in practice [17]. A good initial feasible solution is vital to a nice, fast graph layout and we have found the above approach to be both efficient and satisfactory.

4 Implementation for Trees

Trees are a special type of graph which are widely used in many different application areas. There are a variety of drawing conventions [31] for trees and many different layout algorithms and approaches have been investigated. See for example, [37, 39, 33, 21, 16]. In particular, [22] gives a variant of the spring model to layout trees nicely. Unfortunately, all of these algorithms and approaches are for unconstrained tree layout. In this section we detail a constrained graph layout module for trees. Our model handles arbitrary arithmetic linear equality and inequality constraints and uses the square of the Euclidean distance as the metric over solution values.

Our model is based on viewing unconstrained tree layout as a quadratic optimization problem. Different aesthetic criteria give rise to different optimization problems. As an example of how to capture one particular aesthetic criteria for tree layout as an optimization problem, consider the layout of a *downward rooted tree*. The following constraints, $\mathbf{C}_{tree}$, lay out the tree $T =< V, E >$, where $V = \{1, 2, \ldots, n\}$ is the set of nodes, in an aesthetically pleasing way – By convention, v_1 is the root of the tree:

- for all x_i, x_j, if j is the right neighbor of i, add $x_j - x_i \geq g_x$ into $\mathbf{C}_{tree}$, where g_x is some pre-defined minimal horizontal gap between nodes;
- for any non-leaf node i, if l is i's left-most son and r is i's right-most son, add $x_i = (x_l + x_r)/2$ into $\mathbf{C}_{tree}$;

All vertices' y-coordinates y_i are placed along horizontal lines according to their level. Let $parent(v)$ be the parent of node v. We use the objective function $\phi_{tree} = \sum_{j=2}^{n}(x_j - x_{parent(j)})^2$ to capture the desire to minimize tree width, which is often one of the most important aesthetic criteria in tree-drawing. Thus, we can layout a downward rooted tree by finding the solution to $\mathbf{C}_{tree}$ which minimizes ϕ_{tree}.

Our constrained tree layout module extends this idea. It is based on the following model, *Model C*, which captures the aesthetic layout of trees as a quadratic programming problem. The model is parametric in the choice of $\mathbf{C}_{tree}$ and ϕ_{tree} which are the constraints and objective function which capture the desired aesthetic criteria for a particular type of unconstrained tree layout. The model solves the following optimization problem:

$$minimize \quad \phi_{tree}(\mathbf{v}) + \sum_{v \in \mathbf{v}} w_v \cdot (\psi(v) - v)^2$$

$$subject\ to \quad \mathbf{C} \cup \mathbf{C}_{tree} \tag{7}$$

where: $\mathbf{v} = (x_1, x_2, \ldots, x_n, y_1, y_2, \ldots, y_n)$;
$\mathbf{C}$ is the input set of linear equality and inequality constraints.

As this is a quadratic programming problem there exist many robust and fast techniques for finding the minimum. We use a variant of the active set method.

5 Emprical Evaluation

In this section we detail our empirical evaluation of the three different constrained graph layout modules we have described in the last two sections. The constrained tree layout module, *Model C*, uses the aesthetic criteria given in the example for downward rooted trees. For each model we evaluate both the speed, in seconds, of the layout and the quality of the layout. All programs are implemented in *Borland C++* and run on a *DECpc LPx+ 466d2*. Our first experiment was to compare the quality of layout and speed of layout of *Model A*, *Model B* and Kamada's original algorithm for unconstrained undirected graph layout. Each of the three methods was tried on five sample graphs, of which two, Graph 1 and Graph 2, are Fig. 5.9(b) and Fig. 5.11(c) in [22] respectively; one, Graph 3, is taken from [10]; Graph 4, is given by us;

and the last one is K_{10}. Table 1 shows the time for each method to layout the graph. The layouts using *Model B* of Graph 2 and Graph 4 are shown in Fig. 3 and Fig. 4, respectively. The results show that *Model B* is significantly faster than *Model A* and that *Model B* is not significantly slower than Kamada's original algorithm. This demonstrates that the overhead of a general purpose constraint solving algorithm is not unreasonable even for unconstrained graph layout. The quality of layout produced by *Model B* is satisfactory, and is generally similar to that produced by Kamada's algorithm (and hence *Model A*). The only important difference is for Graph 2. The layout using *Model B* is shown in Fig. 3 and should be compared with the layout given by Kamada's algorithm (shown in [22]). Kamada's algorithm gives a layout with an edge crossing. This is avoided by *Model B* but the layout is still not very good because node 1 is too close to node 2 and node 3 due to the weaker repulsive force, while other nodes are uniformly positioned.

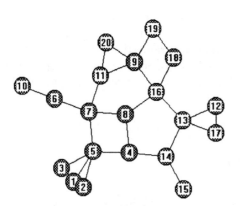

Fig. 3. Graph 2,unconstrained   **Fig. 4.** Graph 4,unconstrained

graph	Kamada's	*Model A*	*Model B*	no. of nodes
Graph 1	0.17	2.08	0.39	10
Graph 2	2.75	33.12	4.89	20
Graph 3	0.77	17.03	1.48	16
Graph 4	0.55	1.59	0.93	11
Graph 5	0.23	4.33	0.16	10

Table 1. Unconstrained Graph Layout

Our second experiment is to compare *Model A* and *Model B* for constrained graph layout. In this experiment we have added several constraints to the example graphs from the first experiment. The constraints require some nodes to be aligned, or to be higher than some other nodes. Table 2 shows the time in seconds for each method to layout the constrained graph. Again, *Model B* is significantly faster than *Model A*. The constrained layouts produced by *Model B* are aesthetically pleasing. As examples of the quality, two of the layed out graphs are shown in Fig. 5 and Fig. 6. Graph 4 has been given constraints requiring that node 4, node 8 and node 9 align vertically, node 4 is lower than node 3, node 5 and node 9, and node 9 is lower than nodes 10 and 11. Note that the layout time for Graph 5 by *Model B* in Table 2 and in Table 1 is quite small. This is because for K_{10}, the initial feasible solution obtained by solving (6) is very close to a minimum. This indicates the importance of an good initial configuration, and *Model B* will be even faster than *Model A* if it is given a better initial feasible solution.

Our third experiments was to evaluate the constrained tree layout module. We took five sample trees: Tree 1 is Fig.2 of [33], Tree 2 and Tree 3 are Fig.10 and Fig.2 of [35] respectively, Tree 4 is Fig.18 of [21] and Tree 5 is a four-level complete binary tree. We lay them out with *Model C* and also with *Model A* and *Model B*, all constrained under $\mathbf{C}_{tree}$. Table 3 shows the time taken to layout the tree with each method. In addition, the *constrained* column in Table 3 gives the time taken to layout each tree, by *Model C*, with constraints which make one of the subtrees of the root an *upward rooted tree* and node gap narrower. Fig. 7 and Fig. 8 are the layout of Tree 1, drawn by *Model B* and *Model C* respectively. Our results demonstrate that *Model C* performs better layout because of width-minimization, and is also significantly faster than the more general *Model B*.

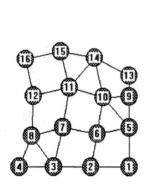

Fig. 5. Graph 3,constrained

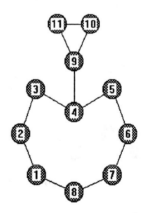

Fig. 6. Graph 4,constrained

Finally, we take Graph 4 as an example to illustrate how suggested values work. As explained above, Fig. 4 is the layout of Graph 4 using *Model B* without

graph	Model A	Model B	no. of constraints
Graph 1	5.78	1.42	4
Graph 2	38.61	14.56	9
Graph 3	18.51	2.86	7
Graph 4	9.67	4.61	7
Graph 5	5.93	0.94	2

Table 2. Constrained Graph Layout

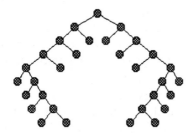

Fig. 7. Tree 1, *Model B* with C_{tree} **Fig. 8.** Tree 1, *Model C* with C_{tree}

graph	Model A	Model B	Model C	constrained	vertices
Tree 1	64.60	37.35	3.41	3.62	31
Tree 2	4.50	2.91	0.82	1.05	16
Tree 3	5.22	1.65	0.55	0.82	13
Tree 4	4.23	1.37	0.55	0.71	10
Tree 5	1.59	1.53	0.61	0.82	15

Table 3. Tree Layout

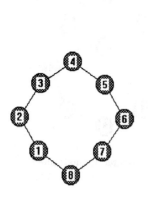
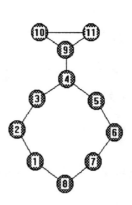
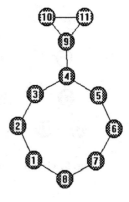

Fig. 9. phase 1   **Fig. 10.** phase 2   **Fig. 11.** phase 3

any constraints and with no suggested values. Now consider a sample interaction in which the suggested values are the old position coordinates. A user draws a graph and gets the unconstrained layout shown in Fig. 9. The user then edits the graph by adding three nodes and four arcs to a graph as displayed in Fig. 10. The user then thinks that this is close to the layout that he feels like and lays it out with the suggested values above mentioned. The layout obtained is shown in Fig. 11. Note that if no *suggested values* were given here, the layout would have been the same as that of Fig. 4.

6 Conclusion

We have introduced a generic model – constrained graph layout – for interactive graph layout and described three implementations of the model, two for undirected graph layout and one for tree layout. The key feature of our new model is that it allows the user to constrain the position of nodes in the graph and to provide suggested values for the node locations. Empirical evaluation of the implementations shows that the greater flexibility of our model does not come at a high computational cost. In particular our second implementation, *Model B*, provides good layout of undirected graphs in the context of arbitrary linear constraints at a reasonable cost, while *Model C*, provides quick and reasonable layout of trees in the context of arbitrary linear constraints.

In practice, *Model A* and *Model B* can be integrated into a two-phase procedure to obtain the benefits of both models. The idea is, in phase one, to use *Model B* to quickly obtain a certain local minimum, then see whether node coincidence occurs or not, if not, layout terminates; otherwise, phase two starts, that is: shift nodes that are coincident slightly along different directions to make every node in a different point, then call *Model A* to finalize the layout. Phase two would not take too long because some local minimum should be closer, and the layout that phase two produces will have no node coincidence.

The main motivation for our work is from work in advanced visual interfaces. First, constrained graph layout can be used in animation in which the new diagram is defined in terms of objects and constraints in the old diagram, and in which remaining objects in the diagram should not be moved unless necessary. The second use is for user interfaces for pen-based computing. One important approach in such user interfaces is to parse the pen-drawn diagram to infer constraints between the components [6]. Constrained graph layout can be used to re-layout or "pretty print" the recognized diagram while preserving its semantics. Constrained graph layout also has numerous other applications in user interfaces – basically whenever the diagram has more semantic structure than a simple graph, then the constrained graph layout model is appropriate.

References

1. G.D. Battisa, P. Eades, R. Tamassia, and I.G. Tollis. Algorithms for drawing graphs: an annotated bibliography. *Computational Geometry: Theory and Applications*, 4:235–282, 1994.

2. A. Borning. The programming language aspects of ThingLab, a constraint-oriented simulation laboratory. *ACM Transactions on Programming Languages and Systems*, 3(4):252–387, 1981.

3. M.J. Box, D. Davies, and W.H. Swann. *Non-linear optimization techniques*. Oliver & Boyd, 1969.

4. F. J. Brandenburg. Designing graph drawings by layout graph grammars. In *Proceedings of DIMACS International Workshop, GD'94, LNCS 894*, Princeton, New Jersey, USA, October 1994. Springer-Verlag.

5. F. J. Brandenburg, M. Himsolt, and C. Rohrer. An experimental comparison of force-directed and randomized graph drawing algorithms. In *Symposium on Graph Drawing, GD'95, LNCS 1027*, Passau, Germany, September 1995. Springer-Verlag.

6. S.S. Chok and K. Marriott. Automatic construction of user interfaces from constraint multiset grammars. In *IEEE Symposium on Visual Languages*, 1995.

7. I. F. Cruz and A. Garg. Drawing graphs by example efficiently: trees and planar acylic digraphs. In *Proceedings of DIMACS International Workshop, GD'94, Princeton, New Jersey, USA, October 1994, LNCS 894*, Princeton, New Jersey, USA, October 1994. Springer-Verlag.

8. R. Davidson and D. Harel. Drawing graphs nicely using simulated annealing. Technical report, Department of Applied Mathematics and Computer Science, 1991.

9. E. Dengler, M. Friedell, and J. Marks. Constraint-driven diagram layout. In *Proceedings of the 1993 IEEE Symposium on Visual Languages*, 1993.

10. P. Eades. A heuristic for graph drawing. *Congressus Numerantium*, 42:149–160, 1984.

11. P. Eades, W. Lai, K. Misue, and K. Sugiyama. Preserving the mental map of a diagram. In *Proceedings of Compugraphics '91*, pages 24–33, 1991.

12. P. Eades and J. Marks. Graph-drawing contest report. In *Symposium on Graph Drawing, GD'95, LNCS 1027*, Passau, Germany, September 1995. Springer-Verlag.

13. R. Fletcher. *Practical Methods of Optimization*. John Wiley & Sons, 1987.

14. A. Frick, A. Ludwig, and H. Mehldau. A fast adaptive layout algorithm for undirected graphs. In *Proceedings of DIMACS International Workshop, GD'94, LNCS 894*, Princeton, New Jersey, USA, October 1994. Springer-Verlag.

15. T. M J Fruchterman and E. M Reingold. Graph drawing by force-directed placement. *Software-Practice and Experience*, 21(11):1129–1164, November 1991.

16. A. Garg, M. T. Goodrich, and R. Tamassia. Area-efficient upward tree drawing. In *Proceedings of the 9th Annual Symposium on Computational Geometry, ACM*, 1994.

17. D Goldfarb and A Idnani. A numerically stable dual method for solving strictly convex quadratic programs. *Math. Prog.*, 27:1–33, 1983.

18. R. Helm and K. Marriott. Declarative graphics. In *Proc. of the 3rd International Conference on Logic Programming, LNCS 225*, pages 513–527, London, England, 1986. Springer-Verlag.

19. R. Helm and K. Marriott. A declarative specification and semantics for visual languages. *Journal of Visual Languages and Computing*, 2:311–331, 1991.

20. R. Helm, K. Marriott, T. Huynh, and J. Vlissides. An object-oriented architecture for constraint-based graphical editing. In *Object-Oriented Programming for Graphics*, pages 217–238. Springer-Verlag, 1995.

21. J. Q. Walker II. A node-position algorithm for general tree. *Software-Practice and Experience*, 20(7):685–705, July 1990.

22. T. Kamada. *Visualizing abstract objects and relations:a constraints-based approach*, volume 5 of *Computer Science*. Singapore, New Jersey:World Scientific, 1989.

23. T. Kamada and S. Kawai. An algorithm for drawing general undirected graphs. *Information Processing Letters*, 31(1):7–15, April 1989.

24. T. Kamps, J. Kleinz, and J. Read. Constraint-based spring-model algorithm for graph layout. In *Symposium on Graph Drawing, GD'95, LNCS 1027*, Passsu, Germany, September 1995. Springer-Verlag.

25. P. Kikusts and P. Rucevskis. Layout algorithm of graph-like diagrams for grade windows graphic editors. In *Symposium on Graph Drawing, GD'95, LNCS 1027*, Passau, Germany, September 1995. Springer-Verlag.

26. T. Lin and P. Eades. Integration of declarative and algorithmic approaches for layout creation. Technical Report TR-HJ-94-10, CSIRO Division of Information Technology, Centre for Spatial Information Systems, 1994.

27. P. Luders, R. Ernst, and S. Stille. An approach to automatic display layout using combinatorial optimization. *Software-Practice and Experience*, 25(11):1183–1202, 1995.

28. K. Misue, P. Eades, W. Lai, and K. Sugiyama. Layout adjustment and the mental map. *Journal of Visual Languages and Computing*, 6:183–210, 1995.

29. B. A. Myers, D. A. Giuse, R. B. Dannenberg, B. V. Zanden, D. S. Kosbie, E. Pervin, A. Mickish, and P. Marchal. Garnet: comprehensive support for graphical highly interactive user interfaces. *Computer*, pages 71–85, November 1990.

30. S. C. North. Incremental layout in dynadag. In *Symposium on Graph Drawing, GD'95, LNCS 1027*, Passau, Germany, September 1995. Springer-Verlag.

31. T. Lin P. Eades and X. Lin. Two tree drawing conventions. Technical Report 174, Key Centre for Software Technology, Department of Computer Science, The University of Queensland, 1990.

32. F. N. Paulisch. *The design of an extendible graph editor*. LNCS 704, Springer-Verlag, 1993.

33. E. M. Reingold and J. S. Tilford. Tidier drawing of trees. *IEEE Trans. on Software Engineering*, SE-7(2):223–228, March 1981.

34. K. Sugiyama and K. Misue. A simple and unified method for drawing graphs: magnetic-spring algorithm. In *Proceedings of DIMACS International Workshop, GD'94, LNCS 894*, Princeton, New Jersey, USA, 1994. Springer-Verlag.

35. K. Tsuchida, Y. Adachi, Y. Oi, Y. Miyadera, and T. Yaku. Constraints and algorithm for drawing tree-structured diagrams. In *Proceedings of the International Workshop on Constraints for Graphics and Visualization, CGV '95*, Cassis, France, September 1995.

36. D. Tunkelang. A practical approach to drawing undirected graphs. Carnegie Mellon University, 1994.

37. J. G. Vaucher. Pretty-printing of trees. *Software-Practice and Experience*, 10:553–561, 1980.

38. L. Weitzman and K. Wittenburg. Relation grammars for interactive design. In *Proceedings of IEEE Visual Languages*, pages 4–11, 1993.

39. C. Wetherell and A. Shannon. Tidy drawing of trees. *IEEE Trans. on Software Engineering*, SE-5(5):514–520, September 1979.

The Graphlet System
(System Demonstration)

Michael Himsolt *

Universität Passau, 94030 Passau, Germany
himsolt@fmi.uni-passau.de

Abstract. Graphlet is a portable, object oriented toolkit for graph editors and graph drawing algorithms, and is the successor of the GraphEd system. Graphlet is based on LEDA and Tcl/Tk. Algorithms can be implemented in C++ and LedaScript, a new scripting language based on Tcl/Tk. The GML format is a portable file format for graphs.

The implementation and visualization of graph algorithms is an important area in research and applications. There is a growing number of systems to support this process. LEDA [8] and GraphBase [7] concentrate on the implementation of a library of graph algorithms. Systems like daVinci [4], DynaDAG [11], D-ABDUCTOR [14] and EDGE [10] concentrate on graph visualization and/or graph editing, while CABRI [3], GD-Workbench [2], GraphEd [6], the Graph Editor Toolkit [9] or VCG [13] combine an editor and a library of algorithms.

The scope of these systems ranges from small and on-purpose systems to large, complex ones. However, the user interfaces of many systems are not easy to customize. Therefore, adding new features or exploring new classes of problems remains difficult. Also, implementing new features usually requires a compiler and a development environment. Furthermore, the application programmer interfaces and file formats of most of these systems are mutually incompatible.

Graphlet is a portable[1], object oriented toolkit for implementing graph editors and graph drawing algorithms. Graphlet is the successor of GraphEd [6]. Besides portability and general overhaul, there are several important improvements from GraphEd to Graphlet. Graphlet is based on three external toolkits, LEDA, Tcl and Tk, which gives us a more powerful design. Graphlet is much more oriented towards rapid prototyping and experiments with algorithms. Graphlet provides a scripting language to make the implementation of tools and customization easier. The GML format provides a powerful and extensible mechanism for writing and exchanging graph files.

---

* This research is partially supported by the Deutsche Forschungsgemeinschaft, Grant Br 835/6-2, research cluster "Efficient Algorithms for Discrete Problems and Their Applications"
[1] Graphlet is available for machines running Sun OS 4 and 5, Linux, Microsoft Windows NT and Microsoft Windows 95.

1 The structure of Graphlet

Graphlet is structured as follows:

Core system Graphlet's core is the platform independent part of Graphlet. The core provides a large set of base data structures, including a parser for the GML file format (see below) and a universal data structure which allows algorithms to add arbitrary attributes to nodes and edges. Most of the algorithm interface resides in the core.

Tcl interface, LedaScript The Tcl interface is a separate layer, which implements the LedaScript language and the graphics interface. It also implements some performance critical user interface operations. The system is designed such that this interface layer can be replaced or augmented with another interface, e.g. for Java.

Editor The main component of Graphlet is a powerful graph editor, implemented in LedaScript. The user interface could be replaced by another interface written in Tcl/Tk. This feature enables us to use Graphlet as the graph visualization subsystem of other Tcl/Tk programs.

Algorithm modules Graphlet uses a modular design where algorithms are separate plugin modules. Algorithms may be written in C++ using LEDA, in LedaScript or in a mixture of both. They can even be separate programs that communicate with Graphlet. Another option is an interface for algorithms that are written with GraphEd's Sgraph data structure, so we can reuse algorithms from GraphEd.

GML Graphlet defines a new, flexible and portable file format for graphs, GML. External applications which understand GML can communicate with Graphlet and use it as an external viewer.

Due to the modular and layered structure, Graphlet can be used in several configurations, It can be used with the core system only, and then is an extension of LEDA. It can also be used as a scripting language, LedaScript, and as a graph editor.

2 Data Structures and Algorithms

Graphlet uses *LEDA* [8] for the implementation of its data structures and algorithms. LEDA is a library of efficient data structures and algorithms. It provides a wide range of basic data structures, such as strings, lists, sets, queues, dictionaries and hash tables. LEDA supports graphs and contains a number of graph algorithms, especially planarity tests and algorithms for computing a planar embedding. These algorithms help implementing graph drawing algorithms more easily. Another important addition are data structures from computational geometry, which help to implement the coordinate assignment in graph drawing algorithms.

Graphlet adds another layer of data structures to LEDA:

- Support for arbitrary attributes for graphs, nodes and edges. Attributes may be added or removed dynamically at runtime. Methods to store these attributes in files (GML format, see below) are also included.
- Support for graphical display of graphs. Most of this is done on a device independent level, which makes the application programmer interface much easier to use.
- Classes for implementing graph algorithms.

3 Tcl and Tk

Tcl [12] is a universal scripting language. *Tk* is a graphical user interface toolkit, which can be used from Tcl and serves for the implementation of Graphlet's user interface. Unlike other toolkits in its class, Tk supports drawing operations.

Tcl and Tk provide several high level functions such as file management, dynamic loading of modules, process communication and, most important, access to the window system. Another advantage of using Tcl/Tk is that there are a large number of software packages, which add various capabilities to Tcl/Tk. These packages can be used in conjunction with Graphlet to provide extended features. In turn, other programs may use Graphlet as a subsystem.

Since LEDA, Tcl and Tk have been ported to UNIX and Microsoft Windows operating systems, Graphlet has a large base of systems to run on. We use Tcl/Tk for all system dependent parts, which minimizes the effort to port Graphlet to other platforms. Actually, taking the code from UNIX to Microsoft Windows systems required only a few changes, most of them due to compiler incompatibilities.

Graphlet provides two programming interfaces: a C++ interface and a scripting language, *LedaScript*. LedaScript is implemented with Tcl and can be used for rapid prototyping and user interface implementation. Except for a few performance critical routines, Graphlet's user interface is written in Tcl/Tk/LedaScript.

4 LedaScript

LedaScript is an extension of Tcl with support for graph operations. Figure 1 shows a LedaScript sample code. Graphs are created in LedaScript with the command **graph**. Each graph is a Tcl command, which is used to change or retrieve parameters such as the label or the coordinates. This is common practice in Tcl/Tk and implements an object oriented programming interface for graphs. Graph algorithms are available as LedaScript commands. Graphlet provides C++ classes which make it easy to add a Tcl interface to algorithms which are implemented in C++.

Of course, algorithms can also be implemented in LedaScript. LedaScript is powerful enough for rapid prototyping of algorithms. Since LedaScript reduces the edit–compile–link–run cycle to edit–run, it is much easier to test variations and debug algorithms. This may even be done without quitting from the program, since LedaScript provides a interactive shell. Also, LedaScript enables users to write batch scripts that run algorithms on graphs. Finally, since Tcl does not have pointers and can continue even after runtime errors, error handling is much easier than in traditional programming languages.

5 User Interface

We designed Graphlet's user interface using our experiences from Graphed. Figure 2 shows two sample screens. Especially, we enhanced the following:

- By default, there is only one graph per window. GraphEd had the ability to display multiple graphs in a window, which turned out to be more confusing than helpful, both for users and programmers. However, the capability to display more than one Graph in a window remains, which may still be useful in non-interactive applications.
- GraphEd's three types of selections were *single node, single edge* or *subgraph.* Graphlet uses a more general model. A selection can be any set of nodes and edges.
- Graphlet offers much wider customization options. For example, self loops or multiple edges can be disabled to avoid undesired structures. All keyboard and mouse bindings are customizable. Graphlet implements modes similar to those in emacs. Also, Graphlet works if only one mouse button is available (e.g. on Macintosh systems).
- Graphlet provides a mode to edit labels.
- Graphlet now offers a much wider scale of graphical representation options for nodes and edges, notably colors and line width.
- Many successful features from GraphEd were kept, as the general window layout with a menubar at the top and a toolbar at the right, the menu layout or a the double click to pop up a window to edit object parameters.

Graphlet's user interface will be continuously enhanced and extended. However, it is obvious that Graphlet cannot satisfy all possible future requirements. One major idea is that it is easy to add features to Graphlet. The easiest way to do that is to add some LedaScript code. LedaScript is efficient enough for small or medium tasks, with only a fraction of the time needed for a C++ implementation. For example, it takes less than an hour to add a routine which outputs a graph in a new format.

6 The GML File Format

We have developed a new file format for Graphlet, GML. Figure 3 shows a simple graph, a circle of three nodes. This example shows several key issues of GML:

ASCII Representation for Simplicity and Portability. A GML file is 7-bit ASCII file. This makes it simple to write files through standard routines.

```
#
# create a canvas ''.c'' (Tcl/Tk drawing area)
#
canvas .c
pack .c

#
# create a graph
#
set g [graph]

#
# draw the graph on canvas .g
#
$g canvases .c

#
# create a node "a" at position (21,21)
#
set n1 [$graph create node]
$g nodeconfigure $n1 -label "a"
$g nodeconfigure $n1 -x 21
$g nodeconfigure $n1 -y 21

#
# create a node "b" at position (42,42)
#
set n2 [$graph create node]
$g nodeconfigure $n2 -label "b"
$g nodeconfigure $n2 -x 42
$g nodeconfigure $n2 -y 42

#
# create an edge from node n1 to node n2
#
set e [$graph create edge source $n1 target $n2]

#
# now draw the graph
#
$g draw
```

Fig. 1. Sample LedaScript code

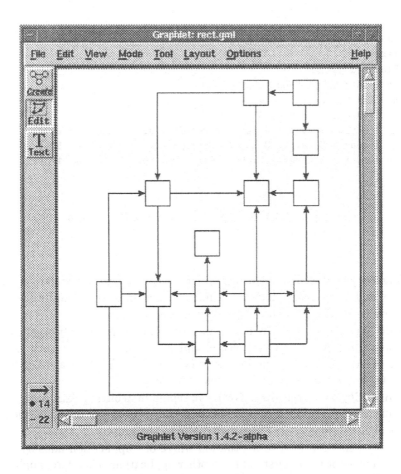

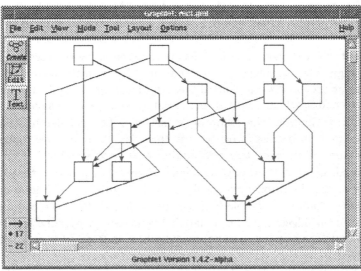

Fig. 2. Graphlet's user interface

```
graph [
  comment "This is a sample graph"
  directed 1
  IsPlanar 1
  node [ id 1 label "Node 1" ]
  node [ id 2 label "node 2" ]
  node [ id 3 label "node 3" ]
  edge [ source 1 target 2 label "Edge from node 1 to node 2" ]
  edge [ source 2 target 3 label "Edge from node 2 to node 3" ]
  edge [ source 3 target 1 label "Edge from node 3 to node 1" ]
]
```

Fig. 3. GML description of a circle of three nodes

Parsers are easy to implement, either by hand or with standard tools like lex and yacc. Also, since GML files are text files, they can be exchanged between platforms without special converters.

Simple Structure. A GML file consists entirely of hierarchically organized tag-value pairs. A tag is a sequence of alphanumeric characters, such as graph or id. A value is either an integer, a floating point number, a string or a list of tag-value pairs enclosed in square brackets.

Extensibility & Flexibility. GML can represent arbitrary data, and it is possible to attach additional information to any object. For example, the graph in Figure 3 adds an IsPlanar attribute to the graph.

Clearly, this can lead to a situation where one application adds some data which cannot be understood by another application. Therefore, applications are free to ignore data they do not understand, except for a basic set which describes graphs. They should, however, save unknown data and write them back.

Representation of Graphs. Graphs are represented by the tags graph, node and edge. The topological structure is modeled with the node's id and the edge's source and target attributes: the id attributes assign unique numbers to nodes, which are then referenced by source and target.

Consistency Consider the following situation: a file includes information of some graph theoretical property, say the existence of a Hamiltonian circle. It is easy to see that this information may become invalid if an edge is removed, but not if an edge is added. However, a program that does not know about Hamilton cycles will not be able to check and guarantee this property. Another example is if a node is moved, then the coordinates of its adjacent edges must be updated. Even more, a hypothetical attribute IsDrawnPlanar might become invalid when node or edge coordinates are changed.

As these examples show, both changes in the structure and in the values of attributes can falsify other attributes. GML therefore specifies that if the name of a tag starts with a capital letter, then it should be considered "unsafe" and discarded whenever changes have occurred in the graph or the attributes, unless the program knows how to deal with the case. A simple strategy could be to not to load unsafe information at all if the program is going to change the graph. This feature gives programs the opportunity to add critical data, and makes at the same time sure that everything remains consistent.

7 Conclusion

The Graphlet system is currently under development at the University of Passau. Versions for several UNIX systems, Microsoft Windows NT and Microsoft Windows 95 are available.

References

1. Di Battista, G., Eades, P., Tamassia, R., Tollis, I.G.: *Algorithms for drawing graphs: an annotated bibliography.* Comput. Geom. Theory Appl. **4** (1994) 235–282
2. Buti, L., Di Battista, G., Liotta, G., Tassinari, E., Vargiu, F., Vismara, L.: GD-Workbench: A System for Prototyping and Testing Graph Drawing Algorithms. In Lecture Notes in Computer Science **1027**, pp. 111–122 (1995)
3. Carbonneaux, Y., Laborde, J.-M., Madani, M.: CABRI-Graph: A Tool for Research and Teaching in Graph Theory. In Lecture Notes in Computer Science **1027**, pp. 409-418 (1995)
4. Fröhlich, M, Werner, M.: Demonstration of the Interactive Graph-Visualization System da Vinci. In Lecture Notes in Computer Science **894**, pp. 266–269. (1994)
5. Information on Graphlet is available on the world wide web at `http://www.fmi.uni-passau.de/Graphlet`.
6. M. Himsolt: GraphEd: A Graphical Platform for the Implementation of Graph Algorithms. In Lecture Notes in Computer Science **894**, pp. 182–193. (1994)
7. Knuth, D.E.: *The Stanford GraphBase: A Platform for Combinatorial Algorithms.* Stanford University (1993)
8. Mehlhorn, K., Näher, S.: LEDA: A library of efficient data types and algorithms, Comm. ACM 38(1), pp. 96–102 (1995).
9. Madden, B., Madden, P., Powers, S., Himsolt, M.: Portable Graph Layout and Editing, In Lecture Notes in Computer Science **1027**, pp. 385–395 (1995)
10. Newberry Paulisch, F.: The Design of an Extendible Graph Editor, Lecture Notes in Computer Science **704** (1993)
11. North, S.: Incremental Layout in DynaDAG, In Lecture Notes in Computer Science **1027**, pp. 409–418 (1995)
12. Ousterhout, J.: Tcl and the Tk Toolkit. Addison-Wesley (1994).
13. Sander, G.: Graph Layout through the VCG Tool, In Lecture Notes in Computer Science **894**, pp. 194–205. (1994)
14. Sugiyama, K., Misue, K.: A Generic Compound Graph Visualizer/Manipulator: D-ABDUCTOR, In Lecture Notes in Computer Science **1027**, pp. 500–503 (1995)

On the Edge Label Placement Problem [*]

Konstantinos G. Kakoulis   and   Ioannis G. Tollis

Department of Computer Science
The University of Texas at Dallas
Richardson, TX 75083-0688
email: kostant@utdallas.edu, tollis@utdallas.edu

Abstract. Let $G(V, E)$ be a graph, and let $f : G \rightarrow R^2$ be a *one to one* function that produces a layout of a graph G on the plane. We consider the problem of assigning text labels to every edge of the graph such that the quality of the labeling assignment is optimal. This problem has been first encountered in automated cartography and has been referred to as the Line Feature Label Placement (LFLP) problem. Even though much effort has been devoted over the last 15 years in the area of automated drawing of maps, the Edge Label Placement (ELP) problem has received little attention. In this paper we investigate computational complexity issues of the ELP problem, which have been open up to the present time. Specifically we prove that the ELP problem is NP-Hard.

1   Introduction

In recent years graph drawing has received increasing attention due to the large number of applications, such as, entity relationship diagrams, software engineering diagrams, CASE tools, debugging tools, communication networks, and database design [3]. The labeling of the graphical features of a drawing in most of these applications is essential, since it gives important information about the relations represented by the drawing.

The problem of labeling graphs can be divided into two subproblems:

- **NLP**: (Node Label Placement) is the problem of placing text labels assigned to particular nodes of the graph.
- **ELP**: (Edge Label Placement) is the problem of placing text labels assigned to particular edges of a graph.

Most of the research addressing the above labeling problem has been done on labeling features of geographical and technical maps. Christensen, Marks and Shieber present a comprehensive survey of algorithms for the labeling problem [2].

The NLP problem has been the subject of extensive research in recent years, and the complexity issues of that problem have been well documented. It has

---

[*] Research supported in part by NIST, Advanced Technology Program grand number 70NANB5H1162.

been proven independently by three different groups [6, 11, 13] that the NLP problem is NP-Complete even in its simplest form.

Even though much effort has been placed in solving the NLP problem, the ELP problem has received little attention [14, 16]. Many heuristics devised to solve the ELP problem [1, 4, 5, 7, 8], are based on exhaustive search algorithms with backtracking. These algorithms do not produce the desirable results, due to the tendency of those methods to get trapped in local optima. They also take exponential time.

Definitely in geographical maps, the labeling of lines (ELP problem) is less restrictive than the labeling of points (NLP problem) due to the fact that lines are usually very long and thus there are many alternative positions to place labels. But the edge density of a drawing can vary, and edge labeling algorithms must perform well even for a high edge density drawing, since unlike in geographical maps the size and density of the edges (lines) of a drawing can and usually are restrictive with respect to label positioning. This makes most of the algorithms in automated cartography for the ELP problem inefficient. Another approach is to adopt techniques used to solve the NLP problem [2] in order to solve the ELP problem.

Up to the present time, the complexity of ELP is an open problem. The NLP and ELP problems have very similar structure. However, as we and other members of the community [12] observed, a direct transformation from NLP to ELP seems very difficult. In this paper we prove that the ELP problem is NP-Hard.

2   The ELP (Edge Labeling Placement) problem

2.1   Labeling quality

The ELP problem is the problem of assigning text labels to any edge of a predefined layout of a graph (the equivalent problem for geographical and technical maps is the LFLP problem) such that the association of the labels to their corresponding edges is clear.

It is very important to note that the visual inspection of a labeling assignment must be sufficient to explain the semantics of any label. Let us consider a labeling assignment of a set of edges where each edge is a street in a city and each label is the name of that street. Then a visual inspection of the labeling assignment must unambiguously reveal the name of each street.

There has been extensive effort especially by cartographers like Imhof [10] and Yoeli [15], to devise rules that measure the semantic clarity of a labeling assignment. Three concepts stand out, and are the basic rules that in general give an accurate assessment of the semantics of labels.

Basic rules for labeling quality [10, 15]:

1. No overlaps of a label with other labels or other graphical features of the layout are allowed.

2. Each label can be easily identified with exactly one graphical feature of the layout (i.e., the assignment is unambiguous).
3. There is a ranking of all potential labels for any particular edge.

A label respects the first rule if it does not overlap any graphical feature, even though it is allowed to touch the edge that it belongs to. Also, a label respects the second rule if it is placed very close or touches the edge that it belongs to.

The ranking of the potential labels for an edge typically captures the aesthetic preference of those labels, which is an essential criterion for the labeling quality of geographical and to some extend technical maps. It also allows to introduce problem specific constraints (i.e., the label of an edge must be closer to the source or destination node). The ranking of a label depends only on its position with respect to its associated edge and is not influenced by overlaps. We associate a penalty for each label according to its ranking.

Since the layout of a graph is fixed, there are instances where even an optimal assignment produces labels that do not strictly follow those rules. In that case we want to have a way of evaluating how good is a given labeling assignment. We can do that by assigning to each label a quantity that evaluates the quality of that label. We name this quantity $COST$. $COST(i, j)$ is a function that gives us the penalty of assigning label j to edge i in the final labeling assignment. The $COST$ for each label j is a linear combination of:

- The penalty with respect to the ranking of label j.
- The penalty which reflects the severity of the violation of the first two basic rules for label j.

Now, let $rank(i, j)$ be the ranking of label j among all potential labels of edge i, where $0 \leq rank(i, j) \leq C$, $C \in \mathbb{R}$. Then $a * rank(i, j)$ is the penalty with respect to the ranking of label j, where a is a constant. Also, let $b * clarity(i, j)$ be the penalty with respect to the second basic rule, if label j is assigned to edge i, where b is a constant. Finally, let $pen(i, j, k, l)$ be the penalty if label[2] j of edge i overlaps label l of edge k. Then:

$$COST(i, j) = a\ rank(i, j) + b\ clarity(i, j) +$$

$$\sum_{k \in E, k \neq i} \sum_{j \cap l \neq \emptyset} pen(i, j, k, l)\ P(k, l)$$

Where:

$$P(k, l) = \begin{cases} 1, & \text{if label } l \text{ is assigned to edge } k \\ 0, & \text{otherwise} \end{cases}$$

The last condition guarantees that only labels assigned to a final labeling assignment affect the cost with respect to the labeling quality.

---

[2] We consider as potential labels only labels that do not overlap any other graphical feature, except other labels and(or) their corresponding edge.

2.2 Label positioning

In order to completely describe the labeling problem we need to define how we construct the set of potential labels for any edge. In $Fig.$ 1(a) labels A, B, C,

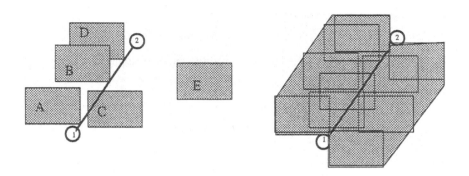

Fig. 1. (a) How to position labels for an edge; (b) Labeling space of an edge.

D and E are potential labels for the edge (1,2). Labels A, B and D follow the first two rules. Instead, label C intersects the edge that belongs to, which is a violation of the first rule. If a drawing has more than one edges, and labels like E floating between edges , it is difficult to have a clear understanding to which edge that label belongs to. So labels like E violate the second rule, and must be avoided.

We can define the set of all potential labels for a given edge in the *discrete* or *continuous* labeling space. In the *discrete* labeling space the set of all potential labels is finite and each label is identified by its position on the layout, like the labels in $Fig.$ 1(a). In the *continuous* labeling space the set of all potential labels is infinite, and for each edge we define a region which is bounded by a closed line, where each potential label for that edge must lay inside that region. Candidate labels are labels that have at least one intersection point with their associated edge. By imposing this restriction we avoid labels like E in $Fig.$ 1(a). The shaded region around the edge (1,2) in $Fig.$ 1(b) is the *continuous* labeling space for that edge, and any label that is placed inside that shaded region is a candidate label for that edge.

2.3 Formulation of the ELP problem

The ELP problem is an optimization problem since the objective is a labeling assignment of minimum cost. Each label which is part of a final assignment carries a penalty that the $COST$ function calculates. The objective is to find a set of labels, one for each edge, that yields minimum total cost. We first consider the problem that optimizes the cost of a labeling assignment.

The Optimal ELP Problem : Let $G(V, E)$ be a graph and let $f : G \to R^2$ be a *one to one* function that produces a drawing for graph G. Let W_i be the set of all potential labels for any edge i.

Question : Find a labeling assignment that minimizes the following function:

$$\sum_{i=1}^{|E|} \sum_{j=1}^{|W_i|} COST(i, j) P(i, j)$$

Where:

$$P(i, j) = \begin{cases} 1, & if\ label\ j\ is\ assigned\ to\ edge\ i \\ 0, & otherwise \end{cases}$$

and

$$\sum_{i=1}^{|E|} \sum_{j=1}^{|W_i|} P(i, j) = |E|$$

and

$$\sum_{j=1}^{|W|} P(i, j) = 1, \qquad 1 \le i \le |E|.$$

The last two conditions guarantee that any edge will have exactly one label assigned to it, since $P(i, j) = 1$ if and only if label j is assigned to edge i. The ELP problem as stated is combinatorial in nature, even though the underlying geometry gives meaning to the cost function, the interpretation of the cost function can be regarded as independent from some particular geometry. In our case we will always interpret the cost function with respect to Euclidean geometry.

Now we will impose some extra constraints in order to obtain a simpler version of the ELP problem. Here we are interested to find if there is an admissible labeling assignment where each label is of zero cost, with respect to the first two basic rules for labeling quality, rather than finding an assignment of optimal cost. This simplification transforms the ELP problem into a decision problem, which enables us to investigate the computational complexity aspects of the problem.

First we redefine the $COST$ function as follows:

$$COST(i, j) = \begin{cases} 0, & if\ rule\ 1\ and\ 2\ are\ followed \\ 1, & otherwise \end{cases}$$

The resulting ELP problem becomes the *Admissible* ELP (AELP) problem. Here the objective is to find a labeling assignment of zero cost with respect to the labeling quality.

The Admissible ELP Problem :

Question : Find a labeling assignment such that:

$$\sum_{i=1}^{|E|} \sum_{j=1}^{|W|} COST(i, j) P(i, j) = 0,$$

subject to the same constraints as above.

Next we further restrict the AELP problem by:

- requiring the labels of each edge to be of the same size.
- restricting the size of the set of potential labels for each edge by not allowing potential labels of the same edge to overlap (i.e., in *Fig.* 1 labels B and D are not considered both as potential labels of edge (1,2)).

The latter constraint guarantees that each edge has a discrete number of potential labels. Hence, the resulting AELP problem becomes the *Discrete* AELP (DAELP) problem.

The Discrete AELP (DAELP) problem : Let $G(V, E)$ be a graph and let: $f : G \rightarrow R^2$ be a *one to one* function that produces a drawing for graph G. Let W_i be the set of all potential labels for any edge i.

Question : Find a labeling assignment such that:

$$\sum_{i=1}^{|E|} \sum_{j=1}^{|W_i|} COST(i,j)P(i,j) = 0$$

Where:

$$P(i, j) = \begin{cases} 1, & \text{if label } j \text{ is assigned to edge } i \\ 0, & \text{otherwise} \end{cases}$$

and

$$\sum_{i=1}^{|E|} \sum_{j=1}^{|W_i|} P(i,j) = |E|$$

and

$$\sum_{j=1}^{|W|} P(i,j) = 1, \qquad 1 \leq i \leq |E|.$$

In the next section we prove that the DAELP problem is NP-Complete.

3 The NP-Completeness of the DAELP problem

We will prove that the DAELP problem is NP-Complete by transforming the 3-SAT problem [9], a well known NP-Complete problem, to it. Recall that 3-SAT is defined as follows:

Instance: Set X of variables, collection U of clauses over X, such that each clause has exactly 3 literals.

Question: Is there a satisfying truth assignment for U?

In order to transform 3-SAT into DAELP problem, we do the following: For each variable in an instance of the 3-SAT problem we transmit to any clause that contains a literal of this variable the information on the status of this variable. This transmission takes place through the construction of a transmition network. The goal is to associate the satisfiability of the 3-SAT problem with the existence of an edge labeling assignment of zero cost for the transmition network. Generally speaking each variable will be linked with all clauses that contain its complement through a route, such that once each variable has been assigned a value, there is only one possible labeling assignment of zero cost for each route. By knowing how each route has been labeled we can conclude the satisfiability of any clause that this route is connected by observing if there is enough room for that clause to have a label of zero cost assigned to it.

To construct the transmition network we need some basic building blocks. The interconnection of these blocks will produce the final transmition network.

The first building block is the **variable_block** shown in $Fig.$ 2. Each variable and its complement will be represented by a *variable_block*. Let a *variable_block* represent X and $\overline{X}$. If $X = TRUE$ then edge $(1,2)$ is assigned label *head*. If $\overline{X} = TRUE$ then edge $(1,2)$ is assigned label *tail*. Every edge in a *variable_block*

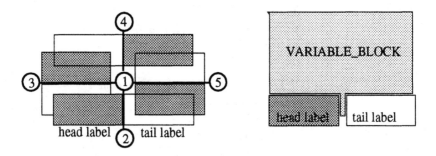

Fig. 2. An example of a variable_block.

has 2 labels of potential zero cost. Because of the cyclic structure and the way labels overlap there can be only two solutions to the DAELP problem for any *variable_block*. One solution contains all the shaded labels and the other all the non-shaded labels, since any shaded label is overlapped by a non-shaded label. Due to space limitations, we present the following results without proofs.

Lemma 1. *For a variable_block we have:*

1. *If a variable has been assigned a value, then the variable_block that represents that variable has only one labeling assignment of zero cost.*
2. *In a labeling assignment of zero cost only one of the head or tail labels is part of this assignment.*

Each clause will be represented by a **clause_block** as shown in $Fig.$ 3. Edge $(1,2)$ has 3 potential labels (X,Y,Z) of zero cost, one for each literal in that clause. If a label of zero cost is assigned to edge $(1,2)$ then the clause represented by *clause_block* is satisfied.

The following building blocks will serve as channels that transfer the truth assignment of the variables (*variable_blocks*) to the clauses (*clause_blocks*).

First we introduce the **pipe_block** shown in $Fig.$ 4. Each edge has only 2 potential labels of zero cost. Also each label overlaps with exactly one other label.

248

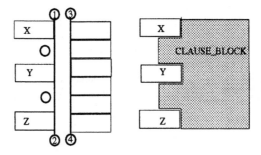

Fig. 3. An example of a clause_block.

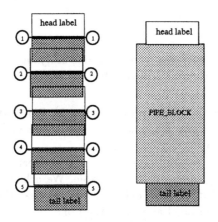

Fig. 4. An example of a pipe_block.

For the *pipe_block* we have a numbering of the edges that reveals the following pattern: Except for the first and last edge in the sequence, the shaded label of an edge overlaps the non-shaded label of the next edge in the sequence.

Lemma 2. *If the head (tail) label for any pipe_block is excluded from a labeling assignment of zero cost, then there exists exactly one labeling assignment of zero cost. That assignment includes the tail(head) label of that pipe_block.*

Next we introduce the **junction_block**, shown in $Fig.$ 5(a). Notice that if one *tail* label is excluded then the *head* label must be included in a labeling assignment of zero cost for a *junction_block*. However this labeling assignment is not unique. A *junction_block* has the following properties:

Lemma 3. *For any junction_block:*

1. *If the head label is excluded then there is exactly one labeling assignment of zero cost, which includes all the tail labels of the block.*

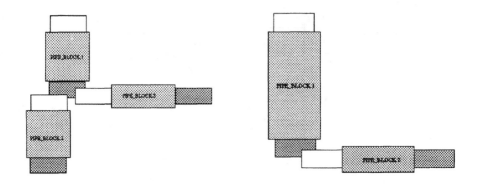

Fig. 5. (a) An example of a junction_block;   (b) An example of a bend_block.

2. *If all tail labels are excluded from a labeling assignment, then there is exactly one labeling assignment of zero cost that includes the head label of the block.*

Next we introduce the **bend_block**, shown in *Fig.* 5(*b*), which behaves exactly the same way as a *pipe_block* does. So what holds for *pipe_blocks* certainly holds for *bend_blocks* also.

The purpose of the transmition network is to connect any variable to clauses that contain the complement of that variable. We achieve that by connecting the building blocks that we have defined so far so that they produce a bigger structure with one *head* label and more than one, if necessary, *tail* labels. The *head* label of this structure will overlap the label in the *variable_block* that represents that variable, and any *tail* label of the structure will overlap a label in some *clause_block* that corresponds to the complement of that variable in every clause that contains it.

To visualize how these connections will be made, one needs to think that for any variable we build a highway that goes from North to South and has exits only towards the east side of the highway, where all the *clause_blocks* are located. We visualize the highway as a vertical line segment and the exits as horizontal line segments. Each variable and the connections to clauses that contain its complement form a set of a highway and its exits. All *variable_blocks* stand on the same horizontal line and all the *clause_blocks* stand on the same vertical line. A highway starts from the *variable_block* and stops at the last exit. Each exit leads to a literal in a *clause_block*.

A highway with its exits is called a *serial interconnection*. *Figure* 6(*a*) illustrates the structure of a *serial interconnection* for variable $\overline{X1}$ that is connected to variable $X1$ in all of the clauses that contain $X1$. To build a serial interconnection one needs to:

1. replace the part of the highway where an exit occurs with a *junction_block*,
2. replace the last exit with a *bend_block*,

3. replace the rest of the highway with an appropriate number of *pipe_blocks*,
4. connect 2 consecutive blocks A1 and A2 in such a way that the *tail* label of block A1 overlaps the *head* label of A2.

By following theses rules the serial interconnection in *Fig.* 6(*b*) which is built by using the building blocks presented so far actually represents the structure in *Fig.* 6(*a*).

Note: *Every serial interconnection has one head label and as many tail labels as the number of exits.*

The *head* label of the structure is the *head* label of the first (top) building block (*Fig.* 6(*b*)). *Tail* labels are the *tail* labels for the last building blocks of any exit. The *serial interconnection* block has the following properties:

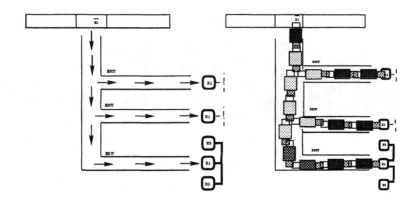

Fig. 6. (a) A serial interconnection. (b) Structure of a serial interconnection.

Lemma 4. *For a serial interconnection:*

1. *If the head label is excluded from a labeling assignment, then there is exactly one edge labeling assignment of zero cost that includes all tail labels of the structure.*
2. *If all tail labels are excluded from a labeling assignment, then there is exactly one labeling assignment of zero cost that includes the head label of the structure.*

For each literal in an instance of a 3-SAT problem there will be a serial interconnection that will represent that literal. The transmition network will be the union of these serial interconnections. *Figure* 8 shows the structure of a transmition network for an instance of a 3-SAT problem of $\{X, Y, Z\}$ variables and $\{\{\overline{X}, \overline{Y}, \overline{Z}\}, \{X, Y, Z\}, \{X, Y, \overline{Z}\}\}$ clauses.

The crossing of highways and exits is unavoidable. This is why we introduce the concept of *over − passes*. If a highway A is to the east of a highway B, and highway A has an exit a, which is south of an exit b of highway B, then exit b will intersect highway A. The building blocks defined so far are sufficient to build the network of these roads but insufficient to built over-passes. In order to be able to build such networks we need to build over-passes.

In *Fig.* 7 the *bridge_block* serves as an over-pass. The *bridge_block* possesses a very interesting property: in a labeling assignment of zero cost if we put pressure on the top (that is the *head1* label is not available) then the pressure comes out at the bottom (that is the *tail1* label must be included in the labeling of the *bridge_block*) regardless of which is the labeling assignment of the rest of the *bridge_block*. The same property is true if the pressure comes from any other direction. This property is essential in the construction of the transmition network, because if we replace any crossing of *serial interconnection* blocks in the transmission network with a *bridge_block* then *Lemma* 4 remains true.

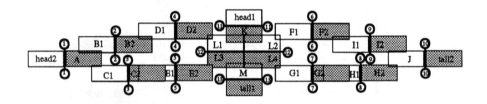

Fig. 7. An example of a bridge_block.

Lemma 5. *If the head1(tail1) or head2(tail2) label for any bridge_block is not available, then label tail1(head1) or tail2(head2) is included in any labeling assignment of zero cost.*

Now we introduce *Algorithm* 1, which given an instance of a 3-SAT problem, it constructs the transmition network for an instance of DAELP problem in polynomial time.

Algorithm 1

Given an instance of the 3-SAT problem, with n variables and m clauses:

1. For each variable we introduce a *variable_block*.
2. For each clause we introduce a *clause_block*.
3. For each *variable_block* we introduce 2 columns in the grid.
4. For each *clause_block* we introduce 3 rows (one row for each literal in the clause) in the grid.
5. For each literal we build a serial interconnection which is a collection of horizontal and vertical segments.
6. For each literal we connect the *variable_block* that corresponds to this literal with the serial interconnection for that literal in such a way that the *head* label of that serial interconnection will overlap the label in *variable_block*

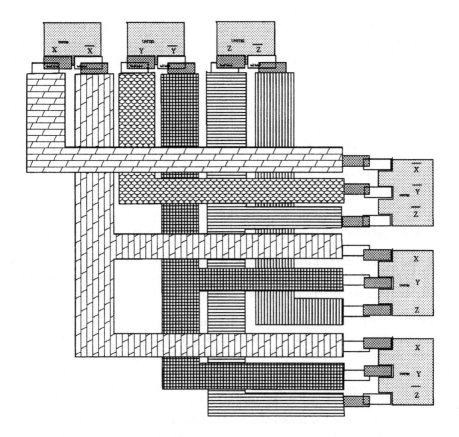

Fig. 8. A transmition network of an instance of a 3-SAT problem

that corresponds to that literal, and all labels corresponding to the complement of that literal in any clause (*clause_block*) will be overlapped by some
tail label of that serial interconnection.

In order to build the transmition network we have to place the serial interconnection of each literal on the plane in order to be able to make the connections
in *Step* 6 of *Algorithm* 1.

Step 5 of *Algorithm* 1 marks the entries of the grid that each serial interconnection occupies. Each grid entry is part of only one serial interconnection
except for the entries where crossings of serial interconnections occur. That entry
is assigned to a *bridge_clause*.

Step 6 of *Algorithm* 1 makes the final connections which complete the transmition network. *Figure* 8 illustrates how these connections take place: First any
pair $(X, \overline{X})$ of columns in the grid is connected with the *variable_block* for variable X. The *head* label of that *variable_block* (see *Fig.* 2) overlaps the *head*
label of the *serial interconnection* for X and the *tail* label of the *variable_block*
overlaps the *head* label of the *serial interconnection* for $\overline{X}$. Secondly, any three

rows in the grid that represent a clause are connected to a *clause_block* for that clause so each of the three labels of the clause (see *Fig.* 3) overlaps the *tail* label of some serial interconnection that ends up in that row and is associated with the complement of the literal that the label in the *clause_block* represents.

Due to space limitations we will not discuss here all the details that show that our transformation takes polynomial time. Briefly the most time consuming step of *Algorithm* 1 is *Step* 5 which can be done in $O(mn^2)$. Also, *Algorithm* 1 requires space proportional to the size of the grid. Hence we have the following:

Lemma 6. *Algorithm* 1 *produces a layout of a graph that describes an instance of 3-SAT, runs in* $O(mn^2)$ *time and requires* $O(n) \times O(m)$ *space.*

In the following theorem we prove that the ELP problem for a transmition network is equivalent to the 3-SAT problem that this network represents.

Theorem 7. *The transmition network* N, *constructed according to Algorithm* 1, *that represents an instance* S *of a 3-SAT problem, has an admissible edge labeling assignment of zero cost if and only if* S *is satisfiable.*

Proof: " ⇒ ": By the hypothesis, N has an edge labeling assignment of zero cost. So there is a set W of labels such that $|W| = |E|$, and each edge has exactly one label assigned to it that does not overlap any other graphical feature of the drawing. Any edge in the *variable_block* has a label of zero cost assigned to it. By *Lemma* 1, either the *head* label or the *tail* label of any *variable_block* is part of the solution but not both. So for each variable represented by a *variable_block* in N we can obtain a truth assignment by examining how each *variable_block* is labeled. We claim that the truth assignment obtained this way satisfies the instance of 3-SAT associated with N.

The *head* or *tail* label of a *variable_block* is part of the solution. Any such label (which is part of the solution) overlaps the *head* label of some serial interconnection. By *Lemma* 4, that label in the *variable_block* puts pressure on that serial interconnection, and all labels in the *clause_blocks* that overlap the *tail* labels of that serial interconnection are excluded from a labeling assignment of zero cost. But any serial interconnection connects a literal (the label that overlaps the *head* label of that serial interconnection) in a *variable_block*, to the complement of that literal in any clause (labels in some *clause_blocks* that overlap some *tail* label of that serial interconnection) of that instance of the 3SAT. This implies that no label of a *clause_block* associated with a literal which has false value is part of the labeling assignment.

But by the hypothesis each edge in N has a label of zero cost assigned to it, which implies that each *clause_block* in N has a label assigned to it. Therefore, for each clause in S there exists at least one literal (the label assigned to the *clause_block* of that clause) with truth value, which implies that S is satisfiable.

" ⇐ ": Let us assume that instance S of the 3-SAT problem is satisfiable. First we construct the transmition network N according to *Algorithm* 1. Since S is satisfiable, there is a truth assignment for every variable. By *Lemma* 1, each *variable_block* has exactly one labeling assignment of zero cost that reflects the truth assignment of S. The labeling assignment of the *variable_blocks* puts pressure (by overlapping their head labels) to those serial interconnections in N that their *tail* labels overlap the labels in any *clause_block* that correspond to literals of false value. By *Lemma* 4 this forces these serial interconnections to include all their *tail* labels to preserve a labeling assignment of zero cost. Consecutively, each label in a *clause_block* of N that corresponds to literals of false value is excluded from a labeling assignment of zero cost.

Since S is satisfiable, for each *clause_block* in N there is at least one label (corresponding to a literal which has true value) that is not overlapped by some *tail* label of the above serial interconnections.

Next, we assign all labels (that correspond to literals of true value in any clause) to their corresponding *clause_blocks*. This assignment puts pressure to all *tail* labels of any serial interconnection that its *head* label intersects a label in a *variable_block* corresponding to false value. But such labels in the *variable_blocks*, by *Lemma* 1, can't be part of a labeling assignment of zero cost. That ensures that those serial interconnections have a labeling assignment of zero cost. Finally, we remove from each *clause_block* all but one labels assigned to it.

Now each *variable_block*, each *clause_block*, and each serial interconnection has a labeling assignment of zero cost, which implies that the transmition network has a labeling assignment of zero cost since any edge in the transmition network belongs to a *variable_block* , a *clause_block* , or a serial interconnection. □

Theorem 8. *The DAELP problem is NP-Complete*

Proof: Since our transformation takes polynomial time, by *Theorem* 7 the DAELP problem is NP-Hard. Now it remains to show that the DAELP problem is in NP. One needs to guess a labeling assignment for each edge and check if it is of zero cost. The checking obviously can be accomplished in polynomial time since for each label we need to check against all the graphical features of the drawing which are of polynomial size. □

Theorem 9. *The AELP problem is NP-hard.*

Sketch of the proof: To prove the NP-Hardness of the Admissible ELP problem we follow the same steps as in the proof of the NP-Hardness of DAELP problem. We must make sure however, that all the properties of the building blocks are preserved. This can be accomplished by adding to the structure of the building blocks nodes of degree zero when edges have more than one labels on each side, as shown in *Fig.*(9). Also we need to adjust the length of each edge that has only one label at each side to be approximately equal to the height or width of the label size, as shown in *Fig.*(10). By applying these two rules we restrict the labeling space essentially down to the discrete labeling space.

As it can be observed from *Figures* 9 and 10, each edge can have an infinite number of potential labels of zero cost for the AELP problem. Also, regardless of

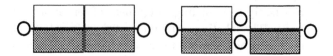

Fig. 9. An edge with more than one labels on each side.

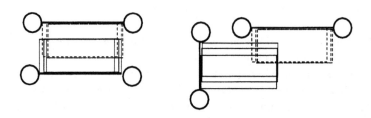

Fig. 10. Edges with one label on each side.

which label we choose for each of these edges, the behavior of these labels with respect to overlapping with other labels is exactly the same as if we had chosen labels that follow the rules of the DAELP problem. Hence, it is clear that the properties that hold for any building block for the DAELP problem, also hold for the AELP problem. □

Finally, we have the following:

Theorem 10. *The Optimal ELP problem is NP-hard.*

4 Extensions and conclusions

We have proven that a restricted version of the ELP problem, namely the AELP problem for discrete labeling space, is NP-Complete. This result implies that efficient heuristics to solve the ELP problem are needed. Even though the ELP and NLP problems are similar, which suggests that some ideas that work on NLP may work on ELP, the challenge remains to design algorithms that will take advantage of the characteristics that are specific to the ELP problem.

The research on the problem of labeling graphical features has been exclusively directed towards predefined drawings. Which is appropriate when the labeling of geographical and technical maps is the objective. But for the graph drawing community, and for anybody that draws graphs to visualize information, the labeling problem can be seen from a different perspective, since the underline geometry of any graph layout can be changed. This presents a possible dual approach. First devise graph layout techniques that reserve space for labels, and secondly devise local improvement techniques that free-up space for labels in the existing layout.

Acknowledgements

We would like to thank Joe Marks for interesting discussions.

References

1. Ahn, J. and H. Freeman, A program for automatic name placement. *Cartographica*, 21(2&3), Summer & Autumn, 1994.
2. J. Christensen, J. Marks and S. Shieber, An empirical study of algorithms for Point Feature Label Placement. *ACM Trans. on Graphics*, 14(3):203-232, July, 1995.
3. Di Battista,G., Eades P., Tamassia R., I. G. Tollis, Algorithms for Drawing Graphs: an Annotated Bibliography. *Computational Geometry, Theory and Applications*, 4(5), pp. 235-282, 1994.
4. Doerschler, J. S and H. Freeman, A rule based system for dense map name placement. *Communications of ACM*, 35(1), pp. 68-79, January, 1992.
5. Ebinger, L. R. and A. M. Goulete, Noninteractive automated names placement for the 1990 decennial census. *Cartography and Geographic Information Systems*, 17(1), pp. 69-78, January, 1990.
6. M. Forman, F. Wagner, A packing problem with applications to lettering of maps . *Proc. of 7-th Annual Symposium on Computational Geometry*, pp. 281-288, 1991.
7. Herbert Freeman, An Expert System for the Automatic Placement of Names on a Geographical Map. *Information Sciences*, 45, pp. 367-378, 1988.
8. Freeman, H. an J. Ahn, On the problem of placing names in a geographical map. *International Journal of Pattern Recognition and Artificial Intelligence*, 1(1), pp. 121-140, 1987.
9. M. R. Garey and D. S. Johnson, *Computers and Intractability: A Guide to the Theory of NP-Completeness*. W.H. Freeman and company, NY, NY, 1979.
10. Eduard Imhof, Positioning names on maps. *The American Cartographer*, 2(2), pp. 128-144, 1975.
11. Kato, T. and H. Imai, The NP-completeness of the character placement problem of 2 or 3 degrees of freedom. *Record of Joint Conference of Electrical and Electronic Engineers in Kyushu*, 1138, 1988. In Japanese.
12. J. Marks, private communication, 1996.
13. J. Marks, S. Shieber, The computational complexity of cartographic label placement. *Technical Report 05-91*, Harvard University, 1991.
14. J. W. van Roessel, An algorithm for locating candidate labeling boxes within a polygon. *The American Cartographer*, 16(3), pp. 201-209, 1989.
15. Pinhas Yoeli, The logic of automated map lettering. *The Cartographic Journal*, 9(2), pp. 99-108, December 1972.
16. S. Zoraster, The solution of large 0-1 integer programming problems encountered in automated cartography, *Operation Research*, 38(5), pp. 752-759, September-October, 1990.

Intersection Graphs of Noncrossing Arc-Connected Sets in the Plane

Jan Kratochvíl[1][*]

Department of Applied Mathematics, Charles University, Prague, Czech Republic

Abstract. Arc-connected sets $A, B \subset E_2$ are called *noncrossing* if both $A - B$ and $B - A$ are arc-connected. A graph is called an *NCAC graph* if it has an intersection representation in which vertices are represented by arc-connected sets in the plane and any two sets of the representation are noncrossing. In particular, disk intersection graphs are NCAC. By a unified reduction we show that recognition of disk intersection and NCAC graphs are NP-hard. A simple observation shows that triangle-free disk intersection and NCAC graphs are planar, and hence recognizable in polynomial time. On the other hand, recognition of triangle-free AC graphs (intersection graphs of arc-connected sets) is still NP-hard.

1 Intersection graphs

Given a family of sets $\mathcal{M}$, a graph G is called an *intersection graph of $\mathcal{M}$* if a mapping $f : V(G) \to \mathcal{M}$ exists so that $f(u) \cap f(v) \neq \emptyset$ iff $uv \in E(G)$ for every pair of distinct vertices $u, v \in V(G)$. The multiset $\{f(u) : u \in V(G)\}$ is called an $\mathcal{M}$ *representation* of G.

Intersection graphs, and namely intersection graphs of geometrical objects (elements of $\mathcal{M}$ are determined by their geometrical shape) were deeply studied both for their interesting graph theoretical properties and for their practical applications. Among these are *interval graphs* (intersection graphs of intervals on a line), *circular arc graphs* (intersection graphs of intervals on a circle), *circle graphs* (intersection graphs of chords of a circle), to mention just a few. These classes of intersection graphs are recognizable in polynomial time and allow polynomial algoritheorems for some otherwise NP-complete optimization problems.

Recently, practical applications led to introduction of more complex classes of intersection graphs, most of which are NP-hard to recognize. We mention *grid intersection graphs* (intersection graphs of vertical and horizontal grid line segments) [5, 11, 3, 1], *box intersection graphs* (intersection graphs of isothetic rectangles) [15, 17, 14], *SEGMENT graphs* (intersection graphs of straight line segments) [4, 12] or *STRING graphs* (intersection graphs of curves) [16, 4, 9, 10]. At Graph Drawing '95, Breu and Kirkpatrick [2] considered *disk intersection graphs* and they presented a proof of NP-hardness of their recognition, both for the case of intersection graphs of unit disks, and of disks of bounded diameters.

---

[*] The author acknowledges partial support of Czech Research grants GA ČR 0194 and GA UK 193 and of Czech-US Science and Technology Research grant No. 94 051.

The unbounded case remained open, and it is the aim of this note to show that recognition of disk intersection graphs (with no bound on the diameters of the disks) is NP-hard. Let us note that the technique of [2] is by its nature of no help for the case of unbounded diameters. Their technique also directly applies to touching graphs of bounded-diameter disks, while unbounded-diamater touching disk graphs are exactly planar graphs (cf. Koebe [8]) and hence easy to recognize. Thus the presented result is the last pebble in the mosaic of the complexity of variants of disk intersection and touching graph recognition problems. We will prove the result in a more general setting of *noncrossing arc-connected sets* in the plane.

2 Noncrossing arc-connected sets

For intersection graphs, it is important to distinguish between *topological connectedness* and *arc-connectedness* of sets in euclidean space. A set is called *topologically connected* if it cannot be split into two disjoint clopen subsets. On the other hand, a set is called *arc-connected* if any two of its points can be connected by a curve within the set (a curve is a homeomorphic image of a closed interval). It is well known [R.Thomas 1988 - unpublished personal communication] that every graph is an intersection graph of topologically connected sets in the plane. On the other hand, intersection graphs of arc-connected sets in the plane are exactly string graphs (intersection graphs of curves in the plane) [16, 10, 4]. These graphs are known to be NP-hard to recognize, but no finite recognition algoritheorem is known.

In some cases, the definitions of intersection graphs depend not only on the shape of the objects used to represent the vertices, but also on their relative position. E.g., grid intersection graphs fulfil the additional requirement that parallel segments of a representation are disjoint (and hence grid intersection graphs are always considered to be bipartite). Touching or contact graphs of segments, curves and disks attracted recently a lot of interest [2, 7]. An important subclass of string graphs are *1-string graphs* where no two curves of a representation are allowed to share more than one intersection point (recognition of such graphs is in NP). In this sense we introduce the following definition: We call two arc-connected sets A, B *noncrossing* if both $A - B$ and $B - A$ are arc-connected. (Figure 1 shows an example of a crossing and noncrossing pair of sets.) In particular, any two disks are noncrossing, and more generally, any two homotopic convex sets are noncrossing. The impact of noncrossingness is illustrated by the following lemma, which will be implicitly used several times in the arguments in the next section.

Lemma 1. *Let A, B be noncrossing arc-connected sets inside a region Ω. Let a, a', b, b' be disjoint curves in Ω such that a, a' start in A, end at the boundary of Ω and are both disjoint with B, while b, b' start in B, end at the boundary of Ω and are both disjoint with A. Then these curves meet the boundary of Ω in the cyclic ordering a, b, b', a' or a', b, b', a.*

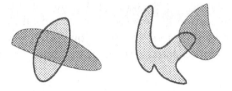

Fig. 1. Crossing and noncrossing pairs of sets.

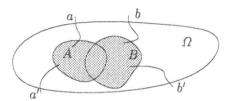

Fig. 2. Illustration to Lemma 1.

Proof. Denote X_a a point in $a \cap A$ (and in a similar way, $X_{a'} \in a' \cap A$, $X_b \in b \cap B$, $X_{b'} \in b' \cap B$). Since $A - B$ is arc-connected, X_a and $X_{a'}$ can be connected by a curve a'' lying inside $A - B$. Similarly, X_b and $X_{b'}$ are connected by a curve b'' lying inside $B - A$. It follows that curves $a \cup a'' \cup a'$ and $b \cup b'' \cup b'$ are disjoint and both connecting two points on the boundary of Ω. The Jordan curve theorem implies that the endpoints of $b \cup b'' \cup b'$ lie both on the same arc of the boundary of Ω determined by the endpoints of $a \cup a'' \cup a'$.

Lemma 2. *Let $\mathcal{A} = \{A_1, A_2, \ldots, A_n\}$ and $\mathcal{B} = \{B_1, B_2, \ldots, B_n\}$ be systems of arc-connected sets such that*
- *every pair A_i, B_j, $i = 1, 2, \ldots, n$, $j = 1, 2, \ldots, m$ is noncrossing,*
- *every set $A_i \in \mathcal{A}$ intersects at most one set of $\mathcal{B}$ and vice versa, and*
- *both $G(\mathcal{A})$, the intersection graph of $\mathcal{A}$, and $G(\mathcal{B})$, the intersection graph of $\mathcal{B}$, are connected.*

Then $A = \bigcup_{i=1}^{n} A_i$ and $B = \bigcup_{i=1}^{m} B_i$ are noncrossing arc-connected sets.

Proof. The union of non-disjoint arc-connected sets is arc-connected, and thus the (graph) connectedness of $G(\mathcal{A})$ implies the arc-connectedness of A (and similarly for B).

Consider a set $A_i \in \mathcal{A}$. Since it is intersected by at most one set of $\mathcal{B}$, we have $A_i - B = A_i$ or $A_i - B = A_i - B_j$ for some $j \in \{1, 2, \ldots, m\}$. In both cases, $A_i - B$ is arc-connected. Since every set of $\mathcal{B}$ intersects at most one set of $\mathcal{A}$, we have $(A_i - B) \cap (A_j - B) = A_i \cap A_j$ for every $i \neq j \in \{1, 2, \ldots, n\}$. Thus the intersection graph of $\{A_i - B | i = 1, 2, \ldots, n\}$ is isomorphic to $G(\mathcal{A})$ and hence connected. It follows that $A - B = \bigcup_{i=1}^{n} (A_i - B)$ is arc-connected. A similar argument shows that $B - A$ is arc-connected, and therefore A and B are noncrossing.

We define *NCAC graphs* as graphs with intersection representations by arc-connected sets in which any two sets of the representation are noncrossing. In particular, disk intersection graphs are NCAC graphs. We will prove in Section 3 that both recognition of disk intersection graphs and of NCAC graphs are NP-hard problems. To get a better feeling about NCAC graphs and their NCAC representations, we state one more lemma here, which is a direct consequence of Lemmas 1 and 2 and which will be important in the proof of Theorem 4 (cf. the ladders in Fig. 4).

Lemma 3. *Let* P_1, P_2 *be two paths such that every vertex of each of them is adjacent to at most one vertex of the other one. Suppose these paths have an NCAC representation in a region* Ω *such that the sets representing the first and last vertex of* P_1 *intersect the boundary of* Ω *and the sets representing the inner vertices of* P_1 *as well as all vertices of* P_2 *lie inside* Ω. *Then all sets representing vertices of* P_2 *lie in the same half-region determined by the chain of arc-connected sets representing* P_1 *and the boundary of* Ω.

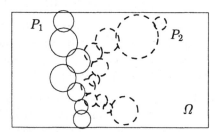

Fig. 3. Illustration to Lemma 3.

In other words, two chains of arc-connected sets cannot cross each other if sets in one of them are each allowed to intersect at most one set of the other chain. To enable a crosssing of chains, one set from one chain must be intersected by at least two sets of the other chain (cf. Fig. 5).

3 Recognition of NCAC graphs

We prove the main result in this section:

Theorem 4. *Recognition of NCAC graphs is NP-hard.*

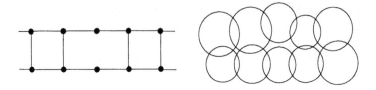

Fig. 4. A ladder.

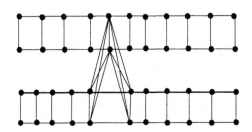

Fig. 5. Crossover component for two ladders.

Proof. We reduce from Not-All-Equal SATISFIABILITY. This is the problem

Instance: A CNF formula $\Phi = \bigwedge_{i=1}^{n} c_i$, where $c_i = (l_{i1} \vee l_{i2} \vee l_{i3})$ is a clause containing literals l_{i1}, l_{i2}, l_{i3} (a literal is a variable or a negation of a variable).

Question: Is there a truth assignment ϕ of the variables such that in every clause c_i, either one or two literals evaluate to TRUE?

Not-All-Equal SATISFIABILITY is known to be NP-complete even if all clauses contain exclusively positive literals (i.e., non-negated variables) and every variable occurs in exactly three clauses (this version is in fact exactly bicolorability of 3-regular 3-uniform hypergraphs) [6]. We assume that we are given such a formula Φ with a set of clauses C over a set of variables X (i.e., $C \subset \binom{X}{3}$).

The reduction follows the guidelines of the reduction used in [9] to show the NP-hardness of recognition of string graphs.

We first embed the incidence graph $G_\Phi = (X \cup C, \{xc : x \in c \in C\})$ of Φ in a grid so that its vertices (both representing variables and clauses) are embedded in the grid vertices, and edges (connecting variables to clauses) are rectilinear, following the grid lines. Here comes the main difference from the construction in [9] – this embedding is not required to be noncrossing, in fact, it better should not be, as Not-All-Equal SAT is known to be solvable in polynomial time for formulas with planar incidence graphs. Note also that for such embedding it is important that G_Φ has maximum degree at most 4 (our G_Φ is cubic). We keep this embedding fixed.

Then we construct a graph G from this embedding by local replacements. Each variable vertex will be replaced by a copy of the variable gadget, which

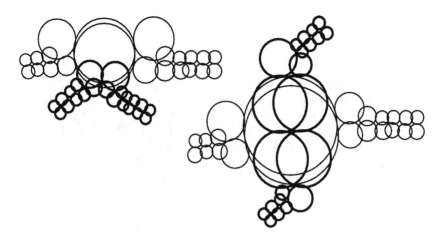

Fig. 6. Noncrossing and crossing realizations of a crossover component.

is simply a cycle of length 8. Egdes of G_Φ are replaced by linking gadgets. In the reduction in [9], the linking gadgets were just pairs of parallel curves. The order of these pairs in which they arrive to the clause gadgets was used to encode whether the variable evaluates to TRUE or FALSE in the clause. In the presented reduction, the truth values will be encoded in the same way. The linking gadgets will be ladders (Fig. 4) of suitable length. A ladder is formed by two disjoint paths such that each vertex of each of them is adjacent to at most one vertex of the other path. Near each variable gadget, two ladders leaving that gadget are connected by a crossover component (Fig. 5). A crossover component allows two ladders to cross each other, if necessary, or to continue in their directions, whatever is necessary in a representation (Fig. 6). This is used when showing that an NCAC representation exists if Φ is Not-All-Equal satisfiable.

Every edge crossing in the drawing of G_Φ is also replaced by a crossover component. An essential argument in [9] was the topological uniqueness of the planar drawing of G_Φ (guaranteed by 3-connectedness of G_Φ). In the present construction we force the topological uniqueness by inserting dummy connectors in G_Φ, which are vertices inserted into faces of G_Φ and connected by chains to new crossover components placed on the edges of G_Φ. (Details can be found in the technical report [13].)

Every clause is replaced by a clause gagdet depicted in Fig. 7. Unlike the gadgets used in [9], our clause gadget is non-symmetric. It requires a more detailed description. The clause gadget is bounded by a cycle of length 15, and every fifth vertex is used for connection with the ladders bringing information from variable gagdets. Let this clause be $\nu = (x \vee y \vee z)$ and let the edges $x\nu, y\nu$ and $z\nu$ arrive to ν in this counterclockwise order in the fixed drawing of G_Φ. We denote by C_x, C_y, C_z the vertices of the clause gadget connected to the ladders, and by $l_u^1, l_u^2, r_u^1, r_u^2, u \in \{x, y, z\}$ the ladder vertices connected to

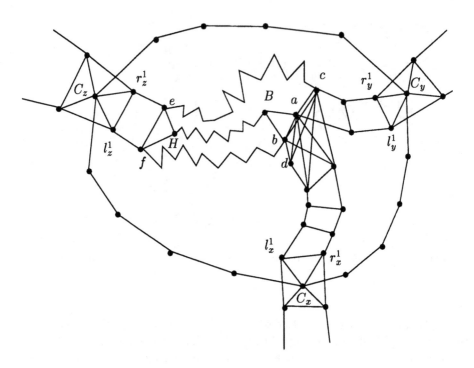

Fig. 7. The clause gadget.

$C_u, u \in \{x, y, z\}$. The ladders continue inside the clause-cycle, now as a part of the clause gadget. The x-ladder ends with vertices b, d, the y-ladder ends with a, c and the z-ladder ends with e, f. The last four vertices of the x-ladder are pairwise adjacent to the last two vertices of the y-ladder. The inner part of the clause gadget is formed by vertices B, H and by paths of suitable lengths (having a representation in mind, we call these paths *chains*): B is adjacent to a and b; H is adjacent to e and f; B is connected by a chain to H; c is connected by a chain to e and d is connected by a chain to f.

If G has a representation by noncrossing arc-connected sets, then it has a representation such that the regions bounded by the cycles in variable gadgets are empty, and the regions bounded by the cycles in clause gadgets contain only the sets representing the vertices of the clause gadgets themselves. Contracting these regions into points and the ladders into edges, we obtain a drawing of G_Φ. The dummy connectors force a topological uniqueness of this drawing and we may assume that this contraction is homotopic to the drawing we started with. In particular, in such a representation the ladders from variables to clauses arrive in the same circular order as in the drawing we started with.

The variable gadget can be represented in two different ways – its cycle being represented clockwise or counterclockwise. These two ways will encode if the variable is TRUE or FALSE. The ladders then transfer this information to

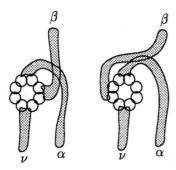

Fig. 8. Schematic picture of ladders leaving a variable occuring in clauses ν, α, β.

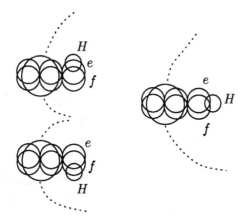

Fig. 9. Disk representations of efH-part of a clause gadget.

the clauses - a variable, say x, is TRUE in a clause ν if and only if the sets l^1_x, r^1_x arrive in this clockwise order to the clause gadget. In the notation of the masterclause $\nu = (x \vee y \vee z)$, x is TRUE in ν iff the sets d, b appear on the boundary of the region bounded by the clause cycle in this clockwise order, y is TRUE in ν iff a, c appear in this clockwise order and z is TRUE in ν if f, e appear in this clockwise order.

Let us have a close look at a representation of the clause gadget. Suppose both variables x, y evaluate to FALSE in ν. Let Ω be the region inside the clause cycle, bounded by the sets representing the x-, y- and z-ladders and the parts of the clause cycle between C_y and C_z, and between C_x and C_z. Since B is connected to H by a chain, the sets representing B and H must penetrate Ω. We attend to the occurrence of c, d, e, f on the boundary of Ω. Since c, a appear in this clockwise order and B intersects a but not c, the sets c, B appear in this order. Similarly for b, d ergo B, d. Hence c, B, d appear in this clockwise order.

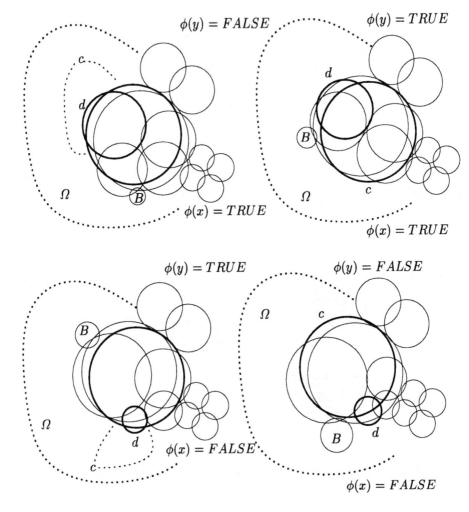

Fig. 10. Disk representations of Bcd-part of the clause gadget.

Now the chains of arc-connected sets from c to e and from d to f satisfy the assumptions of Lemma 3 and therefore f, e must occur in this clockwise order on the boundary of Ω. This means that if x, y are FALSE, z must be TRUE. A similar argument shows that if x, y are TRUE, z must be FALSE. Thus we conclude that Φ is Not-All-Equal satisfiable, provided G is an NCAC graph.

If, on the other hand, Φ is Not-All-Equal satisfied by a truth valuation ϕ, an NCAC representation can be constructed along the same lines. We represent the variable gadgets clockwise or counterclockwise according to their truth values. The ladders will be realized along the edges of G_Φ in the drawing we started with. Near a variable, two ladders either cross or touch each other, in order to lead in the right direction (see schematic Fig. 8 and Fig. 6). So it remains to show that

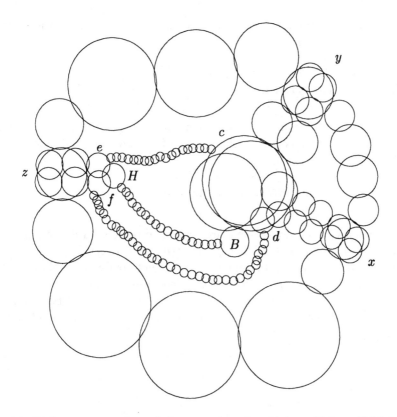

Fig. 11. Disk representation of clause gadget for $\phi(x) = \phi(y) = FALSE$ and $\phi(z) = TRUE$.

the clause gadgets can be realized for all combinations of not-all-equal valuations of their variables. Figure 10 shows disk representations of the x, y-boundary of Ω. It shows realizations of clockwise orders c, a for $\phi(x, y) = $ FF and a, c for $\phi(x, y) = $ TT. It also shows that in case of x, y being evaluated differently, both orders a, c and c, a are realizable. Figure 9 shows the z-boundary of Ω and it shows that all three relative positions of H (efH, eHf, Hef) are realizable. The realization of the entire gadget is then obvious, an example for $\phi(x, y, z) = $ FFT is show in Fig. 11.

Hence $G \in NCAC$ if and only if Φ is Not-All-Equal satisfiable, and G is a disk intersection graph in such a case. Thus we have also proved the following corollary.

Corollary 5. *Recognition of disk intersection graphs is NP-hard.* $\square$

Note also that from the representations of particular gadgets, it would be possible to read a bound on the diameters (independent on Φ) so that G has a bounded-diameter representation whenever Φ is Not-All-Equal satisfiable. However, the bounded-diamater case was completely treated in [2] and therefore we only stress the unbounded case here.

4 Triangle-free graphs

P. Hliněný [personal communication] noted that triangle-free disk intersection graphs are planar. This is true in the setting of noncrossing arc-connected sets as well:

Lemma 6. *Triangle-free NCAC graphs are planar.*

Sketch of proof. Given a noncrossing arc-connected representation of a triangle-free graph G, choose a point in each $f(u) \cap f(v)$, $uv \in E(G)$. Replace every set $f(u)$ by a dendrite (a connected union of curves not containing a closed curve) connecting the points in $f(u)$ so that the dendrites do not cross each other. This is possible since the sets themselves were noncrossing and their pairwise intersections were disjoint. One can easily deduce a planar drawing of G from the dendrites.

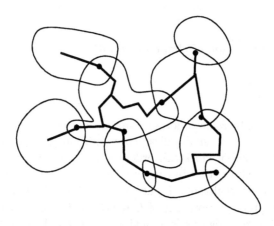

Fig. 12. Illustration to the sketch of proof of Lemma 6.

Since every planar graph is a contact graph of disks [8], and thus a disk intersection graph, we see that triangle-free NCAC (and triangle-free disk intersection graphs) graphs are exactly triangle-free planar graphs. Therefore we conclude:

Corollary 7. *Triangle-free disk intersection graphs, as well as triangle-free NCAC graphs, are recognizable in polynomial time.*

We should note that noncrossingness was crucial for the above observation. In general, traingle-free intersection graphs of arc-connected sets (triangle-free string graphs) are not necessarily planar – e.g., $K_{3,3}$ is a string graph. Also the recognition is more difficult:

Theorem 8. *Recognition of triangle-free string graphs is NP-hard.*

Proof. A reduction from Not-All-Equal SATISFIABILITY similar to the one presented above proves the result. We again start with a CNF formula Φ which has 3 non-negated variables in each clause and in which every variable occurs in 3 clauses. We fix a drawing of the incidence graph of Φ and add dummy connectors to enforce topological uniqueness of the drawing. Then we replace variable vertices by variable gadgets consisting of cycles of length 12, connected to linking gadgets by every other vertex. Linking gadgets are pairs of induced paths (i.e., vertices from different paths are nonadjacent). Crossover components are edges of $K_{2,2}$ between pairs of vertices from two linking gadgets. Clause gadgets are slightly more complicated than in the reduction in [9] and in some sense resemble the gadgets from the previous section. The clause gadget is depicted in Fig. 13. The graph G constructed in this way is triangle-free and we claim that it has an intersection representation by curves (and hence by arc-connected sets) if and only if Φ is Not-All-Equal satisfiable.

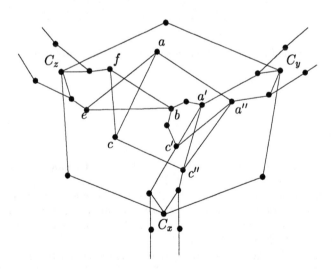

Fig. 13. The clause gadget for triangle-free string graphs.

Again, variable gadgets have two possible ways of representation – the variable-gadget cycle being represented clockwise or counterclockwise. These two ways are used to encode the truth values of the variables. This encoding is such that in the masterclause $\nu = (x \vee y \vee z)$, x (y, z, respectively) is TRUE iff the curves c'', c' (a', a''; f, e, respectively) lie in this clockwise order on the boundary of the region bounded by the clause-gadget cycle. Similarly as in the proof of Theorem 4, Table 1 shows feasible cyclic clockwise orderings of curves a, b, c, e, f in a possible representation by intersections of curves (arc-connected sets) depending on the truth valuations of the variables in a clause $\nu = (x \vee y \vee z)$. It is easy to check (and known from [16, 9]) that the pentagon $acfbe$ has a representation inside a region if and only the cyclic ordering of its curves on the boundary of the region does not coincide with the complementary 5-cycle. Therefore G has an intersection representation by curves if and only if every clause receives at least one TRUE and least one FALSE variable.

$\phi(x, y, z)$	cyclic orders of a, b, c, e, f
FFF	$abcef$
FFT	$abcfe$
FTF	$bacef, bcaef$
FTT	$bacfe, bcafe$
TFF	$acbef, cabef$
TFT	$acbfe, cabfe$
TTF	$cbaef$
TTT	$cbafe$

Table 1. Feasible cyclic orderings for particular truth valuations.

References

1. Bellantoni,S., Ben-Arroyo Hartman,I., Przytycka,T., Whitesides,S.: Grid intersection graphs and boxicity, Discrete Appl. Math. (1994), 41-49
2. Breu, H., Kirkpatrick, D.: On the complexity of recognizing intersection and touching graphs of disks, In: Graph Drawing (F.J.Brandenburg ed.), Proceedings Graph Drawing '95, Passau, September 1995, Lecture Notes in Computer Science 1027, Springer Verlag, Berlin Heidelberg, 1996, pp. 88-98
3. Chandramouli,M., Diwan,A.A.: Upward numbering testing for triconnected graphs, In: Graph Drawing (F.J.Brandenburg ed.), Proceedings Graph Drawing '95, Passau, September 1995, Lecture Notes in Computer Science 1027, Springer Verlag, Berlin Heidelberg, 1995, pp. 140-151
4. Eherlich, G., Even, S., Tarjan, R.E.: Intersection graphs of curves in the plane, J. Combin. Theory Ser. B **21** (1976), 8-20
5. de Fraysseix, H., de Mendez, P.O., Pach, J.: Representation of planar graphs by segments, Intuitive Geometry, Colloquia Mathematica Societatos Janos Bolyai **63** (1991), 109-117

6. M.R. Garey and D.S. Johnson, *Computers and Intractability*, W.H.Freeman and Co., 1978
7. P.Hliněný: Contact graphs of curves, In: Graph Drawing (F.J.Brandenburg ed.), Proceedings Graph Drawing '95, Passau, September 1995, Lecture Notes in Computer Science 1027, Springer Verlag, Berlin Heidelberg, 1996, pp. 312-323
8. Koebe,P.: Kontaktprobleme den konformen Abbildung, Berichte über die Verhandlungen der Sächsischen Akademie der Wissenschaften, Leipzig, Math.-Physische Klasse **88** (1936), 141-164
9. Kratochvíl,J.: String graphs I. There are infinitely many critical nonstring graphs, J. Combin. Theory Ser. B **52** (1991), 53-66
10. Kratochvíl,J.: String graphs II. Recognizing string graphs is NP-hard, J. Combin. Theory Ser. B **52** (1991), 67-78
11. Kratochvíl,J.: A special planar satisfiability problem and some consequences of its NP-completeness, Discrete Appl. Math. **52** (1994), 233-252
12. Kratochvíl,J., Matoušek,J: Intersection graphs of segments, J. Combin. Th. Ser. B **68** (1994), 317-339
13. Kratochvíl,J.: Intersection graphs of noncrossing arc-connected sets in the plane (technical report), KAM Series, Charles University, 1996.
14. Kratochvíl,J., Przytycka,T.:Grid intersection and box intersection graphs on surfaces (extended abstract) In: Graph Drawing (F.J.Brandenburg ed.), Proceedings Graph Drawing '95, Passau, September 1995, Lecture Notes in Computer Science 1027, Springer Verlag, Berlin Heidelberg, 1996, pp. 365-372
15. Roberts,F.S.: On the boxicity and cubicity of a graph, In: W.T. Tutte, ed., Recent Progress in Combinatorics, Academic Press, New York, 1969, pp. 301-310
16. Sinden,F.W.: Topology of thin film RC-circuits, Bell System Tech. J. (1966), 1639-1662
17. Wood,D.: The riches of rectangles, in: Proceedings 5th International Meeting of Young Computer Scientists, Smolenice, 1988, 67-75

Wiring Edge-Disjoint Layouts*

Ruth Kuchem[1] and Dorothea Wagner[2]

[1] RWTH Aachen
[2] Fakultät für Mathematik und Informatik, Universität Konstanz,
78434 Konstanz, Germany

Abstract. We consider the *wiring* or *layer assignment problem* for edge-disjoint layouts. The wiring problem is well understood for the case that the underlying layout graph is a square grid (see [8]). In this paper, we introduce a more general approach to this problem. For an edge-disjoint layout in the plane resp. in an arbitrary planar layout graph, we give equivalent conditions for the k-layer wirability. Based on these conditions, we obtain linear-time algorithms to wire every layout in a tri-hexagonal grid, respectively every layout in a tri-square-hexagonal grid using at most five layers.

1 Introduction

The *wiring problem* consists in converting a two-dimensional edge-disjoint layout into a three-dimensional vertex-disjoint layout. Wiring edge-disjoint layouts is a fundamental and classical problem in VLSI-design. Typically, the general layout problem in VLSI-design consists of two phases, the *placement* and the *routing*. Often, the routing phase is again divided into two steps. First, a two-dimensional layout is constructed satisfying certain conditions induced by the underlying layout model. This layout describes the course of the wires connecting the corresponding terminals. In the second step, the *wiring step* or *layer assignment step*, the edges of the wires are assigned to different layers to avoid physical contacts between different wires. The layout can be viewed as a projection in the plane of this final three-dimensional wiring. In connection with *graph drawing*, the wiring problem is of interest as well. There, an edge-disjoint embedding of a graph is given. The problem consists in a visualization of the graph by a three-dimensional vertex-disjoint embedding whose projection in the plane is again the edge-disjoint embedding of the graph.

Consider an edge-disjoint *layout* in the plane resp. in a planar graph. Such a layout may be an edge-disjoint realization of *nets* or an edge-disjoint embedding of a graph. The construction of edge-disjoint layouts is a fundamental problem. For an overview we refer to [3, 9, 11]. Then the wiring problem can be described as follows. There is a number of graphs isomorphic to the layout, called *layers*. Each path, called *wire*, of the layout is realized by a sequence of subpaths in different

---

* The second author acknowledges the *DFG* for supporting this research under grant *Wa 654/10-2*.

layers such that two different wires are vertex-disjoint. A vertical connection between layers, a *via*, is used at each layer change. The main optimization goal is to minimize the number of layers.

Several results have been obtained for the wiring problem for edge-disjoint layouts in regular grids, especially square grids [1, 2, 4, 5, 6, 10, 12, 15, 14, 16]. Most of these results are based on the combinatorial framework introduced in [8] which applies to edge-disjoint layouts in square grids. Moreover, only layouts where terminals are placed on the boundary of the layout graph are considered. A technique using two-colorable maps is developed, and necessary and sufficient conditions for the wirability in a fixed number of layers are given. For two layers these conditions are easy to test. On the other hand, it is $\mathcal{NP}$-complete to decide if a layout is three-layer wirable [7]. Every layout is wirable in four layers, and such a wiring can be constructed in time linear in the size of the layout [1], [14]. In [14], the concept of two-colorable maps is applied to octo-square grids, but no guarantee for the number of layers required for the wiring is given. If the layout graph is a tri-hexagonal grid, every edge-disjoint layout is wirable in five layers [13].

Nothing is known so far about the wiring problem for layouts in the plane resp. in general planar layout graphs, with arbitrary terminal positions. In this paper, we develop a general approach to this problem. It leads to necessary and sufficient conditions for k-layer wirability of edge-disjoint layouts where at most two different wires meet in a vertex. These conditions generalize the framework given in [8]. Again, two-layer wirability is easy to test. And of course, deciding three-layer wirability is $\mathcal{NP}$-complete as well. For layouts in special planar layout graphs guarantees are given for the number of layers required for the wiring. We prove that every layout in a tri-hexagonal grid is wirable in at most five layers. Moreover, every layout in a tri-square-hexagonal grid is wirable in at most five layers as well. In both cases, such a wiring can be constructed in time linear in the size of the layout. Observe that our approach for layouts in tri-hexagonal grids is different from the approach given in [13].

The wiring theory presented here is restricted to layouts where at most four wire edges are incident to the same vertex. But it is extendable to the case that more than four wire edges belonging to at most two different wires are incident to the same vertex. The case that more than two different wires meet at a vertex seems to be much more involved.

The paper is organized as follows. In Section 2, we introduce the necessary definitions and notations. The general approach to wiring edge-disjoint layouts is developed in Section 3. Finally, in Section 4 we present linear-time algorithms for wiring layouts in tri-hexagonal and tri-square-hexagonal grids using at most five layers.

2 Preliminaries

We consider an edge-disjoint *layout* in the plane or in a planar layout graph. Such a *layout in the plane* consists of vertices, called *terminals*, in the plane and pairwise edge-disjoint Steiner trees connecting specified sets of terminals, called *wires*. A special case of an edge-disjoint layout is an edge-disjoint embedding of a graph, i. e., a mapping of the vertices of that graph into the plane and a realization of its edges by pairwise edge-disjoint paths. Similarly, for an edge-disjoint *layout in a planar graph* consider an undirected graph G, the *layout graph*, with a fixed embedding in the plane. Then the layout consists of *terminals* placed on vertices of G and pairwise edge-disjoint Steiner trees connecting specified sets of terminals, respectively pairwise edge-disjoint paths realizing edges of the embedded graph. A layout in the plane, respectively the vertices and edges of the layout graph G occupied by a layout, induce a planar embedded graph. In the following, we identify a layout L with this induced graph.

A *conducting layer*, or simply *layer* is a graph isomorphic to the layout L. Conducting layers $L_0, \ldots, L_{k-1}$ are assumed to be stacked on top of each other, with L_0 on the bottom and L_{k-1} on the top. A contact between two layers, called a *via*, can be placed only at a vertex of a layer.

A correct *layer assignment* or *wiring* $W(L)$ of a given layout L is a mapping of each edge of L to a layer such that:

1. No two different wires share a vertex on the same layer.
2. If adjacent edges of a wire are assigned to different layers, a via is established between these layers at their common vertex.
3. If a via connects L_h and L_j $(h < j)$, then layers $L_i, h < i < j$, are not used at that vertex by any other wire.

Note that a correct wiring can be interpreted as a three-dimensional configuration of *vertex-disjoint* wires. In the following, we restrict to layouts where at most two different Steiner trees meet at the same vertex. The case that more than two different wires meet at a vertex seems to be much more involved. Because of lack of space, the wiring theory presented here is formulated only for layouts where at most four wire edges are incident to the same vertex. But it is extendable to the case that more than four wire edges belonging to at most two different wires are incident to the same vertex. To determine a correct wiring of a layout L, only those vertices where two different wires meet are of relevance. In the following we call these vertices *non-trivial vertices* of the layout. In Figure 1, all possibilities of non-trivial vertices where at most four wire edges meet are shown.

Definition 1. The subgraph of L induced by all non-trivial vertices and all edges incident to at least one non-trivial vertex of L is called the *core* of L, denoted $core(L)$.

The following lemma shows that we can restrict to the core of a layout L. In [8], it is proved for layouts in grids, where different wires may cross or both bend at the same vertex, but do not meet at terminals. But the proof of the lemma applies to the more general layouts we consider here as well.

Lemma 2. *[8] A layout L is wirable in k layers if and only if each connected component of $core(L)$ is.*

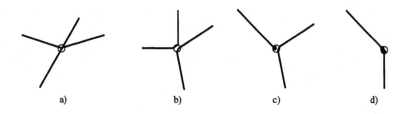

a)   b)   c)   d)

Fig. 1. The different possibilities of non-trivial vertices. Each of the two wires may end at a terminal in that vertex. Possibility c) occurs when one wire ends at a terminal and d) occurs when both wires end at a terminal in that vertex.

Proof. In order to prove the non-trivial part of the lemma, consider a k-layer wiring W of the core $core(L)$. All edges of $L - core(L)$ are assigned to an arbitrary layer, say L_0. If necessary, a via is established on a trivial vertex connecting the wires incident to that vertex.

3   A characterization of k-layer wirability

In this section, we develop a complete characterization of k-layer wirability of planar layouts. First, we observe that we can restrict to the 2-edge-connected components of a layout core.

Lemma 3. *A layout L is wirable in k layers if and only if each 2-edge-connected component of $core(L)$ is wirable in k layers.*

Proof. In order to prove the non-trivial part of the lemma, consider a decomposition of $core(L)$ into its 2-edge-connected components. These 2-edge-connected components can be ordered topologically. Let W be a k-layer wiring of the 2-edge-connected components, and P be a wire going through different components. Then there is an edge $\{u, v\}$ on P such that u and v belong to different components, say $C(u)$ and $C(v)$. Assume w.l.o.g. that $C(u)$ is before $C(v)$ in the topological ordering.

   Now, a correct k-layer wiring W^* of the subgraph of $core(L)$ induced by $C(u)$, $\{u, v\}$ and $C(v)$ is constructed as follows. In case the wirings of $C(u)$ and $C(v)$ are compatible, i. e., P is wired both times above (resp. below) the wire it meets in u and in v, the wiring of $C(v)$ and $C(u)$ remains fixed, $\{u, v\}$ is wired in the same layer as P in u, and an appropriate via is placed on v. Otherwise, only the wiring of $C(u)$ remains fixed and the wiring of $C(v)$ is flipped. That is, if an edge of $C(v)$ belongs to layer $L_i, 0 \le i \le k - 1$ in W, then this edge is assigned to layer L_{k-i} in W^*. Then again, $\{u, v\}$ is wired in the same layer as P in u, and an appropriate via is placed on v. Finally, a sequence of flippings according to an arbitrary linear extension of the topological ordering of the 2-edge-connected components of $core(L)$ induces a correct layer assignment of L in k layers.

In the following, we assume that $core(L)$ is 2-edge-connected. The *dual graph* of a planar graph with respect to a fixed combinatorial embedding is defined as follows. For each face of the graph there is a dual vertex, and there is an edge connecting two dual vertices if and only if their faces are incident with a common primal edge. The subgraph of the dual graph of L corresponding to $core(L)$ is denoted by $core(L)^d$. That is, $core(L)^d$ is the graph induced by all edges dual to edges of $core(L)$. Since $core(L)$ is 2-edge-connected, $core(L)^d$ contains no loops. The edges of $core(L)^d$ that are dual to edges of $core(L)$ incident to a trivial vertex are called the *boundary edges* of $core(L)^d$, or just *the boundary* of $core(L)^d$. Vertices of $core(L)^d$ incident to boundary edges are called *boundary vertices*, all other vertices of $core(L)^d$ are called *inner vertices*. See Figure 3. Obviously, a vertex is an inner vertex if and only if it is dual to a face of $core(L)$ that contains only non-trivial vertices.

Fig. 2. Diagonals (thin) corresponding to a pair of neighbored edges of $core(L)$ that belong to the same wire.

We first give a characterization of two-layer wirable layouts which is of fundamental importance for what follows. It is based on the dual of the layout core. Let us call two layout edges that are incident to the same vertex and have a common face *neighbored*. For a layout L, define *diagonal edges* connecting certain vertices of $core(L)$ with vertices of $core(L)^d$. Precisely, for every pair of neighbored edges that belong to the same wire and whose common vertex is non-trivial, a diagonal edge is introduced conncting that common non-trivial vertex and the vertex of $core(L)^d$ corresponding to the common face. See Figure 2 and Figure 3. For a vertex $v \in core(L)^d$ denote $diag(v)$ the number of diagonals incident to v. Then the *extended degree* of v, $exdeg(v)$, is the sum of the degree of v (denoted $deg(v)$), and the number of diagonal edges incident to v, i. e., $exdeg(v) := deg(v) + diag(v)$. Now, $deg(v)$ is equal to the number of edges on the face F^v dual to v, which is again equal to the number of vertices on F^v. Thus, $exdeg(v) = |F^v| + diag(v)$. A vertex $v \in core(L)^d$ is called *even* if $exdeg(v)$ is an even number, otherwise v is called *odd*.

Lemma 4. *A layout L is two-layer wirable if and only if*

  1. for each inner vertex of $core(L)^d$ the extended degree is even, and

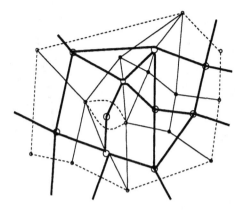

Fig. 3. Part of a layout core (bold), its dual and its diagonals (thin); the dashed edges are the boundary edges of the dual; inner vertices of the dual are black.

2. *for each connected component of the boundary of* $core(L)^d$ *the sum of the extended degrees of the boundary vertices minus the number of boundary edges is even.*

Proof. (**Sketch**) In a two-layer wiring, two neighbored edges incident to a common non-trivial vertex are assigned to different layers if and only if they belong to different wires. Thus, $core(L)$ is two-layer wirable if and only if for every face of $core(L)$ the number of vertices on that face incident to neighbored edges belonging to different wires is even.

For the first part of the proof, consider an inner vertex $v \in core(L)^d$ dual to face F^v of $core(L)$. Let $m(v)$ denote the number of vertices on F^v incident to neighbored edges on F^v belonging to different wires. Since an inner vertex v is dual to a face F^v containing only non-trivial vertices, F^v is two-layer wirable if and only if $m(v)$ is even. Thus F^v is two-layer wirable if and only if $exdeg(v) = |F^v| + diag(v) = m(v) + 2diag(v)$ is even. The proof of the second part is similar, but more technical. It is omitted because of lack of space.

The general wiring theory developed now relies on the construction of an appropriate set of wire edges whose removal leaves a two-layer wirable layout. The declaration what "appropriate" means is our goal now. More precisely, for a k-layer wirable layout we give an equivalent characterization of a set of wire edges whose removal leaves a two-layer wirable layout. This characterization of a set of wire edges consists in giving forbidden patterns for its dual edges.

Definition 5. Consider a subset R of wire edges of $core(L)$. R is called a *removal set* of $core(L)$ if its removal leaves a two-layer wirable layout. The elements of R are also called *removal edges*. Denote R^d the set dual edges of R. A removal set R is called *legal* if R^d contains no cycle and no path connecting two vertices of the same connected component of the boundary of $core(L)^d$.

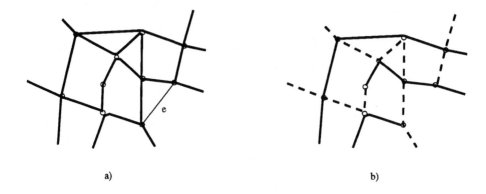

Fig. 4. a) The layout L from Figure 3. Edge e forms a legal removal set (thin). b) A two-layer wiring of $L - \{e\}$

Consider a legal removal set R of $core(L)$. If $core(L)$ is 2-edge-connected, the graph induced by $core(L) - R$ is connected. Thus it has a two-layer wiring that is unique up to the choice of the layers, say in layer *bottom* and layer *top*. Such a two-layer wiring induces a *type classification* of vertices u and v *with respect to* $\{u, v\}$, for every edge $\{u, v\} \in R$.

$$type_{\{u,v\}}(v) := \begin{cases} 1 & \text{if } \{u, v\} \text{ is incident in } v \text{ to a wire edge} \\ & \text{of the same wire in the top layer, or} \\ & \text{of a different wire in the bottom layer;} \\ 2 & \text{otherwise.} \end{cases}$$

Obviously, this classification is well defined. Observe, that for a vertex v incident to edges $\{u, v\}, \{w, v\} \in R$ that belong to different wires, $type_{\{u,v\}}(v) \neq type_{\{w,v\}}(v)$. Lemma 4 implies the following corollary.

Corollary 6. *For a legal removal set R, the edges of R^d form paths connecting odd vertices of $core(L)^d$, respectively an odd vertex to a boundary vertex of $core(L)^d$.*

Since R is legal, these paths form trees whose leaves are odd, respectively lie on the boundary. Note that two different paths containing a vertex of the same boundary component are considered to belong to the same tree.

Lemma 7. *Let R be a legal removal set of $core(L)$. For a type classification induced by a two-layer wiring of $core(L) - R$, $type_{\{u,v\}}(v) \neq type_{\{u,v\}}(u)$ for every edge $\{u, v\} \in R$.*

Proof. **(Sketch)** A tree T of edges of R^d induces a cycle of wire edges not in R, i. e., the cycle of edges of $core(L)$ around T. More precisely, every inner vertex of $core(L)$ belonging to T corresponds to its dual cycle in $core(L)$, and every boundary vertex of $core(L)$ belonging to T corresponds to the cycle in $core(L)$

around the corresponding boundary component. Then the union of these cycles minus the edges of R form the cycle induced by T.

In a two-layer wiring of $core(L) - R$, such a cycle must contain an even number of layer changes. Now, consider an edge $\{u, v\} \in R$ dual to an edge of T. Assume $type_{\{u,v\}}(v) = type_{\{u,v\}}(u)$. Then for all edges $\{x, y\} \in R$ dual to an edge of T we have $type_{\{x,y\}}(x) = type_{\{x,y\}}(y)$. But for an edge of R dual to an edge incident to a leaf of T, the type of the two end vertices must be different with respect to that edge. This is a contradiction.

Lemma 7 guarantees that we can define an orientation on the edges of R.

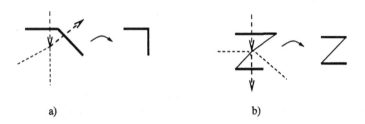

a)   b)

Fig. 5. Removal sets that do not induce 3-layer wirability and the corresponding forbidden patterns. a) Pattern corresponding to neigboring edges. b) Pattern corresponding to edges of the same face that are not neigboring and are both incident to a diagonal ending at the same vertex.

Definition 8. An edge $\{u, v\} \in R$ is oriented from $u \to v$ if $type_{\{u,v\}}(u) = 1$ and $type_{\{u,v\}}(v) = 2$. Let $R^{\to}$ denote the directed graph, called *layer graph*, induced by R and this orientation.

Obviously, two adjacent arcs of $R^{\to}$ that belong to the same wire are oriented towards each other. So, two subsequent edges on a directed path in $R^{\to}$ must belong to different wires. The *length* of a directed path is defined as the number of edges on that path. Now we are ready to prove the main theorem of this section.

Theorem 9. *A layout L is wirable in k layers if and only if there exists a legal removal set R such that the length of any directed path in $R^{\to}$ is at most $k - 2$.*

Proof. "$\Rightarrow$" Consider a wiring of $core(L)$ in layer $L_i, 0 \le i \le k-1$. Let R be the set of all wire edges wired in layer $L_i, 0 \le i \le k - 1$, and $R^{\to}$ the corresponding directed layer graph. For a directed path in $R^{\to}$, two subsequent arcs (u, v) and (v, w) belong to different wires and thus must be wired in different layers. Because of the orientation, the layer of (u, v) must be below the layer of (v, w). Consequently, the length of any directed path in $R^{\to}$ is at most $k - 2$.

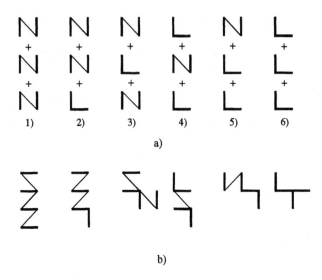

Fig. 6. a) Combinations of forbidden patterns for 3-layer wirability that lead to forbidden patterns for 5-layer wirability; b) examples of forbidden patterns for 5-layer wirability .

"⇐" Assume there exists a legal removal set R such that the length of every directed path in $R^{\rightarrow}$ is at most $k - 2$. For an edge $(u, v) \in R^{\rightarrow}$ denote $l_{max}(u, v)$ the maximum length of a directed path in $R^{\rightarrow}$ terminating with (u, v). A wiring of $core(L)$ in k layers $L_i, 0 \le i \le k - 1$, is constructed as follows.

$$layer(u, v) = \begin{cases} L_0 \text{ resp. } L_{k-1}, & \text{if } (u, v) \notin R^{\rightarrow}; \\ L_i, 1 \le i \le k - 2, & \text{if } (u, v) \in R^{\rightarrow} \text{ and } l_{max}(u, v) = i - 1. \end{cases}$$

The assignment of (u, v) to L_0 resp. L_{k-1} is according to a fixed two-layer wiring of $core(L) - R$.

Remark The proof of Theorem 9 induces an algorithm to construct a k-layer wiring from a legal removal set R where the length of any directed path in $R^{\rightarrow}$ is at most $k - 2$. This algorithm can be easily implemented to run in time linear in the size of the layout.

A characterization of k-layer wirability for $k > 2$ in terms of forbidden patterns in the dual R^d of a legal removal set R is now easily derived from Theorem 9.

Lemma 10. *A layout L is wirable in three layers if and only if there exists a legal removal set R such that R^d contains none of the patterns shown in Figure 5.*

Proof. $core(L)$ is wirable in three layers if and only if there exists a legal removal set R such that $R^{\rightarrow}$ contains no directed path of length two. For a directed path in $R^{\rightarrow}$ two subsequent edges must belong to different wires. These two edges meet at a non-trivial vertex, say v. In case the two corresponding dual edges in R^d are neigboring in v, they form a pattern of type a) shown in Figure 5. Otherwise they form a pattern of type b) shown in Figure 5, i. e., the two corresponding dual edges in R^d belong to the same face, but are not neigboring in v and are each incident to a diagonal ending at v.

Lemma 10 can be easily extended to giving forbidden patterns for wirability in $k \geq 4$ layers. From the patterns shown in Figure 5 we just have to generate patterns dual to directed paths of length $k - 1$ in the layer graph. Forbidden patterns for wirability in four layers are all possible combinations of two forbidden patterns for wirability in three layers,where two patterns are combined by identifying two edges. Accordingly, forbidden patterns for wirability in five layers are all possible combinations of three forbidden patterns for wirability in three layers. See Figure 6.

4 Algorithms

In this section we prove that every layout in a tri-hexagonal grid and every layout in a tri-square-hexagonal grid is wirable in five layers. We present algorithms that construct a removal set for such a layout satisfying Theorem 9 for $k = 5$. That is, the dual R^d corresponding to the constructed removal set R contains none of the patterns illustrated in Figure 6.

4.1 Layouts in Tri-Hexagonal Grid Graphs

A tri-hexagonal grid graph is a grid consisting of grid lines of three different directions, the horizontal direction and two diagonal directions. In every grid point, lines of two different directions meet, either one horizontal and one diagonal line, or two diagonal lines. Consequently, every vertex has degree four. See Figure 7 a) for a tri-hexagonal grid and its dual.

The algorithm to construct a legal removal set R for a layout L in a tri-hexagonal grid works as follows. Firstly, all vertices of the layout are considered to be non-trivial. Then obviously, 5-layer wirability of this layout induces 5-layer wirability of the original layout. The dual of the layout is scanned "row-wise" from bottom to top, and from left to right. For every vertex of the dual of the layout, its extended degree and its vertex class is considered. That is, the vertex set of the grid dual to a tri-hexagonal grid is partitioned into three different classes: vertices of degree six (class 1), vertices of degree three incident to an up-going vertical edge and two diagonal edges (class 2), and vertices of degree three incident to an down-going vertical edge and two diagonal edges (class 3). See Figure 8. Now alternatingly, in every second row only vertices of class 1 and class 2 are visited, or only vertices of class 3. Depending on the extended degree and the class of the visited vertex, an incident edge is added to R^d in such a way, that $exdeg(v)$ is even for all vertices $v \in L^d - R^d$, and R^d contains no forbidden pattern for 5-layer wirability.

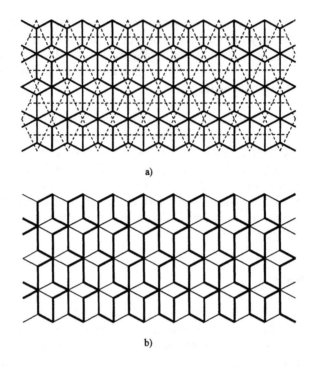

a)

b)

Fig. 7. a) A tri-hexagonal grid graph (dashed) and its dual graph (bold). b) Dual edges that might be added to R^d by the algorithm (bold).

Algorithm 1 Tri-Hexa

Input: *The dual L^d of a layout L in a tri-hexagonal grid.*

Output: *A subset R^d of L^d such that exdeg(v) is even for all vertices $v \in L^d - R^d$ and R^d contains no forbidden pattern for 5-layer wirability.*

begin
$R^d := \emptyset$
for *all rows* **do**
    for *all vertices v of one row* **do**
        if *exdeg(v) is odd in L^d and v*
        belongs to class 2, **then** $R^d := \begin{cases} R^d + e_{right}, & \textit{if right successor is odd;} \\ R^d + e_{up} & \textit{otherwise.} \end{cases}$
        belongs to class 1, **then** $R^d := R^d + e_{up}$
        belongs to class 3, **then** $R^d := R^d + e_{right}$
        $L^d := L^d - R^d$
end

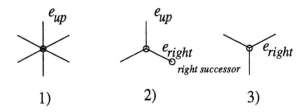

Fig. 8. The three different classes of vertices in the grid dual to a tri-hexagonal grid.

It is easy to prove that Algorithm 1 considers only those edges of L^d shown in Figure 7 b). Then the following theorem can be proved by a detailed case analysis.

Theorem 11. *For a layout L in a tri-hexagonal grid, Algorithm 1 constructs a subset R^d of L^d that contains no forbidden pattern for 5-layer wirability. The running time is linear in the size of the layout.*

Proof is omitted.

From a legal removal set determined by Algorithm 1, a 5-layer wiring of a layout in a tri-hexagonal grid can be constructed in linear time as well according to Theorem 9.

Fig. 9. a) A tri-square-hexagonal grid graph (dashed) and its dual graph (bold). b) Dual edges that might be added to R^d by the algorithm (bold).

4.2 Layouts in Tri-Square-Hexagonal Grid Graphs

A tri-square-hexagonal grid graph is the dual graph of the union of two grids, a hexagonal grid and the dual of a hexagonal grid. Thus it contains only vertices of degree four. Its faces are triangles, squares and hexagons. See Figure 9 for a tri-square-hexagonal grid and its dual.

The algorithm to construct a legal removal set R for a layout L in a tri-square-hexagonal grid again scans the dual of the layout "row-wise" from bottom to top and from left to right. The vertices of the grid dual to a tri-square-hexagonal grid are partitioned into six different classes: vertices of degree six (class 1), three different classes of vertices of degree four (class $2, 3$ and 4), and two different classes of vertices of degree three (class 5 and 6). See Figure 10.

Alternatingly, in every second row only vertices of class 1 and class 2 are visited, or only vertices of class $3, 4, 5$ and 6. Depending on the extended degree and the class of the visited vertex, an edge is added to R^d in such a way that $exdeg(v)$ is even for all vertices $v \in L^d - R^d$, and R^d contains no forbidden pattern for 5-layer wirability.

Algorithm 2 Tri-Square-Hexa

Input: *The dual L^d of a layout L in a tri-square-hexagonal grid.*

Output: *A subset R^d of L^d such that $exdeg(v)$ is even for all vertices $v \in L^d - R^d$, and R^d contains no forbidden pattern for 5-layer wirability.*

> **begin**
> $R^d := \emptyset$
> **for** *all rows* **do**
> > **for** *all vertices v of one row* **do**
> > > **if** *$exdeg(v)$ is odd in L^d and v*
> > > *belongs to class 1*, **then** $R^d := \begin{cases} R^d + e_{right}, & \text{if right successor is odd;} \\ R^d + e_{up} & \text{otherwise.} \end{cases}$
> > > *belongs to class 2*, **then** $R^d := R^d + e_{up}$
> > > *belongs to class 3*, **then** $R^d := R^d + e_{right-up}$
> > > *belongs to class 4*, **then** $R^d := R^d + e_{right-up}$
> > > *belongs to class 5*, **then** $R^d := R^d + e_{right}$
> > > *belongs to class 6*, **then** $R^d := \begin{cases} R^d + e_{right}, & \text{if right successor is odd;} \\ R^d + e_{up} & \text{otherwise.} \end{cases}$
> > > $L^d := L^d - R^d$
>
> **end**

It can be shown that Algorithm 2 considers only those edges of L^d shown in Figure 9 b). Then by a detailed case analysis, the following theorem can be proved.

Theorem 12. *For a layout L in a tri-square-hexagonal grid Algorithm 2 constructs a subset R^d of L^d that contains no forbidden patterns for 5-layer wirability. The running time is linear in the size of the layout.*

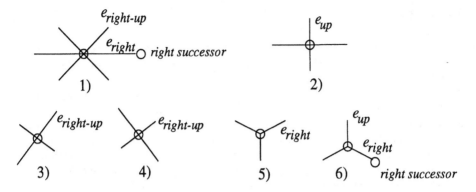

Fig. 10. The six different classes of vertices in the grid dual to a tri-square-hexagonal grid.

Proof is omitted.

From a legal removal set determined by Algorithm 2, a 5-layer wiring of a layout in a tri-square-hexagonal grid can be constructed in linear time as well according to Theorem 9.

5 Concluding Remarks

We presented a general approach to the problem of wiring edge-disjoint layouts. Equivalent conditions for the k-layer wirability of an edge-disjoint layout where at most two wires meet in a vertex are given. Based on these conditions, we obtain linear-time algorithms to wire every layout in a tri-hexagonal grid and every layout in a tri-square-hexagonal grid using at most five layers. Our approach generalizes the framework introduced in [8]. There, equivalent conditions for the k-layer wirability of an edge-disjoint layout in a square grid are given. These conditions are based on a legal partition of the layout grid into a two-colorable map. A legal partition is characterized by "forbidden patterns" for the partition lines. The two color regions of the partition correspond to regions where horizontal edges are wired above vertical edges, resp. to regions where vertical edges are wired above horizontal edges. This framework is not applicable to more general grids resp. planar layout graphs, since there we can have layout edges of more than two different directions. The equivalent conditions for k-layer wirability developed in this paper are based on the characterization of legal removal sets. This leads to "forbidden pattern" as well, which are quite similar to the forbidden patterns given in [8]. Overall, our approach delivers a more general interpretation of the framework introduced there.

Acknowledgements We thank Majid Sarrafzadeh for drawing our attention to the layer assignment problem for general edge-disjoint layouts.

References

1. M. Brady and D. Brown: VLSI routing: Four layers suffice. In F. Preparata, editor, Advances in Computer Research, VOL 2: VLSI Theory, JAI Press Inc. (1984) 245–257

2. M. Brady and M. Sarrafzadeh: Stretching a knock-knee layout for multilayer wiring. IEEE Transactions on Computers, C-39 (1990) 148–152

3. G. Di Battista, P. Eades, R. Tamassia, and I. G. Tollis: Algorithms for Drawing Graphs: an Annotated Bibliography. Computational Geometry 4 (1994) 235–282

4. T. F. Gonzales and S. Zheng: Simple three-layer channel routing algorithms. In J. H. Reif, editor, Proceedings Aegean Workshop on Computing. Springer-Verlag, Lecture Notes in Computer Science, vol. 319 (1988) 237–246

5. M. Kaufmann and P. Molitor: Minimal stretching of a layout to ensure 2-layer wirability. INTEGRATION The VLSI Journal, 12 (1991) 339–352

6. R. Kuchem, D. Wagner, and F. Wagner: Optimizing area for three-layer channel routing. Algorithmica, 15 (1996) 495–519

7. W. Lipski, Jr: On the structure of three-layer wirable layouts. In F. Preparata, editor, Advances in Computer Research, VOL 2: VLSI Theory, JAI Press Inc. (1984) 231–243

8. W. Lipski, Jr and F. P. Preparata: A unified approach to layout wirability. Math. Systems Theory, 19 (1987) 189–203

9. R. H. Möhring, D. Wagner, and F. Wagner: VLSI Network Design: A survey. In M. Ball, T. Magnanti, C. Monma, and G. Nemhauser, editors, Handbooks in Operations Research/Management Science, Volume on Networks, North-Holland, (1995) 625–712

10. P. Molitor: A survey on wiring. J. Inform. Process. Cybernet. EIK, 27 (1991) 3–19

11. H. Ripphausen-Lipa, D. Wagner, and K. Weihe: Efficient algorithms for disjoint paths in planar graphs. In W. Cook, L. Lovász, and P. Seymour, editors, DIMACS, Center for Discr. Math. and Comp. Sc., 20, Springer-Verlag, Berlin (1995) 295–354

12. M. Sarrafzadeh, D. Wagner, F. Wagner, and K. Weihe: Wiring knock-knee layouts: a global approach. IEEE Transactions on Computers 43 (1994) 581–589

13. I. G. Tollis: Wiring Layouts in the Tri-Hexagonal Grid. Intern. J. Computer Math. 37 (1990) 161–171

14. I. G. Tollis: A New Approach to Wiring Layouts. IEEE Transactions on Computer-Aided Design of Integrated Circuits and Systems 10 (1991) 1392–1400

15. I. G. Tollis: Wiring in uniform grids and two-colorable maps. INTEGRATION The VLSI Journal 12 (1991) 189–210

16. D. Wagner: Optimal routing through dense channels. Int. J. on Comp. Geom. and Appl. 3 (1993) 269–289

Proximity Drawings of Outerplanar Graphs [*] (extended abstract)

William Lenhart[1], Giuseppe Liotta[2]

[1] Department of Computer Science, Williams College, Williamstown, MA 01267.
lenhart@cs.williams.edu
[2] Department of Computer Science, Brown University, Providence, RI 02912–1910.
gl@cs.brown.edu

Abstract. A proximity drawing of a graph is one in which pairs of adjacent vertices are drawn relatively close together according to some proximity measure while pairs of non-adjacent vertices are drawn relatively far apart. The fundamental question concerning proximity drawability is: Given a graph G and a definition of proximity, is it possible to construct a proximity drawing of G? We consider this question for outerplanar graphs with respect to an infinite family of proximity drawings called β-drawings. These drawings include as special cases the well-known Gabriel drawings (when $\beta = 1$), and relative neighborhood drawings (when $\beta = 2$). We first show that all biconnected outerplanar graphs are β-drawable for all values of β such that $1 \leq \beta \leq 2$. As a side effect, this result settles in the affirmative a conjecture by Lubiw and Sleumer [15, 17], that any biconnected outerplanar graph admits a Gabriel drawing. We then show that there exist biconnected outerplanar graphs that do not admit any convex β-drawing for $1 \leq \beta \leq 2$. We also provide upper bounds on the maximum number of biconnected components sharing the same cut-vertex in a β-drawable connected outerplanar graph. This last result is generalized to arbitrary connected planar graphs and is the first non-trivial characterization of connected β-drawable graphs. Finally, a weaker definition of proximity drawings is applied and we show that all connected outerplanar graphs are drawable under this definition.

[*] Research supported in part by the National Science Foundation under grant CCR-9423847, by the U.S. Army Research Office under grant 34990–MA–MUR, and by N.A.T.O.- CNR Advanced Fellowships Programme.

1 Introduction and Overview

Proximity drawings have received increasing attention within the graph drawing community (see, e.g., [15, 9, 3, 17, 2, 7, 14]) because of their appealing graphical features: Edges are represented by straight lines, vertices not incident to a certain edge are drawn far apart from that edge, and groups of adjacent vertices tend to cluster together. Also, proximity drawings are of practical interest in several application fields, such as pattern recognition and classification, image processing, geographic variation analysis, GIS, and computational morphology. For a survey on the different application areas of proximity drawings see [12].

A proximity drawing of a graph is has pairs of adjacent vertices relatively close together according to some proximity measure, while pairs of non adjacent vertices are relatively far apart. The choice of proximity measure determines the type of proximity drawing.

One type of proximity drawing is the *visibility drawing*, in which two vertices u and v are close if and only if they are mutually visible along one of k given *directions of visibility*. Application areas such as VLSI routing and circuit layout have stimulated considerable research on both 2-dimensional and 3-dimensional visibility drawings in which vertices are segments or simple non overlapping polygons and the directions of visibility are parallel to the axes. (see, e.g., [1, 10]).

A different type of proximity is based upon the concept of a *region of influence* (also referred to as a *proximity region*). Given two points u and v in the plane, a *proximity region* of u and v is a portion of the plane, determined by u and v, that contains points relatively close to both u and v. A *region of influence based* proximity drawing of a graph G as a *straight-line drawing* (vertices of G are mapped to distinct points in the plane, and edges to straight-line segments) such that: (i) for each edge (u, v) of G, the proximity region of u and v is empty, i.e. it does not contain any other vertex of G; and (ii) for each pair of non-adjacent vertices u, v of G, the proximity region of u and v contains at least one other vertex of G. For example, in a *Gabriel drawing* [11], the proximity region of u and v is the closed disk having u and v as antipodal points; in a *relative neighborhood drawing* [18] the proximity region is the intersection of the two open disks centered at u and at v and with distance $d(u, v)$ as radius. Figure 1 (a) is an example of a Gabriel drawing; the dotted disk is the region of influence of two adjacent vertices. Figure 1 (b) is an example of a relative neighborhood drawing; the dotted lune is the region of influence of two non adjacent vertices. A survey of different types of proximity drawings can be found in [5].

1.1 The Proximity Drawability Problem

Although proximity drawings have many uses, the study of their combinatorial properties is still in its early stages. For example, region of influence based proximity drawings are usually adopted in pattern recognition to associate a *shape* (also called *skeleton*) to a set of points: pairs of points are connected if and only if their proximity region is empty. Clearly, by changing the proximity region one

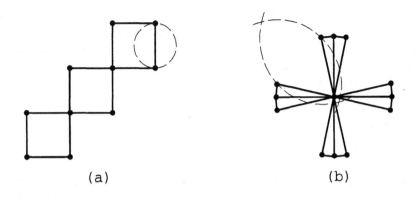

Fig. 1. (a) A Gabriel drawing, and (b) a relative neighborhood drawing.

can obtain different proximity drawings having the same set of vertices. Surprisingly, for most definitions of proximity the basic properties (such as connectivity, planarity, outerplanarity, convexity etc.) of the proximity drawings that one can possibly obtain on a given point set are not well understood. If, for instance, one whishes to give an outerplanar shape to a point set, he/she needs to know what definition of proximity can give rise to such a shape.

The above discussion motivates the study of the *proximity drawability question*: given a graph G and a definition of proximity, is it possible to construct a proximity drawing of G? In this paper we answer the question for biconnected outerplanar graphs with respect to an infinite set of well-known region of influence based proximity drawings called *β-drawings*, which were first introduced by Kirkpatrick and Radke [13] in the context of pattern recognition and computational morphology. β-drawings are based on an infinite family of proximity regions, called *β-regions*, each element of the family being identified by a value of the parameter $β$ ($0 \leq β \leq \infty$). Given a value of $β$, a β-drawing is a proximity drawing in which the proximity region is the β-region. These regions are defined precisely in the next section.

For each value of $β$, the corresponding β-region can be either an open or a closed set. A *[β]-drawing* is a proximity drawing that adopts a closed β-region; similarly, a *(β)-drawing* adopts an open β-region. For example, the 1-region of two points u and v is the disk having u and v as antipodal points; thee 2-region is the intersection of the two disks centered at u and at v and with distance $d(u, v)$ as radius. Hence, a [1]-drawing is a Gabriel drawing and a (2)-drawing is a relative neighborhood drawing (see also Figure 1). In the paper we will use the notation [β]- or (β)-drawing instead of β-drawing only when the distinction between an open and a closed β-region is essential for the discussion. We say that a graph is *β-drawable* if it has a both open and closed β-drawings.

The problem of characterizing β-drawable trees has been studied in [3, 2] in the plane, and in [6] in 3-dimensional space. Lubiw and Sleumer [15, 17] initiated the study of β-drawability of biconnected outerplanar graphs. They focused on the values $\beta = 1$ and $\beta = 2$ and showed that every maximal outerplanar graph admits both a (2)-drawing and a [1]-drawing; furthermore, they proved that all biconnected outerplanar graphs are (2)-drawable and conjectured that such graphs are also [1]-drawable.

1.2 Results

The main results presented in this paper are listed below.

- We settle in the affirmative the conjecture of Lubiw and Sleumer [15, 17], by showing that all biconnected outerplanar graphs are [1]-drawable. We also extend this result and show that all biconnected outerplanar graphs are β-drawable for all values of β such that $1 < \beta < 2$. Observe that the interval $1 \le \beta \le 2$ is the most interesting for studying the β-drawability of planar graphs, since, for $\beta > 2$, no β-drawing can have triangular faces; for $\beta < 1$, β-drawings are not guaranteed to be planar [2].
- We study *convex* β-drawings. A convex drawing of a planar graph is one in which each face is a convex polygon. Convex drawings are an important object of study in graph drawing because of their aesthetic appeal (see, e.g., [8, 4]). We provide here a negative result, by showing that there exist biconnected outerplanar graphs that do not admit any convex β-drawing. We also show that such negative results hold even if the proximity constraint of the drawing is relaxed to the so-called *weak β-proximity* model [7]: a *weak β-drawing* is such that for every pair of adjacent adjacent vertices the corresponding β-region is empty, while for pairs of non adjacent vertices the corresponding β-region may be or may be not empty.
- We extend our investigation to connected graphs. A general theorem is given that provides an upper bound on the number of biconnected components sharing a cut-vertex in a β-drawable graph, for $\beta \in [0, \infty]$. This is of particular interest since it is the first non-trivial result on the connectivity of β-drawable graphs. See, e.g., [19] for more on connectivity, As an application of our result, we prove inclusion relations for connected outerplanar β-drawable graphs with respect to larger classes of graphs.
- We exploit the relationship between β-drawings, minimum spanning trees, and Delaunay triangulations to show classes of *forbidden* outerplanar graphs, i.e. outerplanar graphs that are not β-drawable for $1 \le \beta \le 2$. Our approach generalizes to outerplanar graphs previous techniques developed for studying the β-drawability of trees (see [3, 2]). Also, motivated by the existence of forbidden outerplanar graphs, we consider the representability of outerplanar graphs with the weak β-proximity model and show that all such graphs admit weak β-drawings for any given value of β in the interval $1 \le \beta \le 2$.

Table 1 summarizes the characterization results about the β-drawability of outerplanar graphs. The entries having a bibliographic reference describe previously known results. All other entries describe results from this paper.

$\{\mathcal{CO}\}$, $\{\mathcal{BO}\}$, and $\{\mathcal{MO}\}$ are the set of all connected outerplanar, biconnected outerplanar, and maximal outerplanar graphs, respectively. $\mathcal{G}_{\mathcal{CO}}(\beta)$, $\mathcal{G}_{\mathcal{BO}}(\beta)$, and $\mathcal{G}_{\mathcal{MO}}(\beta)$ are the classes of connected outerplanar, biconnected outerplanar, and maximal outerplanar (β)-drawable graphs, respectively. Similarly, $\mathcal{G}_{\mathcal{CO}}[\beta]$, $\mathcal{G}_{\mathcal{BO}}[\beta]$, and $\mathcal{G}_{\mathcal{MO}}[\beta]$ are the classes of connected outerplanar, biconnected outerplanar, and maximal outerplanar [β]-drawable graphs, respectively. $\mathcal{G}_k$ denotes the class of graphs such that the number of biconnected components sharing a cut-vertex is at most k; $\mathcal{T}_k$ is the class of trees whose vertex degree is at most k; $\overline{\mathcal{T}}$ is the class of forbidden trees described in [3].

β	Trees	Connected	Biconnected	Maximal
$\beta = 1$	$T[1] = T_4 - \overline{T}$ [3]	$\mathcal{G}_2 \not\subseteq \mathcal{G}_{CO}[1] \subset \mathcal{G}_4$	$\mathcal{G}_{BO}[1] = \{BO\}$	$\mathcal{G}_{MO}[1] = \{MO\}$ [15, 17]
$1 < \beta < \dfrac{1}{1-\cos(\frac{2\pi}{5})}$	$T(\beta) = T_4$ [2]	$\mathcal{G}_{CO}(\beta) \subset \mathcal{G}_4$	$\mathcal{G}_{BO}(\beta) = \{BO\}$	$\mathcal{G}_{MO}(\beta) = \{MO\}$
	$T[\beta] = T_4$ [2]	$\mathcal{G}_{CO}[\beta] \subset \mathcal{G}_4$	$\mathcal{G}_{BO}[\beta] = \{BO\}$	$\mathcal{G}_{MO}[\beta] = \{MO\}$
$\beta = \dfrac{1}{1-\cos(\frac{2\pi}{5})}$	$T(\beta) = T_4$ [2]	$\mathcal{G}_{CO}(\beta) \subset \mathcal{G}_4$	$\mathcal{G}_{BO}(\beta) = \{BO\}$	$\mathcal{G}_{MO}(\beta) = \{MO\}$
	$T_4 \subset T[\beta] \subset T_5$ [2]	$\mathcal{G}_4 \not\subseteq \mathcal{G}_{CO}[\beta] \subset \mathcal{G}_5$	$\mathcal{G}_{BO}[\beta] = \{BO\}$	$\mathcal{G}_{MO}[\beta] = \{MO\}$
$\dfrac{1}{1-\cos(\frac{2\pi}{5})} < \beta < 2$	$T_4 \subset T(\beta) \subsetneq T_5$ [2]	$\mathcal{G}_4 \not\subseteq \mathcal{G}_{CO}(\beta) \subset \mathcal{G}_5$	$\mathcal{G}_{BO}(\beta) = \{BO\}$	$\mathcal{G}_{MO}(\beta) = \{MO\}$
	$T_4 \subsetneq T[\beta] \subsetneq T_5$ [2]	$\mathcal{G}_4 \not\subseteq \mathcal{G}_{CO}[\beta] \subset \mathcal{G}_5$	$\mathcal{G}_{BO}[\beta] = \{BO\}$	$\mathcal{G}_{MO}[\beta] = \{MO\}$
$\beta = 2$	$T(2) = T_5$ [3]	$\mathcal{G}_4 \not\subseteq \mathcal{G}_{CO}(2) \subset \mathcal{G}_5$	$\mathcal{G}_{BO}(2) = \{BO\}$ [15, 17]	$\mathcal{G}_{MO}(2) = \{MO\}$ [15, 17]

Table 1. Summarizing the characterization results on the β-drawability of outerplanar graphs for $1 \le \beta \le 2$.

2 Preliminaries

We review first standard definitions on outerplanar graphs. We then formally define the set of β-regions. Finally, we introduce the concept of the β-boundary curve, which will be used for computing β-drawings of biconnected outerplanar graphs.

2.1 Outerplanar Graphs

A planar graph is *outerplanar* if it has a planar embedding such that all vertices lie on a single face.

For a vertex u in a biconnected outerplanar graph G, the *fan of* u, denoted F_u, is the subgraph induced by all vertices which share a face with u. Given an embedding of G, let $u_1, u_2, \ldots, u_k$ denote the neighbors of u in clockwise order, and, for each $i < k$, let $u_{i,1}, u_{i,2}, \ldots, u_{i,m_i}$ denote the vertices of F_u non-adjacent to u on the chain from u_i to u_{i+1}. The edges uu_1 and uu_k are called the *first* and *last* edges of F_u, the edges uu_i are called the *radial* edges, and the remaining edges are called the *fan* edges. Vertex u is the *apex* of fan F_u.

Given a cut-set S of vertices of G, let $G_1, G_2, \ldots, G_n$ be the components of $G - S$. The *S-components of G* are the subgraphs of G induced by the sets

$V(G_i) \cup S$. For a fan edge e of fan F_u, let $G_u(e)$ be the e-component of G not containing u.

2.2 The Set of β-regions

Given a pair x, y of points in the plane, the *open β-region of influence of x and y*, and the *closed β-region of influence of x and y*, denoted by $R(x, y, \beta)$ and $R[x, y, \beta]$ respectively, are defined as follows [13]:

1. For $0 < \beta < 1$, $R(x, y, \beta)$ is the intersection of the two open disks of radius $d(x, y)/(2\beta)$ passing through both x and y. $R[x, y, \beta]$ is the intersection of the two corresponding closed disks.
2. For $1 \le \beta < \infty$, $R(x, y, \beta)$ is the intersection of the two open disks of radius $\beta d(x, y)/2$, centered at the points $(1-\beta/2)x+(\beta/2)y$ and $(\beta/2)x+(1-\beta/2)y$. $R[x, y, \beta]$ is the intersection of the two corresponding closed disks.
3. $R(x, y, \infty)$ is the open infinite strip perpendicular to the line segment $\overline{xy}$ and $R[x, y, \infty]$ is the closed infinite strip perpendicular to the line segment $\overline{xy}$.
4. $R(x, y, 0)$ is the empty set; $R[x, y, 0]$ is the line segment connecting x and y.

Geometric quantities used to analyze β-drawings are the two angles $\alpha(\beta)$ and $\gamma(\beta)$, defined as follows.

1. $\alpha(\beta) = \inf\{\angle xzy \parallel z \in R[x, y, \beta]\}$.
2. $\gamma(\beta)$ is only defined for $\beta \ge 2$, and $\gamma(2) = \frac{\pi}{3}$. For $\beta > 2$, let $z \ne y$ be a point on the boundary of $R[x, y, \beta]$ such that $d(x, y) = d(x, z)$. Then $\gamma(\beta) = \angle zxy$.

In the following property, angles $\alpha(\beta)$ and $\gamma(\beta)$ are denoted as α and γ, for short.

Property 2.1 *[2]*
  The value β is related to angles α and γ as follows.

- $\beta = \sin\alpha$ *for* $0 \le \beta \le 1$ *and* $\frac{\pi}{2} \le \alpha \le \pi$.
- $\beta = \frac{1}{1-\cos\alpha}$ *for* $1 < \beta \le 2$ *and* $0 \le \alpha < \frac{\pi}{2}$.
- $\beta = \frac{1}{\cos\gamma}$ *for* $2 < \beta \le \infty$ *and* $\frac{\pi}{3} \le \gamma \le \frac{\pi}{2}$.

2.3 The β-Boundary Curve

Before we can present our algorithm for producing β-drawings of biconnected outerplanar graphs, we need to introduce and study the following curve.

Definition 1. Given $\beta \in [1, 2]$ and points u and v in the plane, the *β-boundary of u with respect to v* is the curve $z = \mathcal{C}_{u,v,\beta}(\theta)$ where z is the point in the plane such that $\angle vuz = \theta$ and v is on the boundary of $R[u, z, \beta]$. We assume that $-\pi/2 < \theta < \pi/2$ and that a positive angle corresponds to $\triangle uvz$ forming a clockwise cycle.

For example, $\mathcal{C}_{u,v,1}(\theta)$ is the line through v orthogonal to the segment $\overline{uv}$. In this paper, we will be concerned with only those θ in the range $[-\pi/4, \pi/4]$.

Lemma 2. *For* $\beta \in [-\pi/4, \pi/4]$, *if* $u = (0,0)$ *and* $v = (0,1)$, *then*

$$\mathcal{C}_{u,v,\beta}(\theta) = \frac{2(\sqrt{1 - (2/\beta - 1)^2 \sin^2 \theta} - (2/\beta - 1)\cos\theta}{\beta(1 - (2/\beta - 1)^2)}(\sin\theta, \cos\theta).$$

Sketch of proof: Let $z = \mathcal{C}_{u,v,\beta}(\theta)$ and let p be the center of the circle which defines the portion of the boundary of $R[u,z,\beta]$ containing v and z. Since $(\sin\theta, \cos\theta)$ is the unit vector in the direction of z from u, we need only determine $\mid z \mid$ in order to get a formula for z. Let $w = \mid \overline{up} \mid$ and $l = \mid \overline{pz} \mid$; then $w + l = \mid z \mid$ To determine w and l, note that by definition, $w/l = 2/\beta - 1$. Using the fact that $\mid y \mid = 1$ and $\mid \overline{yp} \mid = \mid \overline{pz} \mid$, it is easy to show that

$$w + l = \frac{2}{\beta\sqrt{1 - (2/\beta - 1)^2 \sin^2 \theta} - (2/\beta - 1)\cos\theta}.$$

After this, simple algebra suffices to establish the formula of the lemma. $\square$

The next lemma describes some important properties of $\mathcal{C}_{u,v,\beta}(\theta)$.

Lemma 3. *For* $\beta \in (1,2)$, *the curve* $\mathcal{C}_{u,v,\beta}(\theta)$ *is convex in the range* $-\pi/4 < \theta < \pi/4$ *and the tangent to* $\mathcal{C}_{u,v,\beta}(\theta)$ *at* $\theta = 0$ *is orthogonal to* $\overline{uv}$.

Sketch of proof: It suffices to show the following.

1. $\mathcal{C}_{u,v,\beta}(\theta)$ has non-zero curvature on $(0, \pi/4)$ and $(-\pi/4, 0)$,
2. The tangent to $\mathcal{C}_{u,v,\beta}(\theta)$ for all sufficiently small $\theta > 0$ is negative, and
3. The tangent to $\mathcal{C}_{u,v,\beta}(\theta)$ for all sufficiently small $\theta < 0$ is positive.

The formula for the curvature of a plane curve $(x(\theta), y(\theta))$ is given by

$$\frac{\mid x'y'' - y'x'' \mid}{((x')^2 + (y')^2))^{(3/2)}}.$$

It can be shown by elementary (but tedious) analytic and algebraic methods that the curvature of $\mathcal{C}_{u,v,\beta}(\theta)$ is non-zero for $\theta \in (-\pi/4, 0) \cup (0, \pi/4)$, for all $\beta \in (1,2)$. Similar, but simpler, methods can be used to establish properties 2 and 3. $\square$

3 Proximity Drawings of Biconnected Outerplanar Graphs

We start by showing how to compute a β-drawing of a fan. We then show how to compute a β-drawing of any biconnected outerplanar graph. Finally, we study convexity prperties of β-drawings of outerplanar graphs. For brevity, we will sometimes use the expression "to β-draw a graph G" instead of the expression "to compute a β-drawing of a graph G".

3.1 How to β-draw a Fan

We begin with a technical lemma about β-drawings.

Lemma 4. *Given $\beta, 1 < \beta < 2$, and two graphs G_1 and G_2 whose intersection consists of a single edge uv, let Γ_1 and Γ_2 be β-drawings of graphs G_1 and G_2 both of which have u drawn at point a and v at point b. If Γ_1 and Γ_2 are on opposite sides of the line determined by ab, and the union of the two drawings lies in a convex region which has both a and b on its boundary, then $\Gamma_1 \cup \Gamma_2$ is a β-drawing of $G_1 \cup G_2$.*

Proof omitted in abstract.

Lemma 5. *Let F_u be a fan with apex u such that uu_1 is the first edge of F_u and let β be such that $1 \le \beta < 2$. Then F_u can be β-drawn in any triangle $\triangle abc$ having $\angle abc > \pi/2$, and $\angle bac < \pi/4$, so that vertices u and u_1 are drawn at a and b, respectively, and that the fan edges form a convex chain with the property that given any three vertices v_1, v_2, v_3 on the chain, $\angle v_1v_2v_3 > \pi/2$.*

Sketch of proof: The proof is by induction on the number of neighbors of u. Assume, for the base case, that u has two neighbors u_1 and u_2 connected by a chain $u_{1,1}, \ldots, u_{1,m}$ of non-neighbors of u. Let θ be any angle no greater than $\angle bac$. We describe a β-drawing of the fan in $\triangle abc$ as specified in the statement of the lemma having $\angle u_1uu_2 = \theta$. Moreover, this drawing will have the property that $\angle u_{1,m}u_2, u < \pi/2$.

We first describe how to correctly β-draw vertex u_2, and then how to β-draw the chain of fan vertices from u_1 to u_2. Consider the ray from u at a positive angle θ with the segment uu_1. We guarantee that u_2 is not in $R[u, u_1, \beta]$ by placing it exterior to $R[u, u_1, 2]$, the boundary of which is $C_{u,u_1,2}()$. To ensure that u_1 is not in the region $R[u, u_2, \beta]$, we position u_2 interior to the curve $C_{u,u_1,\beta}()$. Note that this is possible since $C_{u,u_1,\beta}()$ intersects the ray beyond the point at which the ray is intersected by $C_{u,u_1,2}()$.

We show now how to β-draw the chain of fan vertices from u_1 to u_2. We need to ensure that the sequence $u_1, u_{1,1}, u_{1,2}, \ldots, u_{1,m}, u_2$ is convex, that $\angle u_{1,m}u_2u < \pi/2$, and that each region $R[u, u_{1,i}, \beta]$ contains some vertex of the sequence. We will position the vertices of the chain on the curve $C_{uu_2}2$, between the line L determined by u_1c and the curve $C_{uu_1}\beta$. It is easy to see that $u_1, u_{1,1}, u_{1,2}, \ldots, u_{1,m}, u_2$ then form a convex chain in which $\angle u_{1,m}u_2, u < \pi/2$. Since the curve $C_{uu_1}\beta$ intersects each segment $uu_{1,i}$, u_1 is in $R[u, u_{1,i}, \beta]$. Finally, it can be easily checked that $\angle u_1u_{1,i}u_2 > \pi/2$, for every vertex $u_{1,i}$ on the chain; thus the β-region of any two non-consecutive vertices on the chain contains some vertex of the chain. This completes the proof of the base case; we now consider the induction step.

Consider a fan F_u in which u has $k > 2$ neighbors. Let $\theta = \angle bac/(k-1)$. Draw the part of the fan consisting of u and the chain from u_1 to u_2 as in the base case. Now, let c' be the intersection of the line determined by $u_{1,m}u_2$ and the segment ac; let $b' = u_2$. By induction, the fan $F_u - \{u, u_1, u_{1,1}, \ldots, u_{1,m}\}$

can be β-drawn in the (obtuse) triangle $ab'c'$ as specified in the lemma. Now, by Lemma 4, the union of the β-drawings of the two smaller fans is a β-drawing of F_u. Observe that, because of the way $\triangle(ab'c')$ is defined, the construction guarantees that $\angle u_{1,m}u_2u_{2,1} < \pi$. Figure 2(a) shows fan F_u drawn within $\triangle abc$; Figure 2(b) illustrates details of the drawing strategy described above. $\qquad\square$

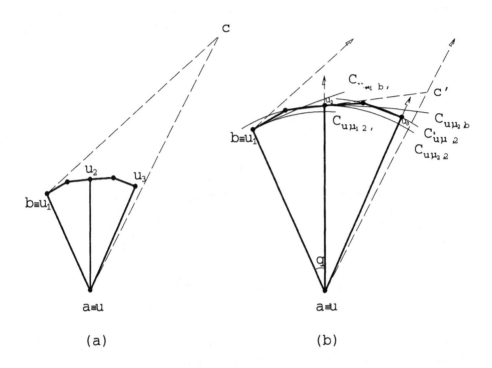

Fig. 2. (a) Fan F_u β-drawn inside $\triangle abc$, and (b) details on how to β-draw F_u.

3.2 How to β-Draw Biconnected Outerplanar Graphs

We are now ready to prove the β-drawability of biconnected outerplanar graphs.

Theorem 6. $\mathcal{G}_{BO}[1] = \{BO\}$, and $\mathcal{G}_{BO}(\beta) = \mathcal{G}_{BO}[\beta] = \{BO\}$ for all values of β such that $1 < \beta < 2$.

Sketch of proof: Let $T = \triangle abc$ be a triangle such that $\angle abc > \pi/2$, let G be a biconnected outerplanar graph, and let uv be an edge of G. We will show that, for any $\beta \in [1, 2)$, G admits a β-drawing inside T such that u is mapped to a and v is mapped to b. Through the proof, we will be using the same notation as in the proof of Lemma 5.

By Lemma 5, F_u can be β-drawn in triangle $\triangle abc$ so that vertices u and v are drawn at a and b, respectively, and that the fan edges form a convex chain. We will show that, for each fan edge e of F_u, an obtuse triangle T_e having e as a side can be defined such that the following two properties hold,

1. for fan edges e, e' such that $e \neq e'$ and any points $x \in T_e$ and $y \in T_{e'}$, some vertex of F_u lies in $R[x, y, \beta]$; and
2. if some T_e contains a β-drawing Γ of a graph such that e is an edge of Γ, then no vertex of Γ lies in $R[u, u_i, \beta]$, for any $i \leq k$.

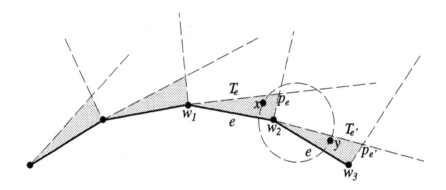

Fig. 3. Illustration for the proof of Theorem 6

This will allow us to β-draw recursively draw each subgraph $G_u(e)$ in T_e, giving a β-drawing of G in the original obtuse triangle T.

The rest of the proof is concerned with showing that the triangles T_e exist. The following two observations can be made about the β-drawing of F_u computed by the algorithm of Lemma 5,

- Since the exterior angles along the chain of fan edges of F_u are all greater than π, it is possible to define, for each fan edge $e = w_1 w_2$ a point p_e in $\triangle abc$ such that $w_1 w_2 p_e$ are vertices of a triangle with $\angle vw p_e > \pi/2$. Further, these points can be positioned so that if w_1, w_2, w_3 are consecutive fan vertices, then letting $e = w_1 w_2$ and $e' = w_2 w_3$, $\angle p_e w_2 p_{e'} > \pi/2$ (see Figure 3).
- The drawing algorithm of Lemma 5 does not make any particular assumption about the angle $\angle u_1 u u_k$ between the first edge uu_1 and the last edge uu_k of F_u, other than $0 < \angle u_1 u u_k < \pi/2$. Also, notice that $\angle uu_1 u_{1,1}$ can be any obtuse angle less than $\angle abc$ and that $\angle u_{k-1,m} u_k u < pi/2$. Thus, by using the algorithm of Lemma 5, it is always possible to β-draw F_u inside $\triangle(abc)$ so that the following is true: given any vertex v ($v \neq u_1$ and $v \neq u_k$) along the convex chain of fan edges, $\angle u_1 v u_k > \pi/2$. This implies that given any three vertices v_1, v_2, v_3 on the convex chain, $\angle v_1 v_2 v_3 > \pi/2$.

The above two observations allow us to prove Property 1.

To prove Property 2, let $e = w_1 w_2$ be a fan edge of F_u, and assume that T_e contains a β-drawing Γ of a graph having e as an edge. Since $R[u, u_i, \beta] \cap T_e$ $(1 \le i \le k)$ is a subset of $R[w_1, w_2, \beta]$, $R[u, u_i, \beta]$ cannot contain any vertex of Γ (other than possibly v or w, if either v or w coincides with u_i). □

3.3 Convexity of Outerplanar β-drawings

In this subsection we present a negative result about the convexity of β-drawings of outerplanar graphs.

We call the outerplanar graph in Figure 4(a) the *octagon* The vertices of degree four are called *corner vertices*; the other vertices are *side vertices*. A [1]-drawing of the octagon is depicted in Figure 4(b).

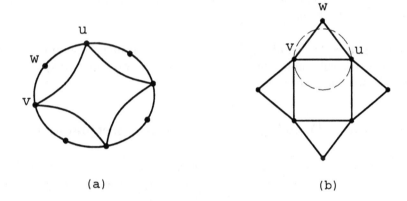

Fig. 4. (a) An octagon graph G, and (b) a [1]-drawing of G.

Lemma 7. *Any β-drawing of the octagon for $1 < \beta \le 2$ is such that the external face of the drawing is non-convex. Furthermore, any [1]-drawing of the octagon has non-convex external face.*

Sketch of proof:

Let G be the octagon and let Γ be any β-drawing of G for some value of β in the interval stated by the theorem. We show that Γ always has two edges incident on a corner vertex that form an angle less than π on the external face.

We can make the following two observations on Γ (see Figure 4).

1. All vertices of Γ must be on the external face. In fact, any 4-cycle in a [1]-drawing must be a face (any point inside a quadrilateral Q is contained in at least one of the closed disks having the edges of Q as diameter). Thus a 4-cycle is a face in any β-drawing with $1 < \beta \le 2$.

2. For every face (v, w, u) of Γ such that v and u are corner and w is side, we have that $\angle vwu < \frac{\pi}{2}$, otherwise $w \in R[v, u, 1]$ (and hence $w \in R[v, u, \beta]$ for larger values of β), contradicting the fact that v and w are adjacent.

Since the outer face of Γ is an octagon (Observation 1) and all side vertices are apices of acute angles (Observation 2), it follows that there must be at least one corner vertex of the polygon that is apex of an interior angle larger than π. □

The reasoning of Lemma 7 can be easily extended to *weak β-drawings* [7]. A *weak β-drawing* is one in which every pair of adjacent vertices has an empty β-region, while pairs of non-adjacent vertices have β-regions which might or might not be empty.

Theorem 8. *There exists an infinite family of outerplanar graphs which do not admit any convex β-drawings or convex weak β-drawings for all values of β such that $1 < \beta \leq 2$. Furthermore, these graphs do not admit any convex [1]-drawing or convex weak [1]-drawing.*

Sketch of proof: The family of graphs consists of all outerplanar graphs having the octagon as a subgraph. The proof of non-convexity of any (weak) β-drawing of such graphs follows from Lemma 7. □

4 Proximity Drawings of Connected Outerplanar Graphs

We first state a general result on the maximum number of biconnected components that can be shared by a cut vertex in any β-drawable graph and then apply the result to outerplanar β-drawable graphs.

4.1 Connectivity of β-drawable Graphs

Theorem 9. *Given β in $[0, \infty]$, let (k) be the maximum vertex degree in the block-cut-vertex tree of a (β)-drawable graph. Similarly, let $[k]$ be the maximum vertex degree in the block-cut-vertex tree of a $[\beta]$-drawable graph. The vertex degrees (k) and $[k]$ are related to the value of β as follows:*

	β	(k)	$[k]$
1	$0 \leq \beta < 2$	$(k) < \lceil \frac{2\pi}{\alpha(\beta)} \rceil$	$[k] \leq \lfloor \frac{2\pi}{\alpha(\beta)} \rfloor$
2	$\beta = 2$	$(k) < \lceil \frac{2\pi}{\alpha(\beta)} \rceil$	$[k] < \lfloor \frac{2\pi}{\gamma(\beta)} \rfloor$
3	$2 < \beta \leq \infty$	$(k) \leq \lfloor \frac{2\pi}{\gamma(\beta)} \rfloor$	$[k] < \lfloor \frac{2\pi}{\gamma(\beta)} \rfloor$

Sketch of proof: The proof generalizes to outerplanar graphs the technique presented in [2] to show upper bounds on the maximum vertex degree of β-drawable trees. In this extended abstract we show the bounds on the value (k). The technique can be easily extended to show also the bounds on $[k]$. We assume that the graphs in question contain at least one cut-vertex (the theorem is trivially verified otherwise) and start by proving the entries of rows 1 and 2.

Let G be a (β)-drawable graph for $0 \leq \beta \leq 2$, and let Γ be a (β)-drawing of G. Let S be the set of points in Γ that represents the vertices of G and let v be a cut-vertex of Γ shared by (k) blocks of G. since $0 \leq \beta \leq 2$, every edge of $MST(S)$ is also an edge of Γ (see. e.g. [2]). Consider any two consecutive edges vx and vy of $MST(S)$ incident with v such that they belong to two distinct blocks of Γ. We show that the minimum angle between vx and vy is always greater than $\alpha(\beta)$, which implies $(k) < \lceil \frac{2\pi}{\alpha(\beta)} \rceil$.

Let T_x and T_y be the subtrees, containing x and y respectively, obtained by removing v from $MST(S)$. Let $x' \in T_x$ and $y' \in T_y$ be the closest pair of vertices in these two subtrees. Notice that these vertices are not adjacent in Γ because they belong to two distinct blocks of G. Thus, there must be some vertex $z \in MST(S)$ such that z is an interior point of $R(x', y', \beta)$. Clearly z is neither a vertex of T_x nor a vertex of T_y, otherwise, since $0 \leq \beta < 2$, we would have that either $d(z, x') < d(x', y')$ or $d(z, y') < d(x'y')$, contradicting the minimality of $d(x', y')$. (Observe that this condition does not preclude z from being either a vertex of the block of G containing x' or of the block of G containing y'.) Since z is a vertex of $MST(S)$ and is not adjacent to either x' or y', we have that $d(z, x') > d(v, x)$ and $d(z, y') > d(v, y)$ (if not, $MST(S)$ would not be a minimum spanning tree). Furthermore, we have that $d(x', y') \leq d(x, y)$ by the definition of points x' and y', and that $\angle x'zy' > \alpha(\beta)$ since z is contained in $R(x', y', \beta)$. Hence, we can conclude that $\angle xvy > \alpha(\beta)$.

We now prove the correctness of the entry in row 3. The fact that $(k) \leq \lfloor \frac{2\pi}{\gamma(\beta)} \rfloor$ for $2 < \beta \leq \infty$ is an immediate consequence of the following claim.

Claim: If a graph is (β)-drawable for $2 < \beta \leq \infty$, then its vertex degree is at most $\lfloor \frac{2\pi}{\gamma(\beta)} \rfloor$.

Proof of the claim:. Let Γ be any (β)-drawing for $2 < \beta \leq \infty$, let v be a vertex of Γ with two consecutive incident edges vx and vy. Assume w.l.o.g. that $d(v, x) \leq d(v, y)$. If the angle $\angle xvy < \gamma(\beta)$ then, by the definition of angle $\gamma(\beta)$, we have that $x \in R(v, y, \beta)$, contradicting the assumption that both edges vx and vy belong to Γ. Thus, the maximum number of edges incident on v cannot be greater than $\lfloor \frac{2\pi}{\gamma(\beta)} \rfloor$. $\square$

Corollary 10. *The following inclusion relations for $\mathcal{G}_{CO}(\beta)$ and $\mathcal{G}_{CO}[\beta]$ are true.*

1. $\mathcal{G}_{CO}[1] \subseteq \mathcal{G}_4$;
2. $\mathcal{G}_{CO}(\beta) \subseteq \mathcal{G}_4$ *and* $\mathcal{G}_{CO}[\beta] \subseteq \mathcal{G}_4$ *for* $1 < \beta < \frac{1}{1-\cos(\frac{2\pi}{5})}$;
3. $\mathcal{G}_{CO}(\beta) \subseteq \mathcal{G}_4$ *and* $\mathcal{G}_{CO}[\beta] \subseteq \mathcal{G}_5$ *for* $\beta = \frac{1}{1-\cos(\frac{2\pi}{5})}$;

4. $\mathcal{G}_{CO}(\beta) \subseteq \mathcal{G}_5$ and $\mathcal{G}_{CO}[\beta] \subseteq \mathcal{G}_5$ for $\frac{1}{1-\cos(\frac{2\pi}{5})} < \beta < 2$;

5. $\mathcal{G}_{CO}(2) \subseteq \mathcal{G}_5$.

Figure 5(a), shows a [1]-drawing of an outerplanar graph of class $\mathcal{G}_4$. The same drawing is also both a β-drawing for $1 < \beta < \frac{1}{1-\cos(\frac{2\pi}{5})}$ and a $(\frac{1}{1-\cos(\frac{2\pi}{5})})$-drawing. Figure 5(b) shows an outerplanar $[\frac{1}{1-\cos(\frac{2\pi}{5})}]$-drawing of an outerplanar graph of class $\mathcal{G}_5$. The same drawing is also both a β-drawing for $\frac{1}{1-\cos(\frac{2\pi}{5})} < \beta < 2$ and a (2)-drawing.

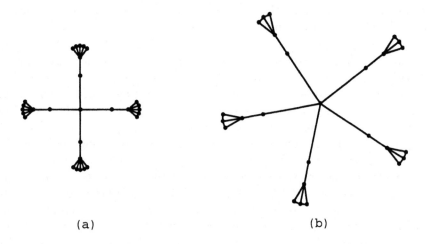

(a)   (b)

Fig. 5. (a) A β-drawing for $1 < \beta < \frac{1}{1-\cos(\frac{2\pi}{5})}$, and (b) a β-drawing for $\frac{1}{1-\cos(\frac{2\pi}{5})} < \beta < 2$.

4.2   Forbidden Outerplanar Graphs and Weak Proximity Drawings

In [16] it is proven that a [1]-drawing is a subgraph of the Delaunay triangulation of its vertex set. In [3], the relationship between [1]-drawings of trees and Delaunay triangulations is exploited to define classes of forbidden trees (i.e., trees that are not [1]-drawable). We generalize the approach of [3] to show outerplanar graphs that do not admit [1]-drawings even if they satisfy the condition of Theorem 9.

The following lemma relates *outerplanar* [1]-*drawings* (i.e. [1]-drawings with all vertices on the external face) to Delaunay triangulations. Its proof is omitted in this extended abstract.

Lemma 11. *Let Γ be an outerplanar* [1]-*drawing, let S be the set of vertices of Γ, and let $DT(S)$ be the Delaunay triangulation of S. Let C be a cycle of $DT(S)$*

that contains at least one vertex of S in its interior. Then there exists an edge $ab \in DT(S)$ and a vertex $p \in S$ such that $ab \notin \Gamma$, $\triangle(apb)$ is a face of $DT(S)$, $\triangle(apb)$ is inside C, and $\angle apb \geq \frac{\pi}{2}$.

Based on the above result, the following lemma can be proved.

Lemma 12. *All exterior angles in an outerplanar* [1]-*drawing of an outerplanar graph are at least* $\frac{\pi}{2}$.

An *extended octagon graph* is a graph constructed by adding k biconnected components to each reflex vertex of the octagon (call this octagon the *root* of the extended octagon graph). The number k is the *degree* of the extended octagon graph. For example, the graph of Figure 6(a) is an extended octagon graph of degree 1, the graph of Figure 6(b) is an extended octagon graph of degree 3.

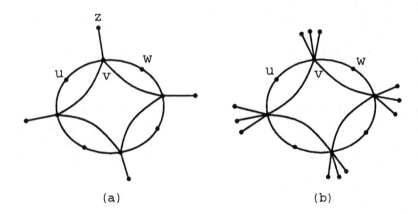

(a)   (b)

Fig. 6. (a) An outerplanar graph that does not admit a [1]-drawing, and (b) an outerplanar graph that does not admit a β-drawing for $1 < \beta \leq 2$.

Lemma 13. *An extended octagon graph of degree 1 does not admit a* [1]-*drawing.*

Sketch of proof: Let G be an extended octagon graph of degree 1 and let G' be its root (see Figure 6 (a)). Suppose G admitted a [1]-drawing Γ. Observe that, by the same reasoning as in the proof of Lemma 7, Γ is an outerplanar [1]-drawing. Also, as in Lemma 7, there exist two consecutive exterior edges vu, vw of G' such that v is a corner vertex and $\angle uvw < \pi$. Let vz be an edge of G not in G'; we have that either $\angle uvz < \frac{\pi}{2}$ or $\angle zvw < \frac{\pi}{2}$ (or both), contradicting Lemma 12. □

Lemma 14. *An extended octagon graph of degree 3 does not admit a* β-*drawing for any value of* β *such that* $1 < \beta \leq 2$.

Sketch of proof: Let G be an extended octagon graph of degree 3 and let G' be its root (see Figure 6 (b)). Suppose G admitted a β-drawing Γ for some $\beta \in [1, 2]$. With the same reasoning as in the proof of the previous lemma, we have that there exist two consecutive exterior edges vu, vw of G' such that v is a reflex vertex and $\angle xvy < \pi$. By the technique in the proof of Theorem 9, it is an easy task to show that the number of blocks that share v (other than G') and that can be drawn between edges uv and vw in Γ is at most $\lfloor \frac{\pi}{\alpha(\beta)} \rfloor < 3$, thus contradicting the assumption that Γ is a β-drawing of G. $\qquad\square$

We are now in a position to prove the main result of this section. The proof of the following theorem directly follows from Lemmas 13, 14, and from the observation that the maximum vertex degree of the block-cut-vertex tree of an extended octagon graph of degree k is $k + 1$.

Theorem 15. *For* $\beta \in [1, 2]$, *the sets* $\mathcal{G}_{CO}(\beta)$ *and* $\mathcal{G}_{CO}[\beta]$ *vary as shown in Table 1.*

Motivated by the existence of forbidden connected outerplanar graphs, we consider the the representability of outerplanar graphs with weak β-drawings.

Theorem 16. *Every connected outerplanar graph* G *admits a weak* [1]*-drawing, and a weak* (2)*-drawing. Furthermore,* G *admits a weak* β*-drawing for any given* β *such that* $1 < \beta < 2$.

Sketch of proof: There are several possible ways for constructing a weak β-drawing of a connected outerplanar graph for $1 \le \beta < 2$. For example, one can add edges to the input graph G to make it biconnected, construct a β-drawing of G by the technique of Theorem 6 and then delete the dummy edges introduced during the make-biconnected step. $\qquad\square$

5   Open Problems

Several questions remain open. For example,

1. Completely characterize the class of β-drawable connected outerplanar graphs for $1 \le \beta \le 2$.
2. The drawing algorithms presented in this paper are based on the real RAM model of computation. It would be interesting to devise numerically robust algorithms for computing proximity drawings of graphs. For example, Theorem 16 shows that weak β-drawings of outerplanar graphs are realizable. However, it would be interesting to study weak β-drawings of graphs where the coordinates of the vertices are integer numbers or rational numbers of limited size.
3. Study other aesthetic criteria of β-drawings, other than convexity. Among them, we find particularly important the area requirement, the aspect ratio, and the ratio of the longest to the shortest edge.
4. Extend the β-drawability question to the study the of non outerplanar graphs. Theorem 9 could be an useful starting point.

References

1. H. Alt, M. Godau, and S. H. Whitesides. Universal 3-dimensional visibility representations for graphs. In *Graph Drawing (Proc. GD '95)*, volume 1027 of *Lecture Notes in Computer Science*, pages 8–19. Springer-Verlag, 1996.

2. P. Bose, G. Di Battista, W. Lenhart, and G. Liotta. Proximity constraints and representable trees. In *Graph Drawing (Proc. GD '95)*, volume 1027 of *Lecture Notes in Computer Science*. Springer-Verlag, 1995.

3. P. Bose, W. Lenhart, and G. Liotta. Characterizing proximity trees. Technical Report TR-SOCS 93.9, School of Computer Science, McGill University, 1993.

4. M. Chroback, M. T. Goodrich, and R. Tamassia. Convex drawings of graphs in two and three dimensions. In *Proc. 12th Annu. ACM Sympos. on Computational Geometry*, pages 319–328, 1996.

5. G. Di Battista, W. Lenhart, and G. Liotta. Proximity drawability: A survey. In *Graph Drawing (Proc. GD '95)*, volume 1027 of *Lecture Notes in Computer Science*. Springer-Verlag, 1995.

6. G. Di Battista and G. Liotta. Computing proximity drawings of trees in 3-d space. In *WADS (Proc. WADS '95)*. Springer-Verlag, 1995.

7. G. Di Battista, G. Liotta, and S. H. Whitesides. The strength of weak proximity. In *Graph Drawing (Proc. GD '95)*, volume 1027 of *Lecture Notes in Computer Science*, pages 178–189. Springer-Verlag, 1996.

8. P. Eades and P. Garvau. Drawing stressed planar graphs in three dimensions. In *Graph Drawing (Proc. GD '95)*, volume 1027 of *Lecture Notes in Computer Science*, pages 212–223. Springer-Verlag, 1996.

9. P. Eades and S. H. Whitesides. The realization problem for euclidean minimum spanning trees is np-hard. In *Proc. ACM Symp. on Comp. Geom.*, 1994.

10. S. Fekete, M. Houle, and S. H. Whitesides. New results on a visibility representation of graphs in 3d. In *Graph Drawing (Proc. GD '95)*, volume 1027 of *Lecture Notes in Computer Science*, pages 234–241. Springer-Verlag, 1996.

11. K. R. Gabriel and R. R. Sokal. A new statistical approach to geographical analysis. *Systematic Zoology*, 18:54–64, 1969.

12. J. W. Jaromczyk and G. T. Toussaint. Relative neighborhood graphs and their relatives. In *Proceedings of the IEEE, 80*, pages 1502–1517, 1992.

13. D. G. Kirkpatrick and J. D. Radke. A framework for computational morphology. In G. T. Toussaint, editor, *Computational Geometry*, pages 217–248. Elsevier, Amsterdam, 1985.

14. W. Lenhart and G. Liotta. Drawing outerplanar minimum weight triangulations. *Inform. Process. Lett.*, 6(12):253–260, 1996.

15. A. Lubiw and N. Sleumer. All maximal outerplanar graphs are relative neighborhood graphs. In *Proc. Fifth CCCG*, pages 198–203, 1993.

16. D. W. Matula and R. R. Sokal. Properties of gabriel graphs relevant to geographic variation research and the clustering of points in the plane. *Geographical Analysis*, 12(3):205–222, 1980.

17. N. Sleumer. Oterplanar graphs as proximity graphs. Master's thesis, University of Waterloo, Watreloo, Canada, 1993.

18. G. Toussaint. The relative neighborhood graph of a finite planar set. *Pattern Recognition*, 12:229–260, 1980.

19. W. T. Tutte. *Connectivity in Graphs*. Oxford University Press, 1972.

Automatic Visualization of Two-Dimensional Cellular Complexes

L. A. P. Lozada,* C. F. X. de Mendonça,* R. M. Rosi,* and J. Stolfi*

Institute of Computing, Unicamp
{lplozada,xavier,marcone,stolfi}@dcc.unicamp.br

Abstract

A *two-dimensional cellular complex* is a partition of a surface into a finite number of elements—faces (open disks), edges (open arcs), and vertices (points). The *topology* of a cellular complex is the abstract incidence and adjacency relations among its elements. Here we describe a program that, given only the topology of a cellular complex, computes a geometric realization of the same—that is, a specific partition of a specific surface in three-space—guided by various aesthetic and presentational criteria.

Keywords: Computer graphics, visualization, graph drawing, solid modeling, minimum-energy surfaces, computational topology.

1 Introduction

1.1 Motivation

In the *boundary representation* technique commonly used in computer graphics and engineering, a solid object is defined indirectly by its surface, which is in turn described as the union of a certain number of *faces*—simple surface patches, flat or curved. Faces are bounded by *edges*—line segments, straight or curved—whose endpoints are the *vertices* of the model.

When manipulating such models, it is advisable to handle separately its *topological* properties (the contacts between faces, edges, and vertices) from its *geometrical* properties (vertex coordinates, face equations, etc.) This separation greatly improves the modularity and versatility of geometric algorithms [12,18].

A *two-dimensional cellular complex* is a mathematical structure that captures the topological aspects of such a boundary model, freed from all geometrical data. Informally, it consists of one or more polygons, whose sides are conceptually identified in pairs, in a specified direction. The pairing is conventionally indicated by labeling each side with a letter and an arrow, as in figure 1(a).

* This research was supported in part by the National Council for Scientific and Technological Development (CNPq) of Brazil and the Foundation for Research Support of the State of São Paulo (FAPESP)

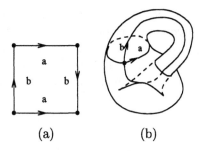

(a)   (b)

Fig. 1. A Klein bottle.

1.2 Topological visualization

The topology of a cellular complex, even a small one, may be hard to understand intuitively, which makes it hard to debug programs that deal with boundary representations. The topological and geometrical aspects are often handled by independent procedures; so one cannot rely on the visual appearance of the object to verify the correctness of its topology. Faces that may look adjacent on the screen may not be adjacent in the topological data structure, and vice-versa.

Motivated by these difficulties, we set out to develop a graphical tool for *automatically visualizing the topology of a cellular complex*: a program which, given only the incidence relations among the elements of a complex, will choose a surface in three-space, partitioned into points, arcs, and disks, that clearly displays those incidence relations. Thus, for example, given a combinatorial description of the complex shown in figure 1(a), the program should ideally output a picture like 1(b).

1.3 What is a good realization?

A cellular complex has infinitely many representations, but not all of them are useful for understanding its structure. Therefore, in order to develop an automatic visualization tool, we must figure out how to recognize a "good" representation.

Obviously we cannot expect a definitive solution to this problem. We can only list some general criteria, suggested by intuition or experiment, which seem to be associated with visual clarity. For example, it seems desirable that the surface be as smooth and flat as possible, in order to minimize self-occlusions and avoid distracting the viewer's attention with graphical artifacts (folds, shadows, silhouette edges, etc.) which have no topological significance. For the same reason, it is generally desirable that the surface be free from self-intersections; or, if that is not possible (as in the case of a Klein bottle), that the extent of self-intersection be somehow minimized.

Not only must the *surface* be easy to understand, but also the edges of the complex must be drawn on that surface in a visually effective way: they must be well-separated, smooth, as straight as possible, and neither too long nor too short. Notice that these requirements may indirectly affect the shape of the surface. For example, the complex shown in figure 2(a) must be drawn on a surface with the topology of a sphere; however, a truly spherical surface, as in figure 2(b), is visually less effective than the sausage-shaped surface shown in figure 2(c).

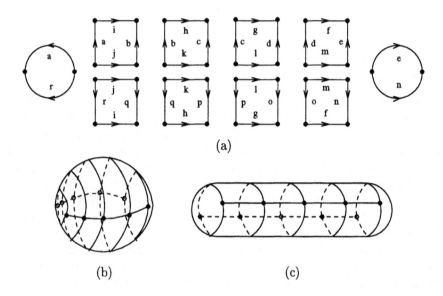

(a)

(b)   (c)

Fig. 2. A complex is more than a surface.

In order to automate the search for a good realization, we borrow a well-established concept from plane graph drawing: that of an *energy function*, a quantitative measure of how badly a solution fails to meet certain visual effectiveness criteria — like the total curvature of the surface, the amount of self-intersection, the variance of edge lengths, and so forth. The problem reduces then to finding, among all possible realizations of the complex, the one which has the minimum energy.

1.4   Related work

This work can be viewed as a three-dimensional extension of the plane graph-drawing problem [8]. Such extensions have been attempted before, but usually by assuming the input to be an ordinary graph, and the output to be only a collection of points and line segments in space, without any surface elements.

Ferguson, Rockwood and Cox [10] addressed the problem of automatically generating a surface with given topology. Since the topology of a surface is completely determined by its genus and orientability, they were able to solve the problem by a direct construction. Our problem differs from theirs because we must not only find a "good" surface with the right topology, but also a "good" drawing of a specified graph on that surface; as we have seen, this requirement influences indirectly the shape of the surface itself.

Another related work is Brakke's Surface Evolver [4,5], a general program to study the evolution of surfaces under arbitrary force laws — such as surface tension, elastic bending, gas pressure, etc. Brakke's evolver has been used to empirically determine the surfaces of minimum energy for various topologies [14]. However, the energies used in those experiments were chosen for their mathematical and physical significance, rather than their visual properties. Furthermore, the energy function depended only on the shape of the surface; there was no graph to be drawn on it.

Energy minimization of surfaces with a given topology has also been proposed as a tool for interpolating a surface over a network of curves [15]. An advanced example of this approach is the work by Moreton and Séquin [21], who consider the problem of realizing a cellular complex with a collection of parametric surface patches, joined with C_1 continuity, given the coordinates and tangent planes at the vertices of the complex. Our approach differs from theirs in the choice of surface model (we use a triangular mesh), and in the amount of geometrical data required from the user (we do not require any).

The general idea of using energy functions to quantify the "ugliness" of a drawing apparently became popular after the graph-drawing paper by Kamada [16,7]. Indeed, some of the energy functions that we use are very similar to his "spring" energies—in spirit, if not in detail. Also, some of our energy functions can be viewed as approximations to the bending energy of a thin elastic membrane (generally assumed to be the square of the surface's curvature), which is often used in minimal surface research [13,15,22,4,14,21].

2 Visualization model

The input to our tool is a purely combinatorial data structure that describes the incidence relationships between the faces, edges and vertices of the complex.

In the literature one can find dozens of data structures that were developed for this purpose [6,9,18,25]. For our work we selected the *quad-edge* data structure [12], a variant of the *winged edge* and *half-edge* structures [2] which are widely used in CAD and computer graphics. The main reason for this choice was that the quad-edge allows the encoding of non-orientable cellular complexes, such as the one of figure 1; as well as degree-1 vertices, loops, bridging edges, and multiple edges with same endpoints.

An obvious way to realize a complex geometrically is to model each face as a polynomial surface patch, implicit or parametric, with suitable continuity constraints between adjacent patches [19,1]. However, in a general complex a face may have any number of sides (including 1 or 2), and may be glued in arbitrarily complex ways. Thus, in general, it is not possible to realize each face as a single polynomial surface patch of bounded degree.

Therefore, instead of using the faces as modeling units, we use what we call the *tiles* of the complex. Each tile is a four-sided surface patch, containing exactly one edge e of the complex and the corresponding edge e' of the dual complex, both running diagonally across the tile. See figure 3.

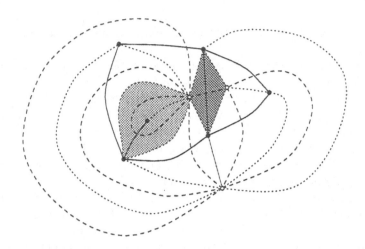

Fig. 3. A two-dimensional complex drawn on the plane (solid), overlaid with its dual complex (dashed), and the tile boundaries (dotted).

2.1 Modeling a tile

Since every tile has only four sides, and therefore only four neighboring tiles, it is possible to realize it as a geometric object of bounded complexity.

A obvious candidate for this role would be a parametric polynomial patch [19,21]. This approach would allow us in principle to obtain a truly smooth surface, with continuity of tangent plane (and possibly curvature) between adjacent tiles. However, it is not at all trivial to enforce those constraints for complexes arbitrary topology. Also, some of the energy functions we use seem hard to evaluate in this model. Given these difficulties, we decided to model each tile as a polyhedral surface; specifically, as a grid of $k \times k$ four-sided *cells*, each consisting of four plane triangles. See figure 4. We thus replace the original cellular complex C, with m edges, by a refined complex T having $6mk^2$ edges and $4mk^2$ faces, all triangles. The elements of C are unions of elements of T; in particular, the edges of C are polygonal paths in T, running diagonally across the corresponding tiles.

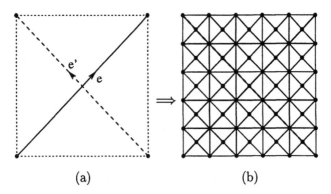

Fig. 4. The triangulated tile model.

Obviously, having adopted a polyhedral model, we must give up any hope of obtaining a really smooth surface, and seek instead to reduce and equalize the external dihedral angles between adjacent triangles. Still, by using a large enough tile order k, we can in principle obtain surfaces that are arbitrarily smooth almost everywhere. Moreover, we can always use standard tricks of computer graphics, such as Gouraud shading [11], to smooth out the corners when rendering the final image.

2.2 Building the triangulation

We use the quad-edge data structure to represent not only the input complex C, but also the refined mesh T. (There are specialized data structures for triangular meshes which use less space than the quad-edge; but the latter is more convenient in the intermediate stages of the tiling procedure, when the mesh T still contains some non-triangular faces.)

After building a triangulated $k \times k$ tile for each edge of the original complex C, we glue the sides of those tiles in pairs, as prescribed by the adjacency relation between the corresponding edges of C.

Note that it is possible for a tile to get glued to itself; that happens, for instance, when the complex C includes faces or vertices of degree 1. In those cases, the resulting mesh may contain pairs of *twin triangles* with the same three vertices, or pairs of *twin edges* having the same endpoints.

Figure 5 illustrates this problem, in this case when the bottom side of a 2×2 triangulated tile (a) gets glued to the right side of the same tile (b,c). Note that in figure 5(c) there are two pairs of twin triangles, highlighted in gray; and two pairs of twin edges, inside and surrounding the gray area..

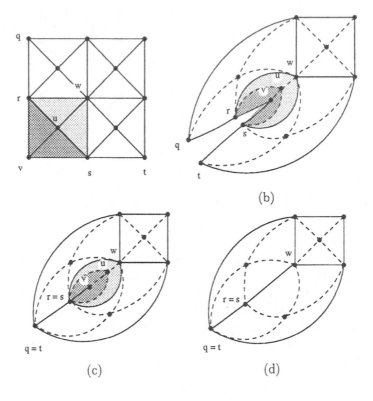

Fig. 5. Gluing a tile to itself.

Obviously, two such triangles will coincide no matter what coordinates we assign to their vertices. If they are left in the mesh, they will look like a loose flap hanging out from the surface. Thus, whenever gluing a tile to itself, we must locate and remove any twin pairs, as shown in figure 5(d).

Fortunately, this cleanup is quite easy. It can be shown [24] that any twin triangles or edges must belong to a grid cell that is adjacent to the two sides of a tile that are being glued together, like the highlighted corner of figure 5(a). As long as k is three or more, the removal of these corner cells does not create any new twin pairs, and original edges of the complex—represented by the tile diagonals—will not be entirely obliterated.

3 Energy functions

Since each tile is modeled by a set of *flat* triangles, the shape of the triangulated mesh is completely determined by the coordinates of its vertices $VT = \{v_1, .. v_n\}$; we say that these n points of $\mathbf{R}^3$ are a *configuration* of the mesh. Our energy function, which measures the "ugliness" of the surface, is therefore a function from $(\mathbf{R}^3)^n$ to $\mathbf{R}$.

The energy functions we have tried were convex combinations $E = \sum_i w_i E_i$ of the functions E_i described below, with fixed weights w_i. In these formulas, $\mathcal{VC}, \mathcal{EC}, \mathcal{DC}$, and $\mathcal{FC}$ denote the vertices, primal edges, dual edges, and faces of a complex C. Also $\vec{\mathcal{E}}C$ and $\vec{\mathcal{D}}C$ denote the set of all directed edges, respectively primal and dual.

Bending energy:

$$E_{\text{bnd}} = \sqrt{|\mathcal{F}T|} \sum_{e \in \mathcal{E}T} l_e \theta_e^2$$

where l_e is the length of edge e, and θ_e is the external dihedral angle at that edge. Minimizing E_{bnd} tends to flatten out the surface, and distribute its curvature evenly among all edges.

Eccentricity energy:

$$E_{\text{ecc}} = |\mathcal{V}T| \sum_{v \in \mathcal{V}T} |v - b_v|^2$$

where b_v is the barycenter of all neighbors of vertex v. Minimizing E_{ecc} also tends to flatten out the surface, and equalize the edge lengths.

Angle variance energy:

$$E_{\text{ang}} = \frac{1}{|\mathcal{V}T|} \sum_{e \in \vec{\mathcal{E}}T} \left| \phi_e - \frac{2\pi}{d_e} \right|^2$$

where ϕ_e is the angle between e and the next edge out of the same vertex, projected onto the tangent plane at that vertex; and d_e is the degree of that vertex. Minimizing E_{ang} tends to equalize the angles between adjacent edges, which helps to unfold the surface and reduce its self-intersections.

Spreading energy:

$$E_{\text{spr}} = \frac{1}{|\mathcal{V}T|} \sum_{v \in \mathcal{V}T} |p_v|^2$$

where p_v is the Cartesian coordinates of vertex v. Minimizing this energy tends to keep all vertices close to the origin of $\mathbf{R}^3$.

Proximity energy:

$$E_{\text{prx}} = \frac{1}{|\mathcal{F}T|^2} \sum_{\substack{r,s \in \mathcal{F}T \\ r \neq s}} \frac{1}{|c_r - c_s|^2 + \rho_r^2 + \rho_s^2}$$

where c_r, c_s are the centroids of triangles r and s, and ρ_r, ρ_s are their average radii (in the root-mean-square sense). This energy can be understood as the electrostatic potential of a set of fuzzy electric charges, located at the triangle centroids. Minimizing E_{prx} tends to spread out the triangles in space, thus avoiding self-intersections (especially grazing ones) and fold-overs.

Patch area energy The two diagonals of a tile divide it into four triangular pieces, each consisting of k^2 triangles. Let $A_1, \ldots A_{4m}$ be the total areas of these quarter-tiles. Then

$$E_{\text{par}} = \frac{1}{4m} \sum_{i=1}^{4m} \left(\frac{A_i}{A_*} + \frac{A_*}{A_i} - 2 \right)$$

where $A_* = \pi/m$ is the "ideal" area of a quarter-tile. Minimizing E_{par} tends to keep the total surface area close to 4π, and to equalize the quarter-tile areas, so that the edges of the original complex are spread out uniformly over the surface.

Note that the energy functions above *must* be used in combination, rather than alone, in order to avoid degenerate configurations. For instance, the proximity energy E_{prx} tends to its minimum value (zero) when the vertices tend to infinity; whereas the bending energy E_{bnd} is minimum when all vertices are at the origin. However, the asymptotic behavior of the two formulas is such that any nontrivial convex combination of them will attain its minimum at configurations of bounded and nonzero radius.

The purpose of the scaling factors in the formulas above, such as $\sqrt{|\mathcal{FT}|}$ in the formula of E_{bnd}, is to make the numerical value of each energy largely insensitive to the number of triangles in the mesh. This scaling allows us to use the same weights for optimizing meshes of different resolutions (different values of k). In particular, it allows us to use the *multi-scale approach*, which consists in running the optimization algorithm simultaneously on several meshes of varying resolution, which are loosely coupled by interpolation and local averaging.

3.1 Selecting the weights

When it comes to selecting the weights, we still do not have any answer much better than trial-and-error, perhaps guided by some physical intuition.

It would be trivial to implement a GUI-based tool that allowed direct adjustment of the weights by the user, with visual feedback. Unfortunately, our optimizer is still too slow for interactive use: it would take tens of minutes to (re)compute the optimum configuration after a change in the weights. (On the other hand, considering our limited expertise in numerical methods, it seems likely that this time can be reduced, by several orders of magnitude, merely by using better optimization algorithms.)

A more speculative solution is to let the computer "learn" the weights from examples, as Eades and Mendonça did for the plane graph drawing problem [20]. The idea is to give the computer a collection of "good" realizations of cellular complexes—generated, say, by ad-hoc programs or manual editing with solid modeling tools—and let the computer find the combination of weights that comes closest to reproducing those shapes.

4 Optimization

Having constructed the triangular mesh, and chosen an energy function, we are faced with problem of finding the configuration of minimum energy. This is a nontrivial optimization problem, since even a small complex determined an energy function with hundreds of independent variables.

We approach this problem at two levels. We use general-purpose non-linear optimization techniques to move from given configuration to a nearby *local* minimum. We repeat this search for a couple dozen random starting points, and (select the local minimum with lowest energy as the answer. It should be possible to use combinatorial heuristics, such as simulated annealing, to extend the search beyond the nearest local minimum; but we haven't been able to get such methods to work fast enough.)

When searching the local minimum, we need a stopping criterion of some sort. Unfortunately, there is no numerical criterion that can tell when the energy is "low enough" to provide a good visualization. So, in practice, we fix a computation budget, and use the best configuration we can find within that limit.

4.1 General optimization methods

To solve this problem, we have tried several general purpose numerical optimization techniques, including Kirkpatrick's *simulated annealing* [17,23], Nadler and Mead's *downhill simplex method* [23, p.289], Powell's *principal directions* method [23, p.298], the naive *single coordinate* optimization, with periodic diagonal steps [23, p.294] and a *gradient descent* method with adaptive step-size.

The tests were performed on various triangulated meshes, ranging from a few tens to a few hundred vertices. The initial guess for the optimization was either a random configuration—where each vertex was chosen independently and uniformly in the unit cube—or a reasonably smooth configuration obtained by the heuristic methods described in the next section. The effectiveness of an algorithm was judged from its energy evolution curve: the energy of the minimum configuration found, as a function of accumulated CPU time.

The ranking of the methods was generally the same on all tests, and consistent over time. Not surprisingly, we found that gradient descent was the most effective. Simulated annealing was so slow that we gave up on it after a few tests. The Nadler-Mead and Brent-Powell algorithms were faster, but not enough to be usable. Moreover, they require $\Omega(n^2)$ storage for a function of n variables, and therefore are restricted to complexes with a few tens of edges.

The naive minimization method consists of adjusting one variable at a time (using, for instance, Brent's univariate minimization algorithm [23, p.283]), while all other variables are held fixed. As recommended in the literature, every $n + 1$ such "axial" steps we perform a "diagonal" step, where we seek the minimum configuration along the line connecting the outcomes of the first and the last of those axial steps.

We implemented this naive method only for the sake of comparison, since textbooks generally claim it is slower than Brent-Powell. To our surprise, it turned out to be much faster. The reason, which was obvious on hindsight, is that all our energy functions are the sum of many terms, each depending on a few vertices only. When varying one coordinate at a time, we could save time by recomputing only the terms that depended on that coordinate. Thus, while the naive algorithm performed somewhat more energy evaluations than Powell's method, each evaluation was faster by one or two orders of magnitude.

In the gradient descent method, we compute the partial derivatives of the energy function by Baur and Strassen's method [3]. The latter, which is basically a systematic use of the chain derivation rule, reduces the cost of computing the gradient of *any* algebraic or transcendental formula to a small constant times the cost of computing the energy itself.

4.2 Heuristic methods

Besides these general-purpose optimization algorithms, we have also used two specialized heuristic methods, the *smoothing heuristic* and the *spreading heuristic*. Both are local operations that are applied to each vertex v in turn, cyclically, up to a prescribed number of passes. The *smoothing heuristic* adjusts the position of vertex v, keeping all other vertices fixed, so as to approximately minimize the bending energy of the edges between the triangles that are incident to v. The *spreading heuristic* keeps the vertex v fixed, but rotates its neighbors $u_1, .. u_k$ around the surface normal r at the vertex so as to better equalize the angles $u_i v u_{i+1}$, when projected on a plane perpendicular to r.

These heuristics are not sensitive to the energy functions or their weights. Therefore, they are useful only as a preprocessing step, when starting from a random configuration, where the goal is merely to untangle the surface and smooth out its largest wrinkles — which are bad under any reasonable energy function. If we want to compare the effect of different energy functions and varying weights, we must eventually follow these heuristics with one of the generic optimization methods above.

Each pass of the heuristic redistributes the "stress" energy of each element among the neighboring elements. It should be noted that, as in any diffusion process, the number of passes needed to achieve a given surface smoothness is expected to scale as the square of the graph-theoretic diameter of the triangulation.

In our test runs, with complexes of a few tens of edges, we found that 20-100 passes of each heuristic already provided a reasonably smooth surface; additional passes were hardly worth their cost.

5 Results

Figure 6 shows the heuristic smoothing of a simple torus-like complex. The original complex had one face, two edges, and one vertex; it was modeled with tiles of order $k = 5$, resulting in a mesh with 300 edges, 200 triangles and 100 vertices. Shown are the initial random configuration (a), and the states after 5, 30, and 100 passes of each heuristic, respectively (b-d).

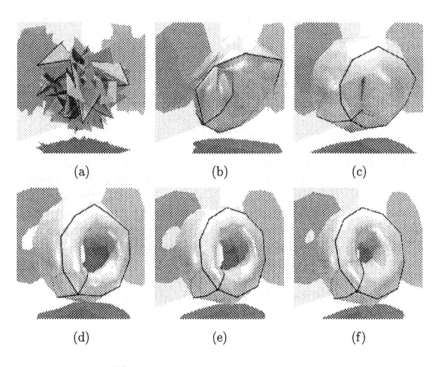

(a)   (b)   (c)

(d)   (e)   (f)

Fig. 6. Smoothing of a torus complex.

The configuration of figure 6(d) was then optimized with the gradient-descent method, for the energy $E = E_{bnd} + E_{par} + E_{ecc} + 5E_{prx}$ Figures 6(e-f) show the state after 100 and 1000 energy evaluations, respectively.

Figure 7 shows renderings of the final stage of optimization for three different complexes with the topology of a sausage, a five-star and a tritorus (a sphere with three handles).

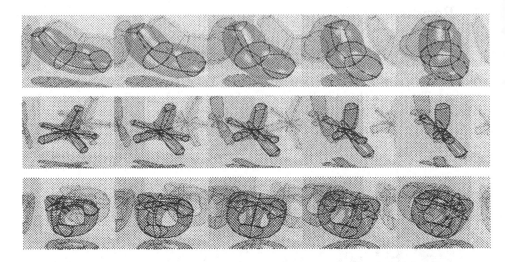

Fig. 7. Realizations of a sausage, a five-star, and a tritorus.

The renderings are further enhanced by a little animation displaying several views of the complexes. The series of pictures are presented as a sequence of stereoscopic pairs. The 3D rendering may be seen by holding the paper about 50 cm away, and converging the eyes toward a point behind it, so that pairs of consecutive drawings are fused into one image. The resulting stereo illusion should compensate for the the small size of the images.

The images were produced with POV-Ray, a freely available ray tracer [26]. Their apparent smoothness is due to Gouraud-shading of the triangles [11]. The edges and vertices are modeled by thin cylinders and small spheres.

6   Conclusions and future work

Our experiments to date are encouraging, but there is still a lot of work to be done here. To begin with, we need to work more on the energy minimization code; speedup several orders of magnitude seem possible, just by using better optimization algorithms.

We still understand very little about the effects of the various energy functions, and the proper weights to use. Among other things, we need to develop energy functions that penalize self-intersections more strongly than the functions we have got. We also need to improve the heuristic methods so as to avoid generating self-intersecting shapes.

Finally, we need to develop combinatorial optimization tools or heuristics to extend the search over several local minima.

References

1. Chandrajit L. Bajaj. Smoothing polyhedra using implicit algebraic splines. In *Proc. SIGGRAPH'92*, pages 79–88, 1992.
2. Bruce G. Baumgart. A polyhedron representation for computer vision. In *Proc. 1975 AFIPS National Computer Conference*, volume 44, pages 589–596, 1975.
3. Walter Baur and Volker Strassen. The complexity of partial derivatives. *Theoretical Computer Science*, 22:317–330, 1983.
4. Kenneth A. Brakke. The Surface Envolver. *Experimental Mathematics*, 1(2):141–165, 1992.
5. Kenneth A. Brakke. *Surface Evolver Manual*. The Geometry Center, University of Minnesota, December 1993. Eletronic Address: brakke@geom.umn.edu.
6. Erik Brisson. Representing geometric structures in d dimensions: Topology and order. *Proc. 5th Annual ACM Symp. on Computational Geometry*, pages 218–227, June 1989.
7. Ron Davidson and David Harel. Drawing graphs nicely using simulated annealing. Technical report, Department of Applied Mathematics and Computer Science, The Weizmann Institute of Science, 1989.
8. G. Di Battista, P. D. Eades, and R. Tamassia. Algorithms for drawing graphs: An annotated bibliography. Technical report, Department of Computer Science, University of Newcastle, 1993.
9. David P. Dobkin and Michel J. Laszlo. Primitives for the manipulations of three-dimensional subdivisions. *Proc. 3rd ACM Symp. on Comp. Geometry*, pages 86–99, June 1987.
10. H. Ferguson, A. Rockwood, and J. Cox. Topological design of sculpted surfaces. In *Proc. SIGGRAPH'92*, pages 149–156, 1992.
11. James D. Foley, Andries van Dam, Steven K. Feiner, and John F. Hughes. *Computer Graphics: Principles and Practice*. Addison-Wesley, second edition, 1990.
12. Leonidas Guibas and Jorge Stolfi. Primitives for the manipulations of general subdivisions and the computation of Voronoi diagrams. *ACM Transactions on Graphics*, 4(2):74–123, April 1985.
13. B. K. P. Horn. The curve of least energy. *ACM Transactions on Mathematical Software*, 9(4):441–460, December 1983.
14. Lucas Hsu, Rob Kusner, and John Sullivan. Minimizing the squared mean curvature integral for surfaces in space forms. *Experimental Mathematics*, 1(3):191–207, 1992.
15. Michael Kallay and Bahram Ravani. Optimal twist vectors as a tool for interpolation a network of curves with a minimum energy surface. *Computer Aided Geometric Design*, pages 465–473, 1990.
16. Tomihisa Kamada and Satoru Kawai. An algorithm for drawing general undirected graphs. *Information Processing Letters*, pages 7–15, April 1989.
17. S. Kirkpatrick, C. D. Gelatt Jr., and M. P. Vecchi. Optimization by simulated annealing. *Science*, 220:671–680, 1983.
18. Pascal Lienhardt. Subdivisions of N-dimensional spaces and N-dimensional generalized maps. *Proc. 5th Annual ACM Symp. on Computational Geometry*, pages 228–236, June 1989.
19. Charles Loop and Tony DeRose. Generalized B-spline surface of arbitrary topology. In *Proc. SIGGRAPH'90*, pages 347–356, August 1990.
20. Cândido X. F. Mendonça and Peter Eades. Learning aesthetics for visualization. In *Anais do XX Seminário Integrado de Software e Hardware*, pages 76–88, Florianópolis, SC (Brazil), 1993.

21. Henry P. Moreton and Carlo H. Séquin. Functional optimization for fair surface design. In *Proc. SIGGRAPH'92*, pages 167–176, 1992.
22. D. B. Parkinson and D. N. Moreton. Optimal biarc-curve fitting. *Computer-Aided Design*, 23(6):411–419, 1991.
23. William H. Press, Brian P. Flannery, Saul A. Teukolsky, and William T. Vetterling. *Numerical Recipes: The Art of Scientific Computing.* Cambridge University Press, 1986.
24. Rober Marcone Rosi and Jorge Stolfi. Automatic visualization of two-dimensional cellular complexes. Technical Report IC-96-02, Institute of Computing, University of Campinas, May 1996.
25. Fujio Yamaguchi and Toshiya Tokieda. A solid modelling system: Freedom-II. *Computers & Graphics*, pages 225–232, 1983.
26. C. Young, D. K. Buck, and A. C. Collins. POV-Ray - Persistence of Vision Raytracer, version 2.0. 1993.

An Alternative Method to Crossing Minimization on Hierarchical Graphs

(Extended Abstract)

Petra Mutzel*

Max-Planck-Institut für Informatik, Im Stadtwald, D-66123 Saarbrücken, Germany

Abstract. A common method for drawing directed graphs is, as a first step, to partition the vertices into a set of k levels and then, as a second step, to permute the vertices within the levels such that the number of crossings is minimized. We suggest an alternative method for the second step, namely, removing the minimal number of edges such that the resulting graph is k-level planar. For the final diagram the removed edges are reinserted into a k-level planar drawing. Hence, instead of considering the k-level crossing minimization problem, we suggest solving the k-level planarization problem. In this paper we address the case $k = 2$. First, we give a motivation for our approach. Then, we address the problem of extracting a 2-level planar subgraph of maximum weight in a given 2-level graph. This problem is NP-hard. Based on a characterization of 2-level planar graphs, we give an integer linear programming formulation for the 2-level planarization problem. Moreover, we define and investigate the polytope $2\mathcal{LPS}(G)$ associated with the set of all 2-level planar subgraphs of a given 2-level graph G. We will see that this polytope has full dimension and that the inequalities occuring in the integer linear description are facet-defining for $2\mathcal{LPS}(G)$. The inequalities in the integer linear programming formulation can be separated in polynomial time, hence they can be used efficiently in a cutting plane method for solving practical instances of the 2-level planarization problem. Furthermore, we derive new inequalities that substantially improve the quality of the obtained solution. We report on first computational results.

1 Introduction

Directed graphs are widely used to represent structures in many fields such as economics, social sciences, mathematical and computer science. A good visualization of structural information allows the reader to focus on the information content of the diagram.

A common method for drawing directed graphs has been introduced by Sugiyama et al. [STT81] and Carpano [Car80]. In the first step, the vertices

* *Partially supported by DFG-Grant Ju204/7-1, Forschungsschwerpunkt "Effiziente Algorithmen für diskrete Probleme und ihre Anwendungen" and by ESPRIT LTR Project No. 20244 – ALCOM-IT*

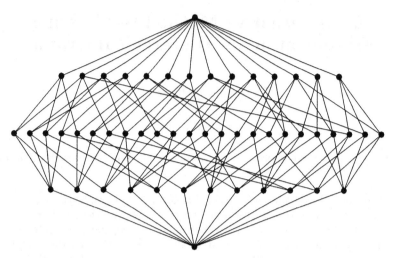

Fig. 1. A real world graph with high crossing number [Fuk96]

are partitioned into a set of k levels, and in the second step, the vertices within each level are permuted in such a way that the number of crossings is small. We suggest an alternative approach for the second step. From now on let us assume that we are given a *k-level hierarchy (k-level graph)*, i.e., a graph $G = (V, E) = (V_1, V_2, \ldots, V_k, E)$ with vertex sets $V_1, \ldots, V_k$, $V = V_1 \cup V_2 \ldots \cup V_k$, $V_i \cap V_j = \emptyset$ for $i \neq j$, and an edge set E connecting vertices in levels V_i and V_j with $i \neq j$ $(1 \leq i, j \leq k)$. V_i is called the *i-th level*. A k-level hierarchy is drawn in such a way that the vertices in each level V_i are drawn on a horizontal line L_i with y-coordinate $k - i$, and the edges are drawn as straight lines. In contrary to the definitions of a hierarchy in [STT81,HP96], we do not care about the direction of the edges, since it is irrelevant for the problem considered here. Essentially, a k-level hierarchy is a k-partite graph that is drawn in a special way.

Even for 2-level graphs the straightline crossing minimization problem is NP-hard. Exact algorithms based on branch and bound have been suggested by various authors (see, e.g., [VML96] and [JM96]). For $k \geq 2$, a vast amount of heuristics has been published in the literature (see, e.g., [War77,STT81,EK86,Mäk90, EW94a] and [Dre94]).

Various authors have already asked the following question: Is a hierarchical drawing with the minimal number of crossings always nicer than a drawing that has many more crossings? They ended up with the following answer: "We merely want to draw a reasonably clear picture which has a "relatively small" number of crossings" [Car80].

For graphs that have a relatively small hierarchical crossing number, this statement goes along with our observation. But in some applications, hierarchical graphs arise that have a relatively high hierarchical crossing number, such as the graph shown in Figure 1. For these graphs we have to find a new method that substantially increases the readability of these diagrams.

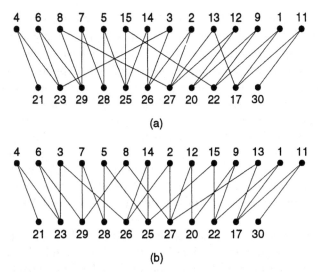

Fig. 2. A graph (a) drawn using k-planarization and (b) drawn with the minimal number of crossings computed by the algorithm in [JM96]

One approach may be to remove a minimal set of edges such that the remaining k-level graph can be drawn without edge crossings. In the final drawing, the removed edges are reinserted. Since the insertion of each edge may produce many crossings, the final drawing may be far from an edge-crossing minimal drawing.

Figure 2(a) shows a drawing of a graph obtained by 2-level planarization, whereas Figure 2(b) shows the same graph drawn with the minimal number of edge crossings (using the exact algorithm given in [JM96]). Although the drawing in Figure 2(a) has 34 crossings, that is 41% more crossings than the drawing in Figure 2(b) (24 crossings), the reader will not recognize this fact. On the contrary, 90% of the colleagues that we have asked thought that the number of crossings in Figure 2(a) is less than in Figure 2(b). This encourages us to study the k-level planarization problem.

Another motivation for studying k-level planarization arises from the fact that the k-level crossing minimization problem is a very hard problem that cannot be solved exactly or approximately (with some reasonable solution guarantees) in practice. Our experiments in [JM96] showed that for sparse graphs, such as they occur in graph drawing, the heuristic methods used in practice are far from the optimum. We believe that the methods of polyhedral combinatorics that have been successfully applied for the maximum planar subgraph problem [JM93a,JM93b,Mut94], and for the straightline crossing minimization problem on two levels where one level is fixed [JM96], may be helpful for getting some better approximation algorithms in practice. But a lot of effort will be needed to get efficient algorithms that will be able to solve the k-level crossing minimization problem for $k > 2$ and $|V_i| \geq 15$ ($i = 1, \ldots, k$) to provable optimality.

The k-level planarization problem, however, may be easier to attack. We build our hope on the fact that there is a linear time algorithm for recognizing

k-level planar graphs (see [HP96] and [BN88]). Moreover, our computational results on 2-level graphs addressed in this paper support our conjecture.

Besides the application in automatic graph drawing, the 2-level planarization problem comes up in Computational Biology. In DNA mapping, small fragments of DNA have to be ordered according to the given overlap data and some additional information. Waterman and Griggs [WG86] have suggested combining the information derived by a digest mapping experiment with the information on the overlap between the DNA fragments. If the overlap data is correct, the maps can be represented as a 2-level planar graph. But, in practice, the overlap data may contain errors. Hence, Waterman and Griggs suggested solving the 2-level planarization problem (see also [VLM96]). Furthermore, the 2-level planarization problem arises in global routing for row-based VLSI layout (see [Len90,Ull84]).

Section 2 reports on previously known results of the 2-level planarization problem. One of the characterizations of 2-level planar graphs leads directly to an integer linear programming formulation for the 2-level planarization problem. In Section 3 we study the polytope associated with the set of all possible 2-level planar subgraphs of a given 2-level graph. From this we obtain new classes of inequalities that tighten the associated LP-relaxation. In order to get practical use out of these inequalities, we have to solve the "separation problem". This question will be addressed in Section 4, where we also discuss a cutting plane algorithm based on those results. First computational results with a cutting plane algorithm are presented in Section 5. In this extended abstract we omit the proofs for some of the theorems.

2   Characterizing 2-Level Planar Graphs

A *2-level graph* is a graph $G = (L, U, E)$ with vertex sets L and U, called lower and upper level, and an edge set E connecting a vertex in L with a vertex in U. There are no edges between two vertices in the same level. A *2-level planar graph* $G = (L, U, E)$ is a graph that can be drawn in such a way that all the vertices in L appear on a line (the lower line), the vertices in U appear on the upper line, and the edges are drawn as straight lines without crossing each other. The difference between a planar bipartite graph and a 2-level planar graph is obvious. For example, the graph shown in Figure 3 is a planar bipartite graph, but not a 2-level planar graph.

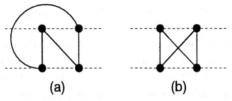

(a)   (b)

Fig. 3. (a) A planar bipartite graph that is (b) not 2-level planar

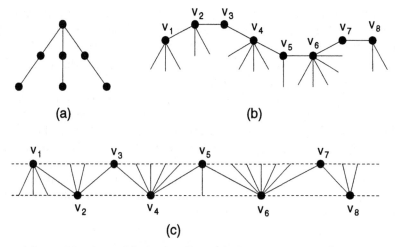

Fig. 4. (a) Double claw. (b) Caterpillar. (c) Caterpillars can be embedded on 2-levels without any crossings.

Given a 2-level graph $G = (L, U, E)$ with weights $w_e > 0$ on the edges, the *2-level planarization problem* (or *maximum 2-level planar subgraph problem*) is to extract a 2-level planar subgraph $G' = (L, U, F)$, $F \subseteq E$, of maximum weight, i.e., the sum $\sum_{e \in F} w_e$ is maximum.

To our knowledge, only the unweighted ($w_e = 1$ for all $e \in E$) 2-level planarization problem has been considered in the literature so far. It was first mentioned in [TKY77]. The authors introduced the problem in the context of graph drawing. They have given the following nice characterization of 2-level planar graphs based on forbidden subgraphs. The characterization was independently given by [EKW86].

We will call the graph shown in Figure 4(a) a *double claw*. A *caterpillar* is a connected graph $G = (V, E)$ having edges on its backbone $(v_1, v_2, \ldots, v_l)$ and single edges (v_i, w), $w \in V \setminus \{v_1, v_2, \ldots, v_l\}$ (see Figure 4(b)).

Theorem 2.1 [TKY77,EKW86]. *A 2-level graph is 2-level planar if and only if it contains no cycle and no double claw.*

Proof. A graph without any cycles is a set of trees. A tree without any double claws is a set of caterpillars. Caterpillars can be embedded on 2-levels without any crossings (see Figure 4(c)). On the other hand, a 2-level planar graph can contain neither a cycle nor a double claw. □

The following alternative characterization leading to a simple linear time algorithm has been given in [TKY77].

Theorem 2.2 [TKY77]. *A 2-level graph G is 2-level planar if and only if the graph G^* that is the remainder of G after deleting all vertices of degree one, is acyclic and contains no vertices of degree at least three.*

However, the 2-level planarization problem is NP-hard even for the case when each vertex in U has degree three and each vertex in L has degree two (by reduction to a Hamiltonian path problem) [EW94b]. Therefore, Eades and Whitesides suggested a heuristic based on the search for a longest path which will be used as a "backbone" of the caterpillar to be constructed.

Tomii et al. suggest an algorithm for acyclic 2-level graphs [TKY77]. The algorithm can be seen as an adaptive greedy algorithm. In each step, the edges are labelled according to some rule and the edge with the highest label will be removed. However, this algorithm does not lead to the optimal solution as shown in Figure 5. The algorithm would remove the edge $(0, 14)$ in a first step. The remaining graph still contains two edge-disjoint double-claws that have to be destroyed by removing two more edges, whereas the optimal solution would be to remove the two edges $(0, 11)$ and $(1, 14)$.

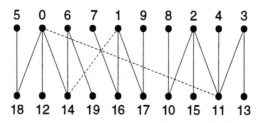

Fig. 5. An acyclic 2-level graph for which the algorithm suggested in [TKY77] leads to a nonoptimal solution

It is an open problem if the 2-level planarization problem can be solved in polynomial time for 2-level acyclic graphs. However, for double claw free graphs, the 2-level planarization problem is equivalent to the maximum forest subgraph problem that can be solved via a simple greedy algorithm.

3 Polyhedral Studies on the 2-Level Planarization Problem

Based on the characterization of 2-level planar graphs in terms of forbidden subgraphs (see Theorem 2.1), it is straightforward to derive an integer linear programming formulation for the maximum 2-level planar subgraph problem. We introduce variables x_e for all edges $e \in E$ of the given 2-level graph $G = (L, U, E)$. We use the following notation: Vectors $\bar{x}$ are column vectors, their transposed vectors $\bar{x}^T$ are row vectors. If $w^T = (w_1, w_2, \ldots, w_m)$ and $\bar{x}^T = (x_1, x_2, \ldots, x_m)$, then $w^T \bar{x} = \sum_{i=1}^m w_i x_i$. We use the notation $x(C) = \sum_{e \in C} x_e$ for $C \subseteq E$.

For any set $P \subseteq E$ of edges we define an incidence vector $\chi^P \in R^{|E|}$ with the i-th component $\chi^P(e_i)$ getting value 1 if $e_i \in P$, and 0 otherwise. Any vector $\bar{x}^T = (x_{e_1}, x_{e_2}, \ldots, x_{e_{|E|}})$, that is the incidence vector of a 2-level planar graph satisfies the following inequalities:

$$0 \leq x_e \leq 1, \qquad \text{for all } e \in E, \qquad (1)$$
$$x(C) \leq |C| - 1, \text{ for all cycles } C \subseteq E \qquad (2)$$
$$x(T) \leq |T| - 1, \text{ for all double claws } T \subseteq E \quad (3)$$
$$x_e \text{ integral}, \qquad \text{for all } e \in E \qquad (4)$$

and vice versa: any vector $\bar{x}^T = (x_{e_1}, x_{e_2}, \ldots, x_{e_{|E|}})$ satisfying inequalities (1), (2), (3) and (4) corresponds to a 2-level planar subgraph of G. Hence, solving the integer linear system $\{\max w^T \bar{x} \mid \text{inequalities (1)-(4) hold for } \bar{x}\}$ will give us the solution of the maximum 2-level planar subgraph problem for a given graph $G = (L, U, E)$ with weights w_e on the edges $e \in E$.

Since solving general integer linear programs is NP-hard, we will have to drop the integrality constraints (4), which gives us a relaxation of the original integer linear programming formulation. In polyhedral combinatorics, we try to substitute the missing integrality constraints by additional inequalities.

We define the polytope $2\mathcal{LPS}(G)$ for a given 2-level graph $G = (L, U, E)$ as the convex hull over all incidence vectors of 2-level planar subgraphs of G. The vertices of this polytope correspond exactly to the 2-level planar subgraphs of G and vice versa. If we can describe the polytope $2\mathcal{LPS}(G)$ as the solution set of linear inequalities, we can optimize any given cost function over the set of all 2-level planar subgraphs of G. Of course, because of the NP-hardness of the problem we cannot expect to find such a description, but in practice a partial description may also suffice.

In an irredundant description only facet-defining inequalities are present. An inequality is said to be *facet-defining* for a polytope $\mathcal{P}$ if it is a face of maximal dimension of $\mathcal{P}$. An inequality $c^T x \leq c_0$ is said to define a *face* of $\mathcal{P}$ if $c^T y \leq c_0$ for all points $y \in \mathcal{P}$ and if there is at least one point y' in $\mathcal{P}$ with $c^T y' = c_0$.

So, our task is to find facet-defining inequalities for the polytope $2\mathcal{LPS}(G)$ for a given 2-level graph G. We will first investigate the inequalities given in the integer linear programming formulation. We will see that the linear inequalities (1) and (3) are facet-defining, but only a part of the inequalities (2). But first we will determine the dimension of $2\mathcal{LPS}(G)$.

Let us consider the set $\mathcal{S}$ of all 2-level planar subgraphs of G. The set $\mathcal{S}$ is a *monotone system* (also called *independence system*), since the empty subgraph is 2-level planar and any subgraph of a 2-level planar graph is also 2-level planar. Hence, we easily get the following theorem using the theory for monotone systems.

Theorem 3.1 *Let $G = (L, U, E)$ be a graph on two levels. The dimension of $2\mathcal{LPS}(G)$, the convex hull of incidence vectors of 2-level planar subgraphs of G, is $|E|$. The trivial inequalities $x_e \geq 0$ and $x_e \leq 1$ are facet-defining for $2\mathcal{LPS}(G)$.*

Proof. It is a well known fact, that for a monotone system $(E, \mathcal{S})$ with ground set E the dimension of the associated polyhedron $P_{\mathcal{S}}$ is $|E| - (|E - \bigcup \mathcal{S}|)$ (a proof is contained, e.g., in [GP85]). Moreover, $x_e \geq 0$ defines a facet of $P_{\mathcal{S}}$ iff $e \in \bigcup \mathcal{S}$. Since every single edge is 2-level planar, we have $\bigcup \mathcal{S} = E$. Hence the dimension of the polyhedron $2\mathcal{LPS}(G)$ is $|E|$ and $x_e \geq 0$ is facet-defining for $2\mathcal{LPS}(G)$.

Let P_i be the 2-level planar graphs induced by the edge sets $\{e \cup e_i\}$ for a given edge $e \in E$ and $e_i \in E \setminus \{e\}$ for $i = 1, \ldots, |E| - 1$. The incidence vectors of the graph P induced by the edge e and the graphs P_i for $i = 1, 2, \ldots, |E| - 1$ are linearly independent and they satisfy $x_e = 1$. Hence we have shown that $x_e \leq 1$ is facet-defining for $2\mathcal{LPS}(G)$. □

Next we will see that not all of the inequalities (2) are facet-defining for $2\mathcal{LPS}(G)$.

Theorem 3.2 *Let $G = (L, U, E)$ be a 2-level graph. The cycle inequalities*

$$x(C) \leq |C| - 1$$

where $C \subseteq E$ is a cycle in G are facet-defining for $2\mathcal{LPS}(G)$ if and only if C is a cycle without chords in G.

Proof. Let $C \subseteq E$ be a cycle without chord in G. We will show that there are $|E|$ incidence vectors of 2-level planar subgraphs induced by the edge set P of G that are linearly independent and that satisfy $\chi^P(C) = |C| - 1$. Consider the graphs P_i induced by the edge sets $C \setminus \{e_i\}$ for $e_i \in C$ for $i = 1, 2, \ldots, |C|$. Moreover, consider the graphs induced by the edge sets $H_j = P_1 \cup f_j$ for $f_j \in E \setminus C$, $j = 1, 2, \ldots, |E| - |C|$. Since the cycle C is chordless, adding any edge $f_j \in E \setminus C$ to P_1 still gives a 2-level planar graph, since neither a cycle nor a double claw destroying 2-level planarity can occur. All the $|E|$ incidence vectors of the 2-level planar graphs induced by P_i for $i = 1, 2, \ldots, |C|$ and H_j for $j = 1, 2, \ldots, |E| - |C|$ are linearly independent and they satisfy $\chi^P(C) = |C| - 1$. Hence the facet-defining property is shown.

Suppose now, $C = (v_1, v_2, \ldots, v_k, v_1)$ is a cycle with a chord $d = (v_h, v_l) \in E$ in G for some $h, l \in \{1, 2, \ldots, k\}$. We will show that there exists a valid inequality $x(D) \leq |D| - 1$ for $2\mathcal{LPS}(G)$ with the properties that $\{x \mid x(C) = |C| - 1\} \subset \{x \mid x(D) = |D| - 1\}$ and $\{x \mid x(C) = |C| - 1\} \neq \{x \mid x(D) = |D| - 1\}$. Hence $x(C) \leq |C| - 1$ cannot define a face of maximal dimension of $2\mathcal{LPS}(G)$. Let $C_1 = (v_1, v_2, \ldots, v_h, v_l, \ldots, v_k, v_1)$ be a cycle consisting of a subset of the cycle C and the chord $d = (v_h, v_l)$ and $C_2 = C \setminus C_1 \cup d$ the remaining part of C together with d. We have $|C| = |C_1| + |C_2| - 2$. Let us assume that $x \in 2\mathcal{LPS}(G)$ with $x(C) = |C| - 1$. We have $x(C) = x(C_1) + x(C_2) - 2x_d < (|C_1| - 1) + (|C_2| - 1) - 2x_d = |C_1| + |C_2| - 2 - 2x_d$. Since $x(C) = |C| - 1 = |C_1| + |C_2| - 2$ and $x_d \geq 0$, we will have $x_d = 0$ and $x(C_1) = |C_1| - 1$ and $x(C_2) = |C_2| - 1$. Hence we have found $D = C_1$ with $\{x \mid x(C) = |C| - 1\} \subset \{x \mid x(C_1) = |C_1| - 1\}$ and obviously $\{x \mid x(C) = |C| - 1\} \neq \{x \mid x(D) = |D| - 1\}$. □

In the following we will see that all the double claws contained in G are present in an irredundant description of $2\mathcal{LPS}(G)$ by linear inequalities.

Theorem 3.3 *Let $G = (L, U, E)$ be a 2-level graph. The double claw inequalities*

$$x(T) \leq |T| - 1$$

where $T \subseteq E$ is a double claw in G are facet-defining for $2\mathcal{LPS}(G)$.

Proof. Let the graphs P_i be induced by the set $T \setminus e_i$ for $i = 1, \ldots, 6$. Obviously, the graphs P_i are 2-level planar graphs and satisfy $x(T) = |T| - 1$. Moreover, consider the graphs $H_j = T \cup f_j$ for $f_j \in E \setminus T$, $j = 1, 2, \ldots, |E| - |T|$. If H_j contains a cycle, we can remove any edge in this cycle in order to get a 2-level planar graph induced by H'_j. In all the other cases there is always an edge we can remove from H_j such that the remaining induced graph H'_j is a set of caterpillars, hence 2-level planar. Clearly, the incidence vectors of the 2-level planar subgraphs P_i, $i = 1, 2, \ldots, 6$, and H'_j, $j = 1, 2, \ldots, |E| - |T|$ of G are linearly independent and satisfy $x(T) \leq |T| - 1$. Hence the facet-defining property is shown. □

We can tighten the LP-relaxation of (1)–(3) by introducing new inequalities that are valid and tight in the sense that they are facet-defining for $2\mathcal{LPS}(G)$. First, we generalize the double-claw inequalities to k-double claw inequalities. Considering a double-claw as a claw having three paths of length two, a *generalized k-double claw* is a claw having k paths of length two (see Figure 6(a)).

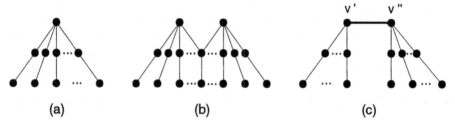

Fig. 6. (a) Generalized k-double claw (b) Combined k-double claw (c) Node-splitted k-double claw

Theorem 3.4 *Let $G = (L, U, E)$ be a 2-level graph. The generalized k-double claw inequalities*

$$x(T) \leq k + 2 \qquad (5)$$

where $T \subseteq E$ is a k-double claw in G $(k \geq 3)$ are facet-defining for $2\mathcal{LPS}(G)$.

Proof. Obviously, the inequality is valid. We denote $x(T) \leq k + 2$ by $c^T x \leq c_0$. Let us assume that there exists an inequality $a^T x \leq a_0$ with $\{x \mid c^T x = c_0\} \subseteq \{x \mid a^T x = a_0\}$. We show that $a_e = \lambda c_e$ and $a_0 = \lambda c_0$ for $\lambda > 0$. Let r be the root of the k-double claw and P denote the subgraph of $G = (V, E)$ induced by the edge set $F := \{(r, w) \mid w \in N(r) \cap V(T)\}$, where $N(r) = \{(r, v) \mid v \in V\}$ is the neighbourhood of r. Adding any two edges $e_1 \neq e_2$ in $T \setminus F$ to P gives a 2-level planar subgraph P' induced by the edge set $F' = \{F \cup e_1 \cup e_2\}$ satisfying $c^T \chi^{F'} = c_0$, hence also $a^T \chi^{F'} = a_0$. Since we can substitute e_1 and e_2 by any of the edges in $T \setminus F'$ we get $a_e = a_f$ for all $e, f \in T \setminus F$. Inserting the edge $e_3 = (w_3, u_3) \in T \setminus F'$ with $w_3 \in N(r) \cap V(T)$ in P' while removing the edge $e'_3 = (r, w_3)$ gives $a_{e_3} = a_{e'_3}$ and finally $a_e = a_f$ for all $e, f \in T$.

For any edge $e \in E \setminus T$ we can find a 2-level planar subgraph induced by the edge set F with $e \in F$ satisfying $c^F \chi^T = c_0$. $\square$

We can prove that the *combined k-double claws* give rise to a class of facet-defining inequalities for our polytope (proof omitted). A *combined k-double claw* consists of two k-double claws (having k_1, and k_2 paths of length two respectively) that share a single edge which has an endnode of degree one (see Figure 6(b)).

Theorem 3.5 *The combined k-double claw inequalities*

$$x(T) \leq k_1 + k_2 + 3 \qquad (6)$$

where $T \subseteq E$ induces a combined k-double claw G with parameters $k_1 \geq 3$ and $k_2 \geq 3$ are facet-defining for $2\mathcal{LPS}(G)$.

The *node-splitting operation* at vertex v in a graph G substitutes the subgraph induced by the edge set $\{(v, w) \mid w \in N(v)\}$ by a new subgraph induced by $\{(v', w') \mid w' \in W'\} \cup \{(v'', w'') \mid w'' \in W''\} \cup \{(v', v'')\}$, where $N(v)$ is the set of adjacent vertices of v in G, $W', W'' \subseteq N(v)$ with $W' \cup W'' = N(v)$ and $W' \cap W'' = \emptyset$. The vertices v' and v'' are the *duplicates* of v. The resulting graph when splitting the root node of a k-double claw is called *node-split k-double claw* with parameters k_1 and k_2 (see Figure 6(c)). The inequalities derived for those graphs contain a coefficient of two.

Theorem 3.6 *Let $G = (L, U, E)$ be a 2-level graph. The node-split k-double claw inequalities*

$$x(T) + 2x_{(v',v'')} \leq k_1 + k_2 + 4 \qquad (7)$$

where $T \subseteq E$ induces a node-split k-double claw G' in G with parameters $k_1 \geq 3$ and $k_2 \geq 3$ are facet-defining for $2\mathcal{LPS}(G)$. Moreover, they are facet-defining for $2\mathcal{LPS}(G')$ for $k_1 \geq 2$ and $k_2 \geq 2$.

Proof. Let $e_0 = (v', v'')$ and $T = T_1 \cup T_2 \cup \{e_0\}$, where T_1 and T_2 are the edge sets inducing the two components of $T \setminus \{e_0\}$. We first show validity. Let us assume that there exists a 2-level planar subgraph P induced by the edge set F violating the inequality (7). We know that $T_1 \cap F$ and $T_2 \cap F$ cannot contain more than $k_1 + 1$ and $k_2 + 2$ edges. If $e_0 \notin F$, the inequality cannot be violated by P. But if $e_0 \in F$, either T_1 contains at most k_1 edges, T_2 contains at most k_2 edges, or T_1 and T_2 contain at most $k_1 + 1$ and $k_2 + 1$ edges in order to ensure 2-level planarity of P. Hence, inequality (7) cannot be violated with P and validity is shown.

Now let us assume that there is an inequality $a^T x \leq a_0$ with $\{x \mid c^T x = c_0\} \subseteq \{x \mid a^T x = a_0\}$, where $c^T x \leq c_0$ denotes inequality (7). Let P be the 2-level planar subgraph induced by $k_1 + 2$ edges in T_1 (edge set F_1) and $k_2 + 2$ edges in T_2 (edge set F_2) not containing e_0. If $k_i \geq 3$, then any edge in F_i can be substituted by an edge $e_i \in T_i \setminus F_i$ maintaining the 2-level planarity. Hence in this case we have shown that $a_e = a_f$ for all $e, f \in T_i$. When inserting the edge e_0 to P, we can remove two edges from F_1 such that the resulting graph P' is still 2-level planar satisfying inequality (7) with equality. Hence, we have that $a_{e_0} = 2a_e$ for some $e \in F$. In the case that $k_1, k_2 \geq 3$, we have shown that inequality (7) is facet-defining for $2\mathcal{LPS}(G')$. Otherwise, let us assume that $k_2 = 2$. Since the degree of vertex v' in P' is exactly one (note that $(v', v'') \in P'$), any edge in $F_2 = T_2$ can be substituted by an edge $(v', w) \in T_1$ without destroying 2-level planarity. Hence, $a_e = a_{(v',w)}$ for all $e \in T_2$ and $(v', w) \in T_1$. If $k_1 = 2$, we apply the procedure symmetrically to T_1. Otherwise, we already know that $a_{(v',w)} = a_f$ for all $f \in T_1$, $(v', w) \in T_1$. Hence, we have shown that $a_{e_1} = a_{e_2}$ for all $e_1 \in T_1$, $e_2 \in T_2$ and $a_{e_0} = 2a_e$ for all $e \in T_1 \cup T_2$ if $k_1, k_2 \geq 2$. Hence, inequality (7) is facet-defining for $2\mathcal{LPS}(G')$.

It remains to show that $a_e = 0$ for all edges $e \in E \setminus E'$ if $G' = (V', E')$ with $E' \subseteq E$ and $V' \subseteq V$. Since zero-lifting is possible for double claw inequalities, we can restrict our attention to edges $e = (v, w)$ with $v \in G_1$ and $w \in G_2$, where G_1 and G_2 denote the graphs induced by the edge sets T_1 and T_2. If the graph $P \cup \{e\}$ is not 2-level planar, we can substitute an edge $e_1 \in F_1$ by an edge $e_1' \in T_1 \setminus F_1$, and an edge $e_2 \in F_2$ by an edge $e_2' \in T_2 \setminus F_2$ such that the resulting graph $P' = P \cup \{e_1', e_2', e\} \setminus \{e_1, e_2\}$ is 2-level planar. We have $0 = a^T \chi^{P'} - a^T \chi^P = a_e$ for all $e \in E \setminus E'$ and the theorem is proven. $\qquad \square$

Complete bipartite subgraphs of a 2-level graph G lead to the so-called *crown inequalities*. The proof of the following theorem is omitted here.

Theorem 3.7 *Let $G = (L, U, E)$ be a 2-level graph containing a complete bipartite subgraph $G' = (L', U', E')$, $E' \subseteq E$. The crown inequalities*

$$x(E') \leq |L'| + |U'| - 1 \qquad (8)$$

with $|L'| \geq 2$ and $|U'| \geq 3$ are facet-defining for $2\mathcal{LPS}(G)$.

In the case that the given 2-level graph contains no double claw, the 2-level planarization problem is equivalent to the maximum forest problem. It is well known that this problem can be solved in polynomial time by a simple greedy algorithm. Moreover, the structure of the associated weighted forest polytope has been well studied (see, e.g., [Edm70]). The inequalities of the weighted forest polytope are still valid for our polytope $2\mathcal{LPS}(G)$, even if the graph G contains double claws. And, as we will see in our computational experiments, they are quite useful in practice.

Theorem 3.8 *Let $G = (L, U, E)$ be a 2-level graph. The forest inequalities*

$$x(F) \leq V(F) - 1 \qquad (9)$$

where $F \subseteq E$ and $V(F)$ is the number of vertices contained in the subgraph induced by F are valid for $2\mathcal{LPS}(G)$.

In the next section we show how the theoretical results obtained in this section can be used in an algorithm for solving practical instances of the 2-level planarization problem.

4 Separation Problems and a Cutting Plane Algorithm

We suggest a branch and cut algorithm for solving practical instances of the maximum 2-level planar subgraph problem. We will explain the reasons why we are confident that branch and cut algorithms will be able to find the optimum solution for moderately sized problem instances in reasonable computation time.

First, 2-level planar subgraphs of any given graph $G = (V, E)$ contain only a linear number of edges, namely, at most $|V| - 1$ edges. In this case, one can use sparse graph techniques, like the ones described in [JRT95] for the Traveling Salesman Problem.

According to results of Grötschel, Lovász, Schrijver [GLS81], Karp and Papadimitriou [KP80], and Padberg and Rao [PR81], we can optimize a linear objective function over a polytope in polynomial time if and only if we can solve the *separation problem* in polynomial time, i.e., given a vector $\bar{x} \in \mathbf{Q}^{|E|}$, decide whether $\bar{x} \in \mathcal{P}$, and, if $\bar{x} \notin \mathcal{P}$, find a vector $d \in \mathbf{Q}^{|E|}$ and a scalar $d_0 \in \mathbf{Q}$ such that the inequality $d^T \bar{x} \leq d_0$ is valid with respect to $\mathcal{P}$ and $d^T \bar{x} > d_0$.

We will see that we can solve the separation problem restricted to the class of inequalities (2) in polynomial time.

Theorem 4.1 *For the cycle inequalities (2) the separation problem can be solved in polynomial time by computing at most $|E|$ shortest path problems.*

Proof. Given a point $\bar{x} \in \mathbf{Q}^{|E|}$, we are searching for a cycle $C \subseteq E$ with $\bar{x}(C) > |C| - 1$. Let us write the inequality in a different way: $|C| - \bar{x}(C) < 1$ which corresponds to $\sum_{e \in C}(1 - x_e) < 1$. For any fixed $e_0 \in E$ we solve a shortest path problem on the graph given by $G - \{e_0\}$ with edge costs $z_e = 1 - x_e$ for $e \in E \setminus \{e_0\}$. Let W be the weight of the shortest path. We then only have to test if $W + z_{e_0}$ is less than one. In this case we have found a cycle C leading to a violated inequality $\bar{x}(C) > |C| - 1$ of $\bar{x}$. If for no $e_0 \in E$ a violated inequality has been found, we have a proof that all the inequalities of type (2) are satisfied at $\bar{x}$. Hence we have solved the separation problem for (2) in polynomial time.

The separation problem can also be solved for the double claw inequalities (3) and their generalization to k-double claw inequalities for fixed k.

Theorem 4.2 *The separation problem for the double claw inequalities and the generalized k-double claw inequalities can be solved in polynomial time for fixed k by computing a series of maximum bipartite matching problems on subgraphs of G.*

Proof. Obviously, all k-double claws for fixed k can simply be enumerated in polynomial time. Faster is the following algorithm that is described for the generalized k-double claw inequalities when k is fixed. Given a point $\bar{x} \in \mathbf{Q}^{|E|}$, we are searching for a k-double claw $T \subseteq E$ with $\bar{x}(T) > k+2$. For any vertex r and any set of k adjacent vertices $w_1, w_2, \ldots, w_k \in N(r)$ let $W := \sum_{i=1}^{k} x_{(r,w_i)}$. We compute a maximum bipartite matching M between the vertex sets $\{w_1, w_2, \ldots, w_k\}$ and $\{N(w_1) \cup N(w_2) \cdots \cup N(w_k)\} \setminus \{r, w_1, w_2, \ldots, w_k\}$. If $W + \sum_{e \in M} x_e \leq k+2$, then no k-double claw inequality rooted at r with neighbours $w_1, w_2, \ldots, w_k$ is violated. Otherwise, M together with $\{(r, w_i) \mid i = 1, 2, \ldots, k\}$ induces a set T for which the inequality $x(T) \leq k + 2$ is violated (T may be only a part of a k-double claw in case that M contains less than k edges). $\qquad\square$

Padberg and Wolsey have already shown that the separation problem for the inequalities occuring in the weighted forest polytope can be solved in polynomial time [PW83].

Theorem 4.3 [PW83]. *The separation problem of the forest inequalities (9) can be solved by computing a minimum cut in a capacitated network G^* constructed from $G = (V, E)$. G^* contains $2(|V| + |E|)$ arcs and $|V| + 2$ vertices.*

We implemented a cutting plane method using the separation routines mentioned above. In a cutting plane algorithm we start with the linear system $\{\max w^T \bar{x} \mid x_e \geq 0, x_e \leq 1 \text{ for all } e \in E\}$. Let x^* denote the optimal solution of the LP-system. We solve the separation problem for inequalities (2), (3) (5) and (9) using Theorems 4.1-4.3. We add all the found inequalities to our system and optimize again. The algorithm stops if no violated inequalities of the above mentioned types are found. If x^* is integer, we know that x^* is the incidence vector of a 2-level planar graph. In this case we have found the optimal solution of the 2-level planarization problem. Otherwise, x^* gives us an upper bound to the value of a maximum 2-level planar subgraph of the given instance G.

In addition, we try to find good solutions to the problem. After each optimization process, we may get new solutions x^* to the problem, most of which are fractional. Fractional solutions x^* may give us a hint about good solutions to the problem. We try to use this information in our heuristics that we apply in each iteration.

5 Computational Results

For our experiments we used the cutting plane algorithm described above. The algorithm stops if either the optimal solution is found or no violated cycle, double claw, generalized double claw or forest inequality can be detected. In any case, the algorithm gives a 2-level planar subgraph together with the solution value of the last linear program that is an upper bound of the optimal solution.

Table 1. Computational Results for graphs on 20 vertices per level

$|V_i|$	$|E|$	Gar	Time	Cycles	2Claw	kClaw	Forest	GarI	TimeI	CyclesI	2ClawI
20	20	0.00	0.01	0.18	0.64	0.00	0.00	0.00	0.01	0.18	0.64
20	25	0.04	0.02	0.71	2.24	0.15	0.00	0.22	0.02	0.71	2.23
20	30	0.35	0.08	1.88	7.52	1.25	0.03	0.84	0.07	1.88	7.52
20	35	0.74	0.28	4.42	18.90	6.15	0.15	2.66	0.20	4.45	18.84
20	40	1.85	0.93	7.92	39.09	21.44	1.02	5.65	0.46	8.00	37.99
20	45	2.69	2.36	11.81	74.82	51.49	3.42	9.81	1.07	12.34	75.31
20	50	2.53	7.35	13.89	93.12	183.98	5.86	10.29	1.58	14.67	94.56
20	55	2.24	11.07	14.89	104.31	245.29	7.10	7.14	1.87	15.83	104.69
20	60	1.22	16.65	16.80	102.68	296.39	7.51	5.69	2.04	17.78	101.29
20	65	0.65	16.69	66.92	66.92	242.31	2.77	2.99	2.10	18.68	89.17
20	70	0.22	12.80	16.62	68.85	178.70	1.46	1.50	1.46	18.53	57.68
20	75	0.00	0.84	17.38	31.23	1.61	0.00	1.83	2.02	19.61	71.45
20	80	0.00	0.83	14.92	33.77	1.35	0.90	0.59	1.10	17.68	33.74
20	85	0.00	0.77	16.15	28.15	2.82	1.43	0.31	0.92	15.19	25.03
20	90	0.00	0.68	13.61	22.54	0.46	0.00	0.28	0.80	13.86	19.79
20	95	0.00	0.37	14.38	7.77	0.00	0.00	0.13	0.52	13.81	10.18
20	100	0.00	0.32	13.92	4.77	0.00	0.00	0.05	0.42	12.00	5.81

Table 1 shows computational results for 100 instances of 2-level graphs with 20 vertices at each level with increasing density. The columns show the number of vertices per level, the number of edges, and the average quality of the solution value, i.e., if Sol denotes the number of edges remaining in a found 2-level planar subgraph and UpBound denotes the value determined by the linear programming relaxation, then the solution guarantee Gar is $\left(\frac{\text{UpBound}-\text{Sol}}{\text{UpBound}}\right) \times 100\%$. Column 4 shows the time on a SUN Ultra 1/170 in seconds. Columns 5 to 8 show the average number of found violated cycle, double claw, generalized k-double claw and forest inequalities.

The results are surprisingly good. On the average, the solution we found is very close (below 3% on average) to the optimal one. If we do not search for violated generalized k-double claw and forest inequalities we get a solution that is worse (up to 11% on average). On some single instances, the obtained solution guarantee was around 20%. Columns 9 to 12 show the average values in this case. Hence, it is really worth studying the associated polytope, i.e., searching for additional inequalities.

Table 2. Computational Results for sparse graphs

$|V_i|$	$|E|$	Gar	Time	Cycles	2Claw	kClaw	Forest	GarI	TimeI	CyclesI	2ClawI
20	40	1.94	0.80	7.85	38.68	18.71	0.54	5.31	0.45	7.88	38.06
30	60	2.47	2.48	11.53	76.36	40.36	0.85	6.86	1.22	11.56	75.25
40	80	3.10	6.49	16.66	128.19	71.22	0.51	8.16	3.67	16.69	127.47
50	100	3.47	11.86	19.63	166.67	114.58	0.52	8.53	5.83	19.68	164.71
60	120	3.83	18.56	24.40	220.36	143.11	0.51	9.18	9.20	24.45	218.48
70	140	4.19	36.60	29.49	273.65	171.79	0.88	9.60	19.55	29.48	270.68
80	160	4.14	48.53	33.18	316.09	200.27	1.07	9.51	25.27	33.16	313.16
90	180	4.36	61.25	37.69	365.75	236.88	0.63	9.99	33.14	37.72	363.14
100	200	4.33	75.99	42.83	408.88	246.42	1.00	9.71	41.02	42.85	405.66

Furthermore, we ran 100 instances on a series of sparse graphs. The results are promising also for these cases (see Table 2). Our solution is at most 5% away from the optimal solution. This confirms our conjecture that when combining our cutting plane algorithm with a branch and bound algorithm ("branch and cut"), we may be able to solve practical instances (of moderate size) to optimality within short computation time.

Consider the graph shown in Figure 2. Our cutting plane algorithm solved the 2-level planarization problem for the given instance provably optimal within 0.05 seconds. During the run 5 violated cycle constraints were found, 10 double claw inequalities, 1 generalized k-double claw inequalities and no forest inequality.

6 Acknowledgements

I started to get interested in the 2-level planarization problem when Peter Eades visited the Max-Planck-Institut für Informatik (Saarbrücken) in June 1995. Thanks to Peter for initiating my interest in this beautiful problem. I am very grateful to René Weiskircher for providing the implementation of the cutting plane algorithm and for running all the computational experiments.

References

[BN88]   G. Di Battista and E. Nardelli: Hierarchies and Planarity Theory. IEEE Transactions on Systems, Man and Cybernetics 18 (1988) 1035 – 1046

[Car80] Carpano, M.J.: Automatic display of hierarchized graphs for computer aided decision analysis. IEEE Transactions on Systems, Man and Cybernetics, SMC-10, no. 11 (1980) 705 – 715

[Edm70] Edmonds, J.: Submodular functions, matroids and certain polyhedra. in: Combinatorial Structures and Their Applications, Gordon and Breach, London (1970) 69 – 87

[Dre94] Dresbach, S.: A New Heuristic Layout Algorithm for DAGs. in: Derigs, Bachem & Drexl (eds.), Operations Research Proceedings 1994, Springer Verlag, Berlin (1994) 121 – 126

[EK86]   Eades, P., and D. Kelly: Heuristics for Reducing Crossings in 2-Layered Networks. Ars Combinatoria 21-A (1986) 89 – 98

[EKW86] Eades, P., B.D. McKay, and N.C. Wormald: On an edge crossing problem. Proc. 9th Australian Computer Science Conference, Australian National University (1986) 327 – 334

[EW94a] Eades, P., and N.C. Wormald: Edge crossings in Drawings of Bipartite Graphs. Algorithmica 10 (1994) 379 – 403

[EW94b] Eades, P. and S. Whitesides: Drawing graphs in two layers. Theoretical Computer Science 131 (1994) 361 – 374

[Fuk96] Fukuda, A.: Face Lattices. Personal Communication (1996)

[GLS81] Grötschel, M., L. Lovász, and A. Schrijver: The Ellipsoid Method and its Consequences in Combinatorial Optimization. Combinatorica 1 (1981) 169 – 197

[GP85]   Grötschel, M. and M.W. Padberg: Polyhedral theory. In E.L. Lawler, J.K. Lenstra, A.H.G. Rinnoy Kan, and D.B. Shmoys (eds.), The Traveling Salesman Problem: A Guided Tour of Combinatorial Optimization, Wiley-Interscience (1985)

[HP96]   Heath, L.S. and S.V. Pemmaraju: Recognizing Leveled-Planar Dags in Linear Time. Lecture Notes in Comp. Sci. 1027, in: F. Brandenburg (ed.), Proceedings on Graph Drawing '95, Passau (1996) 300 – 311

[JM93a]   Jünger, M. und P. Mutzel: Solving the maximum planar subgraph problem by branch and cut. In: L.A. Wolsey and G. Rinaldi (eds.), Proceedings of the 3rd IPCO Conference, Erice (1993) 479 – 492

[JM93b]   Jünger, M. und P. Mutzel: Maximum planar subgraphs and nice embeddings: Practical layout tools. Algorithmica 16, No. 1, Special Issue on Graph Drawing, G. Di Battista and R. Tamassia (eds.), (1996) 33 – 59, also Report No. 93.145, Universität zu Köln, (1993)

[JM96]   Jünger, M. und P. Mutzel: Exact and Heuristic Algorithms for 2-Layer Straightline Crossing Minimization. Lecture Notes in Comp. Sci. 1027, in: F. Brandenburg (ed.), Proc. on Graph Drawing '95, Passau (1996) 337 – 348

[JRT95]   Jünger, M., G. Reinelt, and S. Thienel: Practical Problem Solving with Cutting Plane Algorithms in Combinatorial Optimization. DIMACS Series in Discrete Mathematics and Theoretical Computer Science, 20 (1995) 111 – 152

[KP80]   Karp, R.M. and C.H. Papadimitriou: On Linear Characterizations of Combinatorial Optimization Problems. Proc. of the 21st Annual Symp. on the Foundations of Computer Science IEEE (1980) 1 – 9

[Len90]   Lengauer, T.: Combinatorial algorithms for integrated circuit layout. John Wiley & Sons, Chichester, UK (1990)

[Mäk90]   Mäkinen, E.: Experiments on Drawing 2-Level Hierarchical Graphs. Intern. J. Computer Math. 37 (1990) 129 – 135

[Mut94]   Mutzel, P.: The Maximum Planar Subgraph Problem. Dissertation, Universität zu Köln (1994)

[PR81]   Padberg, M.W. and M.R. Rao: The Russian Method for Linear Inequalities III: Bounded Integer Programming. GBA Working Paper 81 – 39, New York University (1981)

[PW83]   Padberg, M.W. and L.A. Wolsey: Trees and Cuts. Annals of Discrete Mathematics 17 (1983) 511 – 517

[STT81]   Sugiyama, K., S. Tagawa, and M. Toda: Methods for Visual Understanding of Hierarchical System Structures. IEEE Trans. Syst. Man, Cybern., SMC-11 (1981) 109 – 125

[TKY77]   Tomii, N., Y. Kambayashi, and Y. Shuzo: On Planarization Algorithms of 2-Level Graphs. Papers of tech. group on electronic computers, IECEJ, EC77-38 (1977) 1 – 12

[Ull84]   Ullman, J.D.: Computational Aspects of VLSI. Computer Science Press, Rockville, MD (1984)

[VLM96]   Vingron, M., H.-P. Lenhof, and P. Mutzel: Computational Molecular Biology. In: Annotated Bibliographies in Combinatorial Optimization, M. Dell'Amico, F. Maffioli, S. Martello (eds.), Chapter 23, to appear (1996)

[VML96]   Valls, V., R. Marti, and P. Lino: A Branch and Bound Algorithm for Minimizing the Number of Crossing Arcs in Bipartite Graphs. Journal of Operational Research 90 (1996) 303 – 319

[War77]   Warfield, J.N.: Crossing Theory and Hierarchy Mapping. IEEE Trans. Syst. Man, Cybern., SMC-7 (1977) 505 – 523

[WG86]   Waterman, M.S. and J. R. Griggs: Interval graphs and maps of DNA. Bull. Math. Biology 48, no. 2 (1986) 189 – 195

A Linear-Time Algorithm
for Four-Partitioning
Four-Connected Planar Graphs

Shin-ichi Nakano, Md. Saidur Rahman and Takao Nishizeki *

Graduate School of Information Sciences
Tohoku University, Sendai 980-77, Japan.

Abstract. Given a graph $G = (V, E)$, four distinct vertices $u_1, u_2, u_3, u_4 \in V$ and four natural numbers n_1, n_2, n_3, n_4 such that $\sum_{i=1}^{4} n_i = |V|$, we wish to find a partition V_1, V_2, V_3, V_4 of the vertex set V such that $u_i \in V_i$, $|V_i| = n_i$ and V_i induces a connected subgraph of G for each i, $1 \le i \le 4$. In this paper we give a simple linear-time algorithm to find such a partition if G is a 4-connected planar graph and u_1, u_2, u_3, u_4 are located on the same face of a plane embedding of G. Our algorithm is based on a "4-canonical decomposition" of G, which is a generalization of an st-numbering and a "canonical 4-ordering" known in the area of graph drawings.

1 Introduction

Given a graph $G = (V, E)$, k distinct vertices $u_1, u_2, \cdots, u_k \in V$ and k natural numbers $n_1, n_2, \cdots, n_k$ such that $\sum_{i=1}^{k} n_i = |V|$, we wish to find a partition $V_1, V_2, \cdots, V_k$ of the vertex set V such that $u_i \in V_i$, $|V_i| = n_i$, and V_i induces a connected subgraph of G for each i, $1 \le i \le k$. Such a partition is called a *k-partition* of G. A 4-partition of a graph G is depicted in Fig. 1, where the edges of four connected subgraphs are drawn by solid lines and the remaining edges of G are drawn by dotted lines. The problem of finding a k-partition of a given graph often appears in the load distribution among different power plants and the fault-tolerant routing of communication networks [WK94, WTK95]. The problem is NP-hard in general [DF85], and hence it is very unlikely that there is a polynomial-time algorithm to solve the problem. Although not every graph has a k-partition, Györi and Lovász independently proved that every k-connected graph has a k-partition for any $u_1, u_2, \cdots, u_k$ and $n_1, n_2, \cdots, n_k$ [G78, L77]. However, their proofs do not yield any polynomial-time algorithm for actually finding a k-partition of a k-connected graph. For the case $k = 2$ and 3, the following algorithms have been known:

(i) a linear-time algorithm to find a bipartition of a biconnected graph [STN90, STNMU90];

---

* E-mail: nakano@ecei.tohoku.ac.jp, saidur@nishizeki.ecei.tohoku.ac.jp,
nishi@ecei.tohoku.ac.jp

(ii) an algorithm to find a tripartition of a triconnected graph in $O(n^2)$ time, where n is the number of vertices of a graph [STNMU90]; and

(iii) a linear-time algorithm to find a tripartition of a triconnected planar graph [JSN94].

On the other hand, polynomial-time algorithms have not been known for the case $k \geq 4$. **

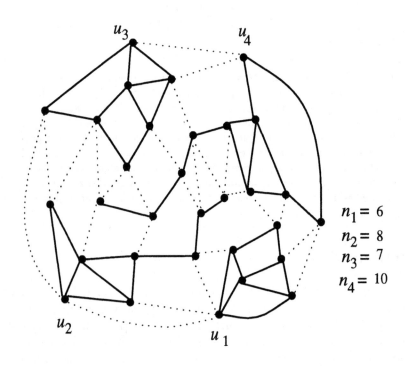

$n_1 = 6$
$n_2 = 8$
$n_3 = 7$
$n_4 = 10$

Fig. 1. A 4-partitioning of a 4-connected plane graph G.

In this paper we give a linear-time algorithm to find a 4-partition of a 4-connected plane graph G if u_1, u_2, u_3, u_4 are located on the same face of G, as illustrated in Fig. 1. We first bipartition the 4-connected graph G into two biconnected graphs having about $n_1 + n_2$ and $n_3 + n_4$ vertices respectively, we then bipartition each of them to two connected graphs, and, by adjusting the numbers of vertices in the resulting four graphs, we finally obtain a required 4-partition of G. To bipartition G into two biconnected graphs, we will newly define and use a "4-canonical decomposition" of G, which is a generalization of an st-numbering and a "canonical 4-ordering" known in the area of graph drawings [E79, K94, KH94].

** A polynomial-time algorithm for any k is claimed in [MM94], but is not correct [G96].

The remainder of the paper is organized as follows. In Section 2 we introduce our notations and give a linear-time algorithm to find a 4-canonical decomposition of a 4-connected planar graph. In Section 3 we present a linear-time algorithm to find a 4-partition of a 4-connected planar graph. Finally we put our discussions in Section 4.

2 4-Canonical Decomposition

In this section we introduce some definitions and prove that every 4-connected plane graph has a 4-canonical decomposition and it can be found in linear time.

Let $G = (V, E)$ be a connected simple graph with vertex set V and edge set E. Throughout the paper we denote by n the number of vertices in G, that is, $n = |V|$. An edge joining vertices u and v is denoted by (u, v). The *degree* of a vertex v is the number of neighbors of v in G. The *connectivity* $\kappa(G)$ of a graph G is the minimum number of vertices whose removal results in a disconnected graph or a single-vertex graph K_1. G is called a *k-connected* graph if $\kappa(G) \geq k$. We call a vertex of G a *cut vertex* if its removal results in a disconnected or single-vertex graph. For $W \subseteq V$, we denote by $G - W$ the graph obtained from G by deleting all vertices in W and all edges incident to them.

A graph is *planar* if it can be embedded in the plane so that no two edges intersect geometrically except at a vertex to which they are both incident. A *plane* graph is a planar graph with a fixed embedding. The *contour* $C(G)$ of a biconnected plane graph G is the clockwise (simple) cycle on the outer face. We write $C(G) = w_1, w_2, \cdots, w_h, w_1$ if the vertices $w_1, w_2, \cdots, w_h$ on $C(G)$ appear in this order. A *chord* in a biconnected plane graph G is a path P in G satisfying the following (a) - (d):

(a) P connects two vertices w_p and w_q, $p < q$, on $C(G)$;
(b) P does not pass through any vertices on $C(G)$ except the ends w_p and w_q;
(c) P lies on an inner face; and
(d) there is no edge e on $C(G)$ such that P together with e forms an inner face.

The chord is said to be *minimal* if none of $w_{p+1}, w_{p+2}, \cdots, w_{q-1}$ is an end of a chord. Thus the definition of a minimal chord depends on which vertex is considered as the starting vertex w_1 of $C(G)$. Let $\{v_1, v_2, \cdots, v_{p-1}, v_p\}$ be a set of three or more consecutive vertices on $C(G)$ such that the degrees of the first vertex v_1 and the last one v_p are at least three and the degrees of all intermediate vertices $v_2, v_3, \cdots, v_{p-1}$ are two. Then we call the set $\{v_2, v_3, \cdots, v_{p-1}\}$ an *outer chain* of G.

For a cycle C in a plane graph G, we denote by $I(C, G)$ the subgraph of G inside C, that is, the plane subgraph of G induced by the set of vertices inside (or on) the cycle C. Clearly $I(C, G)$ is biconnected if G is biconnected. We have the following lemma.

Lemma 1. *Assume that G is a 4-connected plane graph and that a cycle $C = w_1, w_2, \cdots, w_h, w_1$ in G is not a face of G. Let w_p and w_q be the two ends of any* *minimal chord in* $I(C,G)$ *if* $I(C,G)$ *has a chord, and let* $w_p = w_1$ *and* $w_q = w_h$ *if* $I(C,G)$ *has no chord. Then the following* (a) *and* (b) *hold:*

(a) *If* $W = \{w_{p+1}, w_{p+2}, \cdots, w_{q-1}\}$ *is an outer chain of* $I(C,G)$, *then* $I(C,G) - W$ *is biconnected.*

(b) *Otherwise, there is a set* $W' = \{w_{p'}, w_{p'+1}, \cdots, w_{q'}\}$ *of one or more consecutive vertices on* C *such that*
   (i) $p < p' \leq q' < q$, *and*
   (ii) *none of the vertices in* W' *except the first vertex* $w_{p'}$ *and the last one* $w_{q'}$ *has a neighbor in the proper outside of* C.
   Moreover, for any of sets W' *satisfying* (i) *and* (ii), $I(C,G) - W'$ *is biconnected.*

Proof. (a) Assume that $W = \{w_{p+1}, w_{p+2}, \cdots, w_{q-1}\}$ is an outer chain of $I(C,G)$. Then $I(C,G)$ has a minimal chord with ends w_p and w_q. Suppose for a contradiction that $I(C,G) - W$ is not biconnected. Then $I(C,G) - W$ has a cut vertex v. Since G is 4-connected, v must be on C. However, the chord above passes through v, and $v \neq w_p, w_q$, contradicting to the condition (b) of the definition of a chord.

(b) Assume that W is not an outer chain of $I(C,G)$. Then $q \geq p+2$ since G is 4-connected. Obviously any singleton set $W' = \{w_{p'}\}$, $p < p' < q$, satisfies (i) and (ii). Therefore it suffices to prove that $I(C,G) - W'$ is biconnected for any of sets W' satisfying (i) and (ii). Suppose for a contradiction that $I(C,G) - W'$ is not biconnected for a set W' satisfying (i) and (ii). Then $I(C,G) - W'$ has a cut vertex v. If v is not on C, then the removal of v and one or two appropriate vertices in W' disconnects G and hence G would not be 4-connected, a contradiction. If v is on C, then either G would not be 4-connected or a chord with ends w_p and w_q would not be minimal, a contradiction. □

Let $G = (V, E)$ be a connected graph, and let $(s, t) \in E$. We say that an ordering $\pi = v_1, v_2, \cdots, v_n$ of the vertices of G is an *st-numbering* of G if the following conditions are satisfied:

(st1) $v_1 = s$ and $v_n = t$; and
(st2) each $v_i \in V - \{v_1, v_n\}$ has two neighbors v_p and v_q such that $p < i < q$.

Not every connected graph has an *st*-numbering, but the following lemma holds.

Lemma 2. [E79] *Let* G *be a biconnected graph, and let* (s, t) *be any edge of* G. *Then* G *has an st-numbering* $\pi = v_1, v_2, \cdots, v_n$ *such that* $v_1 = s$ *and* $v_n = t$, *and* π *can be found in linear time.*

A bipartition of a biconnected graph can be found by an *st*-numbering as follows [STNMU90, STN90]. Let $G = (V, E)$ be a biconnected graph, let $u_1, u_2 \in V$ be two designated distinct vertices, and let n_1, n_2 be two natural numbers such that $n_1 + n_2 = n$. We may assume without loss of generality that $(u_1, u_2) \in E$; otherwise, consider as G the graph obtained from G by adding a new edge (u_1, u_2). Since G is biconnected, by Lemma 2 G has an *st*-numbering $v_1(= u_1), v_2, \cdots, v_n(= u_2)$. Clearly the following fact holds:

(st3) both the subgraphs of G induced by $\{v_1, v_2, \cdots, v_i\}$ and $\{v_{i+1}, v_{i+2}, \cdots, v_n\}$ are connected for each i, $1 \le i < n$.

Thus, choosing $i = n_1$, one can find a required bipartition of G in linear time.

Generalizing an *st*-numbering in a sense, we define a "4-canonical decomposition" of a 4-connected plane graph G and in the succeeding section we give an algorithm to find a 4-partition of G by using the "4-canonical decomposition." We now give the definition of a 4-canonical decomposition.

Assume that $G = (V, E)$ is a 4-connected plane graph with four designated distinct vertices u_1, u_2, u_3, u_4 on the same face of G. We may assume that u_1, u_2, u_3, u_4 lie on the contour $C(G)$ of G, since, for any face F of G, we can re-embed G so that F becomes the outer face. We may furthermore assume that the four vertices u_1, u_2, u_3, u_4 appear on $C(G)$ of G in this order. Moreover we may assume that $(u_1, u_2), (u_3, u_4) \in E$; otherwise, consider as G the new graph obtained from G by adding edges (u_1, u_2) and (u_3, u_4). For a set $W_1, W_2, \cdots, W_i$ of pairwise disjoint subsets of V, we denote by G_i the subgraph of G induced by $W_1 \cup W_2 \cup \cdots \cup W_i$, and by $\overline{G_i}$ the subgraph of G induced by $V - W_1 \cup W_2 \cup \cdots \cup W_i$, that is, $\overline{G_i} = G - W_1 \cup W_2 \cup \cdots \cup W_i$. We say that a partition $\Pi = W_1, W_2, \cdots, W_l$ of V is a *4-canonical decomposition* of G if the following three conditions (co1)–(co3) are satisfied:

(co1) W_1 is the set of vertices on the inner face containing edge (u_1, u_2), and W_l is the set of vertices on the inner face containing edge (u_3, u_4);
(co2) for each i, $1 \le i < l$, both G_i and $\overline{G_i}$ are biconnected; and
(co3) for each i, $1 < i < l$, either W_i consists of exactly one vertex on both $C(G_i)$ and $C(\overline{G_{i-1}})$ or W_i is an outer chain of G_i or $\overline{G_{i-1}}$.

Fig. 2 illustrates the condition (co3); (a) for the case $|W_i| = 1$, and (b) and (c) for the cases w_i is an outer chain of G_i and $\overline{G_{i-1}}$ respectively, where G_i and $\overline{G_{i-1}}$ are indicated by different shading and the vertices in W_i are drawn in black dots.

The 4-canonical decomposition defined for 4-connected plane graphs is a generalization of the "canonical 4-ordering" defined for internally triangulated 4-connected plane graphs [K94, KH94].

We have the following two lemmas.

Lemma 3. *Let $G = (V, E)$ be a 4-connected plane graph with four designated distinct vertices u_1, u_2, u_3, u_4 appearing on $C(G)$ in this order. Then G has a 4-canonical decomposition $\Pi = W_1, W_2, \cdots, W_l$. Furthermore Π can be found in linear time.*

Proof. Let W_1 be the set of vertices on the inner face containing edge (u_1, u_2). Clearly G_1 is biconnected, and $u_3, u_4 \notin W_1$. We now claim that $\overline{G_1} = G - W_1$ is also biconnected. Let C be the contour of the biconnected plane graph obtained from G by deleting edge (u_1, u_2). Clearly $I(C, G)$ has neither a chord nor an outer chain; otherwise, G would not be 4-connected. Let cycle C start with u_4,

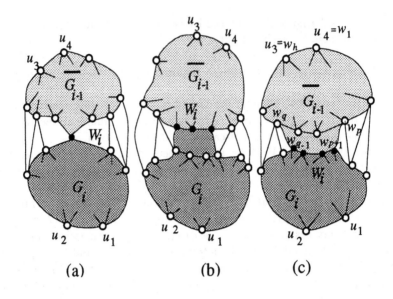

Fig. 2. Illustration of the condition (co3).

then the set W_1 of vertices are consecutive on C and satisfies (i) and (ii) in Lemma 1(b). Therefore $\overline{G_1} = I(C, G) - W_1$ is biconnected.

Assume that we have chosen $W_1, W_2, \cdots . W_{i-1}$, $i \geq 2$, such that the conditions (co2) and (co3) hold for each j, $1 \leq j \leq i-1$, and that $u_3, u_4 \notin W_1 \cup W_2 \cup \cdots \cup W_{i-1}$. Then we show that there is a set W_i ($\subseteq V - W_1 \cup W_2 \cup \cdots \cup W_{i-1}$) such that

(1) G_i is biconnected,
(2) either $u_3, u_4 \notin W_i$ or $u_3, u_4 \in W_i$;
(3) if $u_3, u_4 \notin W_i$, then $\overline{G_i}$ is biconnected and W_i satisfies the condition (co3); and
(4) if $u_3, u_4 \in W_i$, then $l = i$, that is, $V = W_1 \cup W_2 \cup \cdots \cup W_l$, and W_l is the set of vertices on the inner face containing edge (u_3, u_4).

There are the following two cases.

Case 1: graph $\overline{G_{i-1}} = G - W_1 \cup W_2 \cup \cdots \cup W_{i-1}$ is a cycle.

In this case $\overline{G_{i-1}}$ is the inner face of G containing edge (u_3, u_4). We set $l = i$ and $W_l = V - W_1 \cup W_2 \cup \cdots \cup W_{i-1}$. Then $u_3, u_4 \in W_l$, and $V = W_1 \cup W_2 \cup \cdots \cup W_l$. Since $G_i = G$, G_i is biconnected.

Case 2: otherwise.

Let $C(\overline{G_{i-1}}) = w_1, w_2, \cdots, w_h, w_1$ be the contour of $\overline{G_{i-1}}$ with the starting vertex $w_1 = u_4$. Then $\overline{G_{i-1}} = I(C(\overline{G_{i-1}}), G)$. If $\overline{G_{i-1}}$ has a chord then let w_p and w_q be the two ends of a minimal chord, otherwise let $w_p = w_1 = u_4$ and $w_q = w_h = u_3$. Let $W = \{w_{p+1}, w_{p+2}, \cdots, w_{q-1}\}$. We now have the following three subcases.

Subcase 2(a): W is an outer chain of $\overline{G_{i-1}}$.

In this subcase we set $W_i = W$. Then $u_3, u_4 \notin W_i$, and W_i satisfies (co3). Since G is 4-connected, each vertex in W_i has at least two neighbors in the biconnected graph G_{i-1} induced by $W_1 \cup W_2 \cup \cdots \cup W_{i-1}$. Therefore the graph G_i induced by $(W_1 \cup W_2 \cup \cdots \cup W_{i-1}) \cup W_i$ is biconnected. By Lemma 1(a), $\overline{G_i} = \overline{G_{i-1}} - W_i$ is biconnected too.

Subcase 2(b): W is not an outer chain of $\overline{G_{i-1}}$, but a vertex w_r in W has two or more neighbors in G_{i-1}.

In this subcase we set $W_i = \{w_r\}$. Then $u_3, u_4 \notin W_i$, and w_r lies on both $C(G_i)$ and $C(\overline{G_{i-1}})$ and hence W_i satisfies (co3). Since w_r has two or more neighbors in G_{i-1}, G_i is biconnected. Since $W_i = \{w_r\}$ satisfies (i) and (ii) in Lemma 1(b), $\overline{G_i} = \overline{G_{i-1}} - W_i$ is biconnected.

Subcase 2(c): otherwise.

In this subcase, W is not an outer chain of $\overline{G_{i-1}}$, and every vertex in W has at most one neighbor in G_{i-1}. Since G is 4-connected, W contains two vertices $w_{p'}$ and $w_{q'}$ such that

(1) $p < p' < q' < q$,
(2) each of $w_{p'}$ and $w_{q'}$ has exactly one neighbor in G_{i-1} and these neighbors are different from each other, and
(3) none of $w_{p'+1}, w_{p'+2}, \cdots, w_{q'-1}$ has a neighbor in G_{i-1}.

We now set $W_i = \{w_{p'}, w_{p'+1}, \cdots, w_{q'}\}$. Clearly $u_3, u_4 \notin W_i$, G_i is biconnected, and W_i is an outer chain of G_i and hence satisfies (co3). Since W_i satisfies (i) and (ii) in Lemma 1(b), $\overline{G_i} = \overline{G_{i-1}} - W_i$ is biconnected.

Thus we have proved that there exists a 4-canonical decomposition.

One can implement an algorithm for finding a 4-canonical decomposition, based on the proof. It maintains a data-structure to keep the outer chains and minimal chords of $\overline{G_i}$. The algorithm traverses every face at most a constant times, and runs in linear time. □

Lemma 4. Let $W_1, W_2, \cdots, W_l$ be a 4-canonical decomposition of a 4-connected plane graph G. Then the following (a) and (b) hold for any i, $1 < i < l$:

(a) If W_i is an outer chain of G_i as illustrated in Fig. 2(b), then, for any $W_i' \subseteq W_i$, $\overline{G_{i-1}} - W_i'$ is biconnected.
(b) If W_i is an outer chain of $\overline{G_{i-1}}$ as illustrated in Fig. 2(c), then, for any $W_i' \subseteq W_i$, $G_i - W_i'$ is biconnected.

Proof. We give only a proof for (b) since the proof for (a) is similar. Let W_i be an outer chain of $\overline{G_{i-1}}$. The graph G_{i-1} is biconnected. Since G is 4-connected, each vertex in W_i has at least two neighbors in G_{i-1}. Therefore the graph $G_i - W_i'$ induced by $W_1 \cup W_2 \cup \cdots \cup W_{i-1} \cup (W_i - W_i')$ is also biconnected. □

3   4-Partition of 4-Connected Plane Graph

In this section we give our algorithm to find a 4-partition of a 4-connected plane graph G. Assume that the four designated distinct vertices u_1, u_2, u_3, u_4

appear on $C(G)$ in this order and n_1, n_2, n_3, n_4 are natural numbers such that $\sum_{i=1}^{4} n_i = n$.

Algorithm Four-Partition
Find a 4-canonical decomposition $\Pi = W_1, W_2, \cdots, W_l$ of G;
Let i be the minimum integer such that $\sum_{j=1}^{i} |W_j| \geq n_1 + n_2$;
Let $r = \sum_{j=1}^{i} |W_j| - (n_1 + n_2)$, that is, r is the excess of the number of vertices in $W_1 \cup W_2 \cup \cdots \cup W_i$ over $n_1 + n_2$;
There are the following two cases (1) $r = 0$, and (2) $r \geq 1$;
Case 1: $r = 0$.
{In this case, G_i contains $n_1 + n_2$ vertices, and $\overline{G_i}$ contains $n_3 + n_4$ vertices.}
Find a bipartition V_1, V_2 of the biconnected graph G_i such that $u_1 \in V_1$, $u_2 \in V_2$, $|V_1| = n_1$, $|V_2| = n_2$, and both V_1 and V_2 induce connected subgraphs;
Find a bipartition V_3, V_4 of the biconnected graph $\overline{G_i}$ such that $u_3 \in V_3$, $u_4 \in V_4$, $|V_3| = n_3$, $|V_4| = n_4$, and both V_3 and V_4 induce connected subgraphs;
Return V_1, V_2, V_3, V_4 as a 4-partition of G.
Case 2: $r \geq 1$.
{ In this case, G_i contains $n_1 + n_2 + r$ vertices, and $\overline{G_i} = \overline{G_{i-1}} - W_i$ contains $n_3 + n_4 - r$ vertices. Since $r \geq 1$, $|W_i| \geq 2$ and hence W_i is an outer chain of either $\overline{G_{i-1}}$ or G_i.}
Let $C(\overline{G_{i-1}}) = w_1, w_2, ..., w_h, w_1$ where $w_1 = u_4$;
Assume that $W_i = \{w_{p+1}, w_{p+2}, \cdots, w_{q-1}\}$ is an outer chain of $\overline{G_{i-1}}$ as illustrated in Fig 2(c), otherwise, interchange the roles of u_1, u_2 and u_3, u_4;
Find an st-numbering $v_1, v_2, \cdots, v_{n_3+n_4-r}$ of $\overline{G_i}$ such that $s = v_1 = u_4$ and $t = v_{n_3+n_4-r} = u_3$;
Let $w_p = v_{p'}$ and $w_q = v_{q'}$;
Assume that $p' < q'$, otherwise, interchange the roles of u_3 and u_4;
There are the following three subcases (a) $n_4 \leq p'$, (b) $p' + r \leq n_4$, and (c) $p' < n_4 < p' + r$;
Subcase 2(a): $n_4 \leq p'$.
{In this subcase, the last r vertices in the outer chain W_i are added to $\overline{G_i}$ as the deficient r vertices.}
Let $V_4 = \{v_1, v_2, \cdots, v_{n_4}\}$ be the first n_4 vertices in the st-numbering of $\overline{G_i}$;
Let $V_3' = \{v_{n_4+1}, v_{n_4+2}, \cdots, v_{n_4+n_3-r}\}$ be the remaining $n_3 - r$ vertices in $\overline{G_i}$;
{By the fact (st3) of an st-numbering both V_4 and V_3' induce connected graphs.}
Let $W_i' = \{w_{q-1}, w_{q-2}, \cdots, w_{q-r}\}$ be the set of the last r vertices in W_i;
Let $V_3 = V_3' \cup W_i'$;
{Since w_{q-1} is adjacent to w_q in V_3', V_3 induces a connected graph of n_3 vertices.}
Let $G_{12} = G_i - W_i'$;
{ G_{12} is biconnected by Lemma 4(b), and has $n_1 + n_2$ vertices.}
Find a bipartition V_1, V_2 of G_{12} such that $u_1 \in V_1$, $u_2 \in V_2$, $|V_1| = n_1$, $|V_2| = n_2$, and both V_1 and V_2 induce connected subgraphs;
Return V_1, V_2, V_3, V_4 as a 4-partition of G.
Subcase 2(b): $p' + r \leq n_4$.
{In this subcase, the first r vertices in W_i are added to $\overline{G_i}$ as the deficient r vertices.}

Let $V_4' = \{v_1, v_2, \cdots, v_{n_4-r}\}$ be the set of the first $n_4 - r$ vertices of $\overline{G_i}$, where $w_p = v_{p'} \in V_4'$;

Let $V_3 = \{v_{n_4-r+1}, v_{n_4-r+2}, \cdots, v_{n_4+n_3-r}\}$ be the remaining n_3 vertices of $\overline{G_i}$;

Let $W_i' = \{w_{p+1}, w_{p+2}, \cdots, w_{p+r}\}$ be the set of first r vertices in W_i;

Let $V_4 = V_4' \cup W_i'$;

{V_3 and V_4 induce connected graphs having n_3 and n_4 vertices, respectively.}

Let $G_{12} = G_i - W_i'$;

Find a bipartition V_1, V_2 of the biconnected graph G_{12} such that $u_1 \in V_1$, $u_2 \in V_2$, $|V_1| = n_1$, $|V_2| = n_2$, and both V_1 and V_2 induce connected subgraphs;

Return V_1, V_2, V_3, V_4 as a 4-partition of G.

Subcase 2(c): $p' < n_4 < p' + r$.

{In this subcase, the first $n_4 - p'$ and the last $r - (n_4 - p')$ vertices in W_i are added to $\overline{G_i}$ as the deficient r vertices.}

Let $W_{i4}' = \{w_{p+1}, w_{p+2}, \cdots, w_{p+n_4-p'}\}$ be the set of the first $n_4 - p'$ vertices in W_i;

Let $W_{i3}' = \{w_{q-1}, w_{q-2}, \cdots, w_{q-(r-n_4+p')}\}$ be the set of the last $r - (n_4 - p')$ vertices in W_i;

{Since $|W_{i4}'| + |W_{i3}'| = r \leq |W_i|$, $W_{i4}' \cap W_{i3}' = \phi$ and $|W_{i4}' \cup W_{i3}'| = r$};

Let $V_4 = \{v_1, v_2, \cdots, v_{p'}\} \cup W_{i4}'$;

Let $V_3 = \{v_{p'+1}, v_{p'+2}, \cdots, v_{n_4+n_3-r}\} \cup W_{i3}'$;

{ $|V_4| = n_4$, $|V_3| = n_3$, $w_p = v_{p'} \in V_4$, $w_q \in V_3$, and hence both V_4 and V_3 induce connected graphs.}

Let $G_{12} = G_i - W_{i4}' \cup W_{i3}'$;

Find a bipartition V_1, V_2 of the biconnected graph G_{12} such that $u_1 \in V_1$, $u_2 \in V_2$, $|V_1| = n_1$, $|V_2| = n_2$, and both V_1 and V_2 induce connected subgraphs;

Return V_1, V_2, V_3, V_4 as a 4-partition of G.

□

Clearly the running time of the above algorithm is $O(n)$. Thus we have the following theorem.

Theorem 5. *A 4-partition of any 4-connected plane graph G can be found in linear time if the four vertices u_1, u_2, u_3, u_4 are located on the same face of G.*

One can easily derive the following fact from Lemma 3 or directly from the canonical 4-ordering by Kant and He [K94, KH94].

Fact 6. *For any given internally triangulated 4-connected plane graph $G = (V, E)$, two distinct edges (u_1, u_2) and (u_3, u_4) on $C(G)$, and two numbers n_1, n_2 such that $n_1 + n_2 = n$ and $n_1, n_2 \geq 3$, there exists a partition V_1, V_2 of V such that $u_1, u_2 \in V_1$, $u_3, u_4 \in V_2$, $|V_1| = n_1, |V_2| = n_2$, and both V_1 and V_2 induce biconnected subgraphs of G.*

Proof. By Lemma 3, G has a 4-canonical decomposition $\Pi = W_1, W_2, \cdots, W_l$. Since G is internally triangulated, each W_i, $i = 2, 3, \cdots, l-1$, is not an outer chain of G_i or $\overline{G_{i-1}}$ and hence each W_i consists of exactly one vertex on both $C(G_i)$ and $C(\overline{G_{i-1}})$. Thus all W_i's except W_1 and W_l are singleton sets, each of W_1 and W_l contains exactly three vertices, and hence $l = n - 4$. For j, $1 < j < l$, the vertex in W_j has four or more neighbors, two of which are in $W_1 \cup W_2 \cup \cdots \cup W_{j-1}$ and other two of which are in $W_{j+1} \cup W_{j+2} \cdots \cup W_l$. Thus, for j, $1 < j < l$, both $V_1 = W_1 \cup W_2 \cup \cdots \cup W_j$ and $V_2 = V - (W_1 \cup W_2 \cup \cdots \cup W_j)$ induce biconnected graphs. Hence, it suffices to choose $j = n_1 - 2$. $\qquad\square$

4 Conclusion

In this paper we give a linear-time algorithm to find a 4-partition of a 4-connected plane graph G in the case four vertices u_1, u_2, u_3, u_4 are located on the same face of G. It is remained as future work to find efficient algorithms for finding a k-partition of a k-connected (not always planar) graph for $k \geq 4$.

References

[DF85]   M.E. Dyer and A.M. Frieze, *On the complexity of partitioning graphs into connected subgraphs*, Discrete Applied Mathematics, 10 (1985) 139-153.

[E79]   S. Even, *Graph Algorithms*, Computer Science Press, Potomac (1979).

[G78]   E. Győri, *On division of connected subgraphs*, Proc. 5th Hungarian Combinational Coll., (1978) 485-494.

[G96]   E. Győri, *Private communication*, March 21, 1996.

[JSN94] L. Jou, H. Suzuki and T. Nishizeki, *A linear algorithm for finding a non-separating ear decomposition of triconnected planar graphs*, Tech. Rep. of Information Processing Society of Japan, AL40-3 (1994).

[K94]   G. Kant, *A more compact visibility representation*, Proc. of the 19th International Workshop on Graph Theoretic Concepts in Computer Science (WG'93), LNCS 790 (1994) 411-424.

[KH94]   G. Kant and X. He, *Two algorithms for finding rectangular duals of planar graphs*, Proc. of the 19th International Workshop on Graph Theoretic Concepts in Computer Science (WG'93), LNCS 790 (1994) 396-410.

[L77]   L. Lovász, *A homology theory for spanning trees of a graph*, Acta Math. Acad. Sci. Hunger, 30 (1977) 241-251.

[MM94]   J. Ma and S. H. Ma, *An $O(k^2 n^2)$ algorithm to find a k-partition in a k-connected graph*, J. of Computer Sci. & Technol., 9, 1 (1994) 86-91.

[STN90] H. Suzuki, N. Takahashi and T. Nishizeki, *A linear algorithm for bipartition of biconnected graphs*, Information Processing Letters 33, 5 (1990) 227-232.

[STNMU90] H. Suzuki, N. Takahashi, T. Nishizeki, H. Miyano and S. Ueno, *An algorithm for tripartitioning 3-connected graphs*, Journal of Information Processing Society of Japan 31, 5 (1990) 584-592.

[WK94]   K. Wada and K. Kawaguchi, *Efficient algorithms for triconnected graphs and 3-edge-connected graphs*, Proc. of the 19th International Workshop on Graph Theoretic Concepts in Computer Science (WG'93), LNCS 790 (1994) 132-143.

[WTK95] K. Wada, A. Takaki and K. Kawaguchi, *Efficient algorithms for a mixed k-partition problem of graphs without specifying bases*, Proc. of the 20th International Workshop on Graph Theoretic Concepts in Computer Science (WG'94), LNCS 903 (1995) 319-330.

Graphs Drawn with Few Crossings per Edge

János Pach[*] and Géza Tóth[**]

Courant Institute, New York University
251 Mercer Street, New York, NY 10012

Abstract. We show that if a graph of v vertices can be drawn in the plane so that every edge crosses at most $k > 0$ others, then its number of edges cannot exceed $4.108\sqrt{k}v$. For $k \leq 4$, we establish a better bound, $(k+3)(v-2)$, which is tight for $k = 1$ and 2. We apply these estimates to improve a result of Ajtai et al. and Leighton, providing a general lower bound for the crossing number of a graph in terms of its number of vertices and edges.

1 Introduction

Given a simple graph G, let $v(G)$ and $e(G)$ denote its number of vertices and edges, respectively. We say that G is *drawn* in the plane if its vertices are represented by distinct points of the plane and its edges are represented by Jordan arcs connecting the corresponding point pairs but not passing through any other vertex. Throughout this paper, we only consider *drawings* with the property that any two arcs have at most one point in common. This is either a common endpoint or a common interior point where the two arcs properly cross each other. We will not make any notational distinction between vertices of G and the corresponding points in the plane, or between edges of G and the corresponding Jordan arcs.

We address the following question. What is the maximum number of edges that a simple graph of v vertices can have if it can be drawn in the plane so that every edge crosses at most k others? For $k = 0$, i.e. for *planar* graphs, the answer is $3v - 6$. Our first theorem generalizes this result to $k \leq 4$. The case $k = 1$ has been discovered independently by Bernd Gärtner, Torsten Thiele, and Günter Ziegler (personal communication).

Theorem 1. *Let G be a simple graph drawn in the plane so that every edge is crossed by at most k others. If $0 \leq k \leq 4$, then we have*

$$e(G) \leq (k+3)(v(G) - 2).$$

For $k = 0, 1, 2$, the above bound cannot be improved (see Remark 2.3 at the end of the next section.)

The *crossing number* $\mathrm{cr}(G)$ of a graph G is the minimum number of crossing pairs of edges, over all drawings of G in the plane.

---

[*] Supported by NSF grant CCR-94-24398 and PSC-CUNY Research Award 667339
[**] Supported by OTKA-T-020914 and OTKA-14220.

Ajtai et al. [AC82] and, independently, Leighton [L83] obtained a general lower bound for the crossing number of a graph, which found many applications in combinatorial geometry and in VLSI design (see [PA95], [PS96], [S95]). Our next result, whose proof is based on Theorem 1, improves the bound of Ajtai et al. by roughly a factor of 2.

Theorem 2. *The crossing number of any simple graph G satisfies*

$$\operatorname{cr}(G) \geq \frac{1}{33.75} \frac{e^3(G)}{v^2(G)} - 0.9v(G) > 0.029 \frac{e^3(G)}{v^2(G)} - 0.9v(G).$$

Theorem 3. *Let G be a simple graph drawn in the plane so that every edge is crossed by at most k others, for some $k \geq 1$. Then we have*

$$e(G) \leq \sqrt{16.875k}v(G) \approx 4.108\sqrt{k}v(G).$$

Theorems 2 and 3 do not remain true if we replace the constants 0.029 and 4.108 by 0.06 and 1.92, respectively (see Remarks 3.2 and 3.3).

In the last section, we use the ideas of Székely [S95] to deduce some consequences of Theorem 2.

2 Proof of Theorem 1

First we need a lemma for *multigraphs*, i.e., for graphs that may have multiple edges. In a *drawing* of a multigraph, any two non-disjoint edges either share only endpoints or have precisely one point in common, at which they properly cross.

Let M be a multigraph drawn in the plane so that every edge crosses at most k other edges. Let M' be a sub-multigraph of M with the largest number of edges such that in the drawing of M' (inherited from the drawing of M), no two edges cross each other. We say that M' is a *maximal plane sub-multigraph* of M, and its faces will be denoted by $\Phi_1, \Phi_2, \ldots, \Phi_m$. Let $|\Phi_i|$ denote the number of edges of M' along the boundary of Φ_i, where every edge whose both sides belong to the interior of Φ_i is counted *twice*. It follows from the maximality of M' that every edge e of $M - M'$ crosses at least one edge of M'. The closed portion between an endpoint of e and the nearest crossing of e with an edge of M' is called a *half-edge*. Thus, every edge of $M - M'$ contains two half-edges. Every half-edge lies in a face Φ and intersects at most $k - 1$ other half-edges and an edge of Φ_i (not counting the incidences at the vertices of M). Let $h(\Phi_i)$ denote the number of half-edges in Φ_i.

Lemma 2.1. *Let $0 \leq k \leq 4$ and let M be a multigraph drawn in the plane so that every edge crosses at most k others. Let M' be a maximal plane sub-multigraph of M, and let Φ denote a face with $|\Phi| = s \geq 3$ sides in M', whose bondary is connected.*

Then the number of half-edges in Φ satisfies

$$h(\Phi) \leq (s - 2)(k + 1) - 1.$$

Proof: We proceed by induction on s. First, let $s = 3$ and denote the vertices of Φ by A, B, and C. Let a, b, and c denote the number of half-edges in Φ emanating from A, B, and C, respectively. We have to show that $a + b + c \leq k$. For $k = 0$, there is nothing to prove. We check the cases $k = 1, 2, 3, 4$, separately.

• **$k = 1$:** If $a = b = c = 0$, we are done. Assume without loss of generality that $a \geq 1$. But then $a = 1$, because all half-edges in Φ emanating from A intersect the edge BC. Since any half-edge in Φ emanating from B or C would create another intersection on the half-edge starting from A, we obtain $b = c = 0$. Hence, $a + b + c = 1$.

• **$k = 2$:** Suppose without loss of generality that $a \geq 1$. Clearly, $a \leq 2$. If $a = 1$, the unique half-edge in Φ emanating from A intersects all half-edges coming from B and C. So $1 + b + c = a + b + c \leq 2$. If $a = 2$, any half-edge from B would intersect both half-edges emanating from A and the edge AC, which is impossible. Hence, $b = 0$. Similarly, $c = 0$, and $a + b + c = 2$.

• **$k = 3$:** Just like before, we can exclude all cases when $a + b + c > 3$, except for the case $a = b = 2$ and $c = 0$. Now let e_1 and e_2 denote the edges containing the two half-edges in Φ emanating from A. Both of them intersect the two half-edges starting from B and the edge BC. So they cannot cross any other edge. Removing BC from M' and adding e_1 and e_2, we would obtain a larger plane sub-multigraph of M, contradicting the maximality of M'.

• **$k = 4$:** We can again exclude all cases when $a + b + c > 4$, with the exception of the case $a = 2$, $b = 3$, $c = 0$. As before, let e_1 and e_2 denote the edges containing the two half-edges in Φ emanating from A. Now both e_1 and e_2 are intersected by the three half-edges emanating from B and by the edge BC. Hence, there are no other edges crossing them, and the number of edges of M' can be increased by replacing BC with e_1 and e_2. Contradiction.

Now let $s > 3$, and suppose that the lemma has already been proved for faces with fewer than s sides. Let $A_1, A_2, \ldots, A_s$ denote the sequence of vertices of Φ, listed in clockwise order. In this sequence, the same vertex may occur several times (as many times as it is visited during a full clockwise tour around the boundary of Φ). For simplicity, let $A_0 = A_s$ and $A_{s+1} = A_1$.

We call an open arc *empty* if it does not intersect any half-edge in Φ.

Case 1. Assume that there is a half-edge $e = A_i E$ in Φ, where E is an interior point of the side $A_j A_{j+1}$, and either
(i) the arc $A_j E \subseteq A_j A_{j+1}$ is empty and $i \neq j - 1$, or
(ii) the arc $E A_{j+1} \subseteq A_j A_{j+1}$ is empty and $i \neq j + 2$.

By symmetry, we can suppose that e satisfies (i). Let M^* denote the (multigraph) drawing obtained from M by replacing the edge of M containing e with a new edge $e' = A_i A_j$ lying in Φ and running very close to e and the arc $A_j E$.

Since $A_j E$ was empty in M, e' crosses exactly the same half-edges as e. Thus, M^* also satisfies the condition that each of its edges crosses at most k other edges. Clearly, $M' \cup e'$ is a maximal plane sub-multigraph of M^*. In M^*, e' divides Φ into two faces Φ' and Φ'' having s' and s'' sides, respectively, where $3 \leq s', s'' < s$, $s' + s'' = s + 2$.

Each half-edge in Φ, except e, corresponds to a half-edge in Φ' or Φ''. By the induction hypothesis,

$$h(\Phi) = h(\Phi')+h(\Phi'')+1 \leq (s'-2)(k+1)+(s''-2)(k+1)-2+1 = (s-2)(k+1)-1.$$

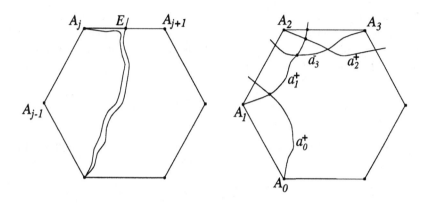

Figure 1.

Case 2. Assume that there is no half-edge in Φ that satisfies the condition of Case 1.

Then, for any non-empty side A_iA_{i+1} of Φ, the half-edge a_{i-1}^+ (resp. a_{i+2}^-) whose intersection with A_iA_{i+1} is closest to A_i (resp. closest to A_{i+1}) starts at the vertex A_{i-1} (resp. A_{i+2}).

Since any side of Φ intersects at most k half-edges, if there are two empty sides of Φ, then $h(\Phi) \leq (s-2)k \leq (s-2)(k+1) - 1$. So we can suppose that Φ has at most one empty side. Since $s > 3$, there are three consecutive non-empty sides, say, A_1A_2, A_2A_3, and A_3A_4.

Then a_1^+ must intersect a_0^+, a_2^+, a_3^-, and the side A_2A_3. Similarly, a_4^- must intersect a_5^-, a_3^+, a_2^+, and the side A_2A_3. This is clearly impossible if $k = 1, 2$ or 3.

For $k = 4$, let e_1 and e_2 denote the edges of M containing a_1^+ and a_4^-, respectively. Both of these edges cross three half-edges and the side A_2A_3 of Φ, so neither of them can cross any further edges. Removing the edge A_2A_3 from M' and adding e_1 and e_2, we would obtain a plane sub-multigraph of M, whose number of edges is larger than the number of edges of M'. This contradicts the maximality of M', completing the proof of Lemma 2.1. $\square$

For any face Φ with at least 3 sides, let $t(\Phi)$ denote the number of triangles in a triangulation of Φ.

Lemma 2.2 *Let Φ be any face of M' with $|\Phi| \geq 3$ sides. Then the number of half-edges of Φ satisfies*

$$h(\Phi) \leq t(\Phi)k + |\Phi| - 3.$$

Proof: If the boundary of Φ is connected, then $t(\Phi) = |\Phi| - 2$. Hence, by Lemma 2.1, $h(\Phi) \leq (|\Phi| - 2)(k+1) - 1 = t(\Phi)k + |\Phi| - 3$.

For any face Φ, the number of half-edges in Φ is at most $|\Phi|k$, because every side of Φ intersects at most k half-edges. If the boundary of Φ is not connected, then $t(\Phi) \geq |\Phi|$. Therefore, in this case, we have $h(\Phi) \leq |\Phi|k \leq t(\Phi)k + |\Phi| - 3$. □

Now we are ready to prove Theorem 1. Suppose that a simple graph G is drawn in the plane with at most k crossings on each edge. Let G' be a maximal plane subgraph of G. Denote the faces of G' by $\Phi_1, \Phi_2, \ldots \Phi_m$. To triangulate Φ_i, we need at least $|\Phi_i| - 3$ edges. Therefore,

$$e(G') \leq 3v - 6 - \sum_{i=1}^{m}(|\Phi_i| - 3).$$

Every edge of $G - G'$ gives rise to two half-edges. So, Lemma 2.2 yields that

$$e(G - G') \leq \frac{1}{2}\sum_{i=1}^{m}\left(t(\Phi_i)k + |\Phi_i| - 3\right).$$

Summing up the last two inequalities and noticing that the total number of triangles satisfies $\sum_i t(\Phi_i) = 2v(G) - 4$, we obtain

$$e(G) \leq 3v(G) - 6 + \frac{1}{2}\sum_{i=1}^{m}\left(t(\Phi_i)k - (|\Phi_i| - 3)\right) \leq 3v(G) - 6 + (v(G) - 2)k$$

$$= (k + 3)(v(G) - 2),$$

as desired. □ □

Remark 2.3. For $k = 0$, the bound $e \leq 3v - 6$ is tight for any triangulation.

For $k = 1$, the obtained bound $e \leq 4v - 8$ is also tight, provided that $v \geq 12$. First we show that for every $v \geq 12$ there is a planar graph with v vertices, all of whose faces are quadrilaterals, and no two faces share more than one edge. Indeed, Figure 2 illustrates that such graphs exist for $v = 8, 13, 14, 15$. Once we have an example G with v vertices, we can construct another one with $v + 4$ vertices, by replacing some face of G by the 8-point example in Figure 2. Notice that if we add both diagonals of each face (including the external face), then we obtain a graph with $4v - 8$ edges such that along each edge there is at most one crossing.

Figure 2.

For $k = 2$, the bound $e \leq 5v - 10$ is sharp for all $v \geq 50$ such that $v \equiv 2 \pmod 3$. For simplicity, we only exhibit a construction for $v \equiv 5 \pmod{15}$. First we construct a planar graph whose faces are pentagons and two faces have at most one edge in common. For $v = 20$, such a graph is shown in Figure 3. The number of vertices of such an example G can be increased by 15, by replacing some face of G with the graph depicted in Figure 3. Notice that if we add all 5 diagonals of each face to G, then we obtain a graph with v vertices and $5v - 10$ edges, in which every edge crosses at most 2 others.

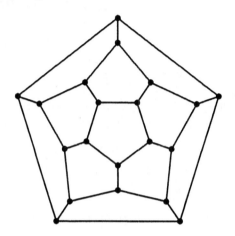

Figure 3.

3   Proofs of Theorems 2 and 3

In this section, we slightly improve the best known general lower bound on the crossing number of a graph, due to Ajtai et al. and Leighton. Our proof is based on the following consequence of Theorem 1.

Corollary 3.1. *The crossing number of any simple graph G with at least 3 vertices satisfies*

$$\mathrm{cr}(G) \geq 5e(G) - 25v(G) + 50.$$

Proof: If $e(G) \leq 3v(G) - 6$, then the statement is void. Assume $e(G) > 3v(G) - 6$.

It follows from Theorem 1 that if $e(G) > (k + 3)(v(G) - 2)$, then G has an edge crossed by at least $k + 1$ other edges ($k \leq 4$). Deleting such an edge, we obtain by induction on $e(G)$ that the number of crossings is at least

$$\sum_{k=0}^{4} [e(G) - (k + 3)(v(G) - 2)] = 5e(G) - 25v(G) + 50.$$

□

Proof of Theorem 2. Let G be a simple graph drawn in the plane with $\mathrm{cr}(G)$ crossings, and suppose that $e(G) \geq 7.5v(G)$.

Construct a *random* subgraph $G' \subseteq G$ by selecting each vertex of G independently with probability $p = 7.5v(G)/e(G) \leq 1$, and letting G' be the subgraph induced by the selected vertices. The expected number of vertices of G', $E[v(G')] = pv(G)$. Similarly, $E[e(G')] = p^2 e(G)$. The expected number of crossings in the drawing of G' inherited from G is $p^4 \mathrm{cr}(G)$, and the expected value of the crossing number of G' is even smaller.

By Corollary 3.1, $\mathrm{cr}(G') \geq 5e(G') - 25v(G')$ for every G'. Taking expectations,

$$p^4 \mathrm{cr}(G) \geq E[\mathrm{cr}(G')] \geq 5E[e(G')] - 25E[v(G')] = 5p^2 e(G) - 25pv(G).$$

This implies that

$$\mathrm{cr}(G) \geq \frac{1}{33.75} \frac{e^3(G)}{v^2(G)}, \tag{1}$$

whenever $e(G) \geq 7.5v(G)$. With a slight modification, we can extend this bound to the range $e(G) < 7.5v(G)$. In fact, using Corollary 3.1, it is easy to check that

$$\mathrm{cr}(G) \geq \frac{1}{33.75} \frac{e^3(G)}{v^2(G)} - 0.9v(G)$$

is valid for every simple graph G. $\square$

Proof of Theorem 3. For $k \leq 4$, the result is weaker than the bounds given in Theorem 1.

So let $k > 4$, and consider a drawing of G such that every edge crosses at most k others. Let C denote the number of crossings in this drawing. If $e(G) < 7.5v(G)$, then there is nothing to prove. If $e(G) \geq 7.5v(G)$, then using the stronger form (1) of Theorem 2, we obtain

$$\frac{1}{33.75} \frac{e^3(G)}{v^2(G)} \leq \mathrm{cr}(G) \leq C \leq \frac{e(G)k}{2}.$$

Consequently,

$$e(G) \leq \sqrt{16.875}\sqrt{k}v(G). \ \square$$

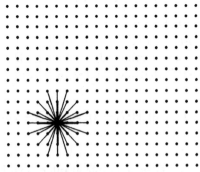

Figure 4.

Remark 3.2. The bound given in Theorem 2 is asymptotically tight, apart from the values of the constants. The best construction we found is the following.

Let $v \ll e \ll v^2$. Let $V(G)$ be a set of v points arranged in a slightly perturbed unit square grid of size $\sqrt{v} \times \sqrt{v}$, so that the points are in general position. Let $d = \sqrt{2e/\pi v}$, so that $d^2\pi = e/2v$.

Connect two points by a straight-line segment if and only if their distance is at most d. Then $v(G) = v$, $e(G) \approx vd^2\pi/2 = e$.

To count the number of crossings in G, let $S(a) = \{(x,y)|1 \le x, y \le a\}$, and for any two segments (u_1, u_2), (v_1, v_2), $(u_1, u_2) \otimes (v_1, v_2)$ means that the two segments cross each other. Then the number of crossings in G is

$$\frac{1}{8} | \{(u_1, u_2, v_1, v_2) \in [V(G)]^4 \mid \|u_1 - u_2\|, \|v_1 - v_2\| \le d, \ (u_1, u_2) \otimes (v_1, v_2)\} |$$

$$\approx \frac{1}{8} \int_{u_1 \in S(\sqrt{v})} \int_{\substack{u_2 \in S(\sqrt{v}) \\ \|u_1 - u_2\| \le d}} \int_{v_1 \in S(\sqrt{v})} \int_{\substack{v_2 \in S(\sqrt{v}) \\ \|v_1 - v_2\| \le d \\ (u_1, u_2) \otimes (v_1, v_2)}} 1 \ dv_2 dv_1 du_2 du_1$$

$$= \frac{2\pi}{27} vd^6 (1 + o(1)).$$

Thus,

$$\mathrm{cr}(G) \le \frac{2\pi}{27} vd^6 (1 + o(1)) \approx \frac{16}{27\pi} \frac{e^3}{v^2} \approx .06 \frac{e^3}{v^2}.$$

J. Spencer [S96] showed that the limit

$$c = \lim \frac{v^2}{e^3} \min_{\substack{|V(G)| = v \\ |E(G)| = e}} \mathrm{cr}(G)$$

exists, as $v \to \infty$ and $v \ll e \ll v^2$. By our results, $.06 \ge c \ge .029$.

Remark 3.3. To see that the bound obtained in Theorem 3 is also asymptotically tight. Consider the same construction as in Remark 3.2, but now set $d = \sqrt[4]{3k/2}(1-o(1))$, as k tends to infinity. Just like above, it can be shown that no edge crosses more than k other edges. The number of edges

$$e(G) = v(G) \frac{d^2\pi}{2}(1 - o(1)) = v(G)\sqrt{k}\frac{\sqrt{3}\pi}{\sqrt{8}}.$$

Thus, we have

$$1.92(1 - o(1))\sqrt{k}v(G) < \max e(G) < 4.108\sqrt{k}v(G),$$

where the maximum is taken over all simple graphs with $v(G)$ vertices that have a drawing with at most k crossings per edge.

4 Two further applications

Using Székely's method (see [S95]) and Theorem 2, we can improve the constant in the Szemerédi–Trotter theorem [ST83], [CE90].

Theorem 4.1. *Given m points and n lines in the Euclidean plane, the number of incidences between them is at most $2.57n^{2/3}m^{2/3} + 2.07(m + n)$.*

Proof. Define a graph G drawn in the plane such that the vertex set of G is the given set of m points, and join two points with an edge drawn as a straight line segment if the two points are consecutive along one of the lines. Let I denote the total number of incidences between the given m points and n lines. Then $v(G) = m$ and $e(G) = I - n$. Since every edge belongs to one of the n lines, $\mathrm{cr}(G) \leq \binom{n}{2}$. Applying Theorem 2 to G, we obtain that

$$\binom{n}{2} \geq \mathrm{cr}(G) \geq \frac{1}{33.75}\frac{(I - n)^3}{m^2} - 0.9m,$$

which implies, by Jensen's inequality, that

$$I \leq 2.57n^{2/3}m^{2/3} + 3.13m + n.$$

By symmetry we also have that

$$I \leq 2.57n^{2/3}m^{2/3} + m + 3.13n.$$

Taking the average of the last two inequalities, the result follows. □

Remark 4.2. As Erdős pointed out fifty years ago, the order of magnitude of the bound in Theorem 4.1 cannot be improved. To see this, one can take n points arranged in a unit square grid of size $\sqrt{n} \times \sqrt{n}$ and consider the m most "popular" lines.

More precisely, for any $1 > \varepsilon > 0$, take all lines which contain at least $\varepsilon\sqrt{n}$ of the points. Then, for the number of lines m we have

$$m \approx 4\sqrt{n}\sum_{r=1}^{1/\varepsilon}\sum_{\substack{s < r \\ (r,\, s) = 1}} r + s - 2rs\varepsilon = 6\sqrt{n}\sum_{r=1}^{1/\varepsilon} r\phi(r) - 4\sqrt{n}\varepsilon\sum_{r=1}^{1/\varepsilon} r^2\phi(r) \approx \frac{6\sqrt{n}}{\pi^2\varepsilon^3}.$$

Here $\phi(n)$ denotes Euler's function and we used the formula $\sum_{r=1}^{N} \phi(r) \approx 3N^2/\pi^2$ (see e. g. [HW54]). By similar calculations, for the number of incidences I we get

$$I \approx 4n\sum_{r=1}^{1/\varepsilon}\sum_{\substack{s < r \\ (r,\, s) = 1}} 1 - rs\varepsilon^2 = 4n\sum_{r=1}^{1/\varepsilon} \phi(r) - 2n\varepsilon^2\sum_{r=1}^{1/\varepsilon} r^2\phi(r) \approx \frac{3n}{\pi^2\varepsilon^2}.$$

Comparing the last two expressions, we obtain

$$I \approx cn^{2/3}m^{2/3} \quad \text{with} \quad c = \sqrt[3]{\frac{3}{4\pi^2}} \approx 0.42.$$

We can also generalize Theorem 2 for multigraphs with bounded edge-multiplicity, improving the constant in Székely's result [S95].

Theorem 4.3. *Let G be a multigraph with maximum edge-multiplicity m. Then*

$$\text{cr}(G) \geq \frac{1}{33.75} \frac{e^3(G)}{mv^2(G)} - 0.9m^2 v(G).$$

Proof. Define a random simple subgraph G' of G on the same vertex set. For each pair of vertices v_1, v_2 of G, let $e_1, e_2, \ldots e_k$ be the edges connecting them. With probability $1 - k/m$, G' will not contain any edge between v_1 and v_2. With probability k/m, G' contains precisely one such edge, and the probability that this edge is e_i is $1/m$ $(1 \leq i \leq k)$.

Applying Theorem 2 to G', the result follows. □

References

[AC82] M. Ajtai, V. Chvátal, M. Newborn, E. Szemerédi, Crossing-free subgraphs, *Ann. Discrete Mathematics* **12** (1982), 9–12.

[CE90] K. Clarkson, H. Edelsbrunner, L. Guibas, M. Sharir, E. Welzl, Combinatorial complexity bounds for arrangements of curves and surfaces, *Discrete and Computational Geometry* **5** (1990), 99-160.

[HW54] G. H. Hardy, E. M. Wright, *An Introduction to the Theory of Numbers*, University Press, Oxford, 1954.

[L83] T. Leighton, *Complexity Issues in VLSI*, Foundations of Computing Series, MIT Press, Cambridge, MA, 1983.

[PA95] J. Pach and P.K. Agarwal, *Combinatorial Geometry*, John Wiley, New York, 1995.

[PS96] J. Pach and M. Sharir, On the number of incidences between points and curves, *Combinatorics, Probability, and Computing*, submitted.

[S96] J. Spencer, personal communication.

[S95] L. Székely, Crossing numbers and hard Erdős problems in discrete geometry, *Combinatorics, Probability, and Computing*, to appear.

[ST83] E. Szemerédi and W. T. Trotter, Extremal problems in discrete geometry, *Combinatorica* **3** (1983), 381-392.

A Pairing Technique for Area-Efficient Orthogonal Drawings* (Extended Abstract)

Achilleas Papakostas   and   Ioannis G. Tollis

Dept. of Computer Science
The University of Texas at Dallas
Richardson, TX 75083-0688
email: papakost@utdallas.edu, tollis@utdallas.edu

Abstract. An orthogonal drawing of a graph is a drawing such that vertices are placed on grid points and edges are drawn as sequences of vertical and horizontal segments. In this paper we present linear time algorithms that produce orthogonal drawings of graphs with n nodes. If the maximum degree is four, then the drawing produced by our first algorithm needs area no greater than $0.76n^2$, and introduces no more than $2n + 2$ bends. Also, every edge of such a drawing has at most two bends. Our algorithm is based on forming and placing pairs of vertices of the graph. If the maximum degree is three, then the drawing produced by our second algorithm needs at most $\frac{1}{4}n^2$ area, and at most $\lfloor \frac{n}{2} + 2l + 1 \rfloor$ bends ($\lfloor \frac{n}{2} \rfloor + 3$ bends, if the graph is biconnected), where l is the number of biconnected components that are leaves in the block tree. For biconnected graphs, this algorithm produces optimal drawings with respect to the number of bends (within a constant of two), since there is a lower bound of $\frac{n}{2} + 1$ in the number of bends for orthogonal drawings of degree 3 graphs.

1 Introduction and Preliminaries

Research on algorithms for drawing graphs has received increasing attention recently. For a survey of graph drawing algorithms and other related results see the annotated bibliography of Di Battista, Eades, Tamassia and Tollis [4]. In this paper we focus on the problem of *orthogonal* drawings of graphs, that is drawings in which each edge of the graph is a polygonal chain consisting of horizontal and vertical segments. Orthogonal drawings are grid drawings, such that vertices and bends correspond to grid points (i.e., they have integer coordinates), and edges correspond to grid paths. A graph admits an orthogonal drawing if it has maximum degree 4. There are various aesthetic criteria for orthogonal drawings that look "nice"; minimizing the area, bends and crossings are some of these criteria. Our goal, here, is to minimize the area and to obtain drawings with a small number of bends.

* Research supported in part by NIST, Advanced Technology Program grant number 70NANB5H1162.

Several results have appeared in the literature regarding planar orthogonal drawings of graphs. In [17] and [19] it is shown that every biconnected planar graph of maximum degree 4 can be embedded in an $n \times n$ grid with $2n+4$ bends. If the graph is not biconnected then the total number of bends rises to $2.4n+2$. In all cases, no more than 4 bends per edge are required. The algorithms of [19] take linear time and produce drawings, such that at most one edge may have 4 bends. Kant [10] shows that if the graph is triconnected of maximum degree 4, then it can be drawn on an $n \times n$ grid with at most 3 bends per edge. The total number of bends is no more than $\lceil \frac{3}{2}n \rceil + 3$. Even and Granot [7] present an algorithm for obtaining an orthogonal drawing of a 4-planar graph with at most 3 bends per edge. If the embedding of a planar graph is fixed, then an orthogonal drawing with the minimum number of bends can be computed in $O(n^2 \log n)$ time [18]. If the planar embedding is not given, the problem is polynomially solvable for 3-planar graphs [6], and NP-hard for 4-planar graphs [9].

It is interesting to note that there is a lower bound of $2n - 2$ bends for biconnected planar graphs [20]. There are also examples of biconnected graphs [20], for which every bend-optimal planar drawing introduces a single edge with length $\Omega(n^2)$ having $\Omega(n)$ bends along it. Although these drawings achieve optimality in terms of the total number of bends, they are not aesthetically pleasing. This suggests that research in this area should concentrate on deriving orthogonal drawings of graphs with $O(1)$ bends per edge (usually 2 or 3) and $O(n)$ maximum edge length.

Upper and lower bounds have been proved in the case when the orthogonal drawing of a graph is not necessarily planar. Leighton [11] presented an infinite family of planar graphs which require area $\Omega(n \log n)$. Independently, Leiserson [12] and Valiant [21] showed that every planar graph of degree 3 or 4 has an orthogonal drawing with area $O(n \log^2 n)$. Valiant [21] showed that the orthogonal drawing of a general (nonplanar) graph of degree 3 or 4 requires area no more than $9n^2$, and described families of graphs that require area $\Omega(n^2)$. Valiant was not concerned about minimizing the total number of bends. In fact, an analysis of his construction shows that each edge can have up to 4 bends. Recently, Biedl [2] has presented a number of lower bounds on the area and number of bends for orthogonal drawings of different families of graphs of maximum degree 4.

Some algorithms for drawing general graphs of degree 4 in an orthogonal but not necessarily planar fashion have appeared in the literature. Schäffter [16] presented such an algorithm which constructs orthogonal drawings of graphs with at most two bends per edge. The area required was $2n \times 2n$. A better algorithm was presented in [1, 3], which draws the graph within an $n \times n$ grid with no more than 2 bends per edge. This algorithm introduced at most $2n + 2$ bends. In [14] we presented an algorithm that produces orthogonal drawings of graphs of maximum degree 4 in less than $n \times n$ area. The bounds depend on the number of balanced vertices (i.e., vertices that have 2 incoming and 2 outgoing edges) produced by the st-numbering. It turns out that for many graphs the number of balanced vertices is very low. So, although an upper bound of $0.81n^2$ can be shown for graphs with high enough number of balanced vertices, this is not a general upper bound. Also, the bounds cannot be extended easily to one-connected graphs. An extensive experimental work appeared in [5] where four general purpose orthogonal graph drawing algorithms where implemented and compared.

In this paper we present an algorithm that produces an orthogonal drawing of an n-vertex graph of maximum degree 4 that needs at most $0.76n^2$ area and at most $2n + 2$ bends. The number of bends that appear on each edge is no more than two. Our algorithm is based on pairing the vertices of the graph, and placing the pairs in the plane. If the maximum degree is 3, then we present another algorithm which produces an orthogonal drawing which needs area at most $\frac{1}{4}n^2$ and $\lfloor \frac{n}{2} + 2l + 1 \rfloor$ bends ($\lfloor \frac{n}{2} \rfloor + 3$ bends, if the graph is biconnected). In this drawing, no more than one bend appears on each edge except for one edge, which may have at most two bends. Note that l is the number of the biconnected components of G that are leaves in G's block tree. A preliminary version of this algorithm appeared in [14]. Due to space limitations, we omit some details and proofs, but they can be found in [13].

We use n (resp. m) to denote the number of vertices (resp. edges) of a graph. An st-ordering is an ordering $v_1, v_2, \cdots, v_n$ of the vertices such that every v_j ($2 \leq j \leq n - 1$) has at least one predecessor and at least one successor, that is neighbors v_i, v_k with $i < j < k$. It is known that:

Theorem 1. *[8] Let G be biconnected and s, t be two vertices of G. Then there exists an st-ordering such that s is the first and t is the last vertex, and it can be computed in $O(m)$ time.*

An incremental linear time algorithm for producing orthogonal drawings of bi-connected graphs of maximum degree 4 is presented in [1, 3]. The main result is given by the following theorem:

Theorem 2. *[1, 3] Let G be a biconnected graph of maximum degree 4. Then there exists a linear time algorithm which embeds G on an $n \times n$ grid with at most $2n + 2$ bends. Each edge is bent at most twice.*

2 Using Pairing to Improve the Area Bounds

In this section we present an algorithm for obtaining orthogonal drawings of general (nonplanar) biconnected graphs of maximum degree 4. Our algorithm achieves better results in terms of area than any previously known algorithm for orthogonal drawings. Let G be a general (nonplanar) biconnected graph of maximum degree 4. We obtain an st-numbering for G, with s as the source and t as the sink. We call a vertex with a incoming edges and b outgoing edges an a-b vertex ($1 \leq a, b \leq 4$). For example, a vertex with one incoming edge and two outgoing, is a 1-2 vertex.

After the st-numbering is complete, we scan the graph G looking for those 1-1 vertices whose outgoing edge enters a 1-2 or a 1-3 vertex, if there are any. In order to simplify the description of our algorithm, we "absorb" these vertices into a single edge until no 1-2 or 1-3 vertex has a 1-1 vertex as its unique immediate predecessor. Notice that no double edge is introduced when these 1-1 vertices are (temporarily) removed from G. Also notice that if a vertex was an a-b vertex in G and it was not removed as a result of the above procedure, it will still be an a-b vertex in the *reduced* graph. Let us use the notation G' for the reduced graph, and n' for its number of vertices. We then modify the st-numbering of G' so that there are no gaps in the st-number sequence assigned to the vertices of G' as a result of the removal of some 1-1 vertices from G.

The main idea of the algorithm is to create pairs of vertices of G' so that every 1-2, 1-3 and 2-2 vertex is a member of exactly one such pair. We distinguish between two different kinds of pairs:

- *row pairs.* The two vertices of such a pair share the same row in the final drawing of G. In other words, we save a row.
- *column pairs.* The two vertices of such a pair are placed in such a way so that a column is *reused* in the final drawing of G. A column is reused when at least two different edges use it. In other words, we save a column.

Algorithm Form_pairs considers the vertices of G' in reverse order of the st-numbering starting with the vertex which is right before the sink t. If a vertex already belongs to a pair, the vertex is called *assigned*, otherwise it is called *unassigned*. The next unassigned vertex we consider is always either a 1-2, 1-3, or 2-2 vertex and we pair it with some other lower numbered vertex in G'. The assignment of the 1-2, 1-3, and 2-2 vertices of G' to pairs is called *pairing* of G'. The vertex of a pair with the lowest st-number of the two is called the *first* vertex of the pair, and the other is called the *second* vertex of the pair.

In the rest of this paper, when we talk about a predecessor of a vertex in G or G' with respect to the st-numbering, we mean the immediate predecessor of this vertex.

Algorithm: Form_pairs
Input: A reduced graph G' of maximum degree 4 along with an st-numbering
Output: A pairing of G'

1. $i := n' - 1$;
2. While $i > 2$ do
   (a) Consider the next vertex v_i according to a decreasing order of the st-numbering.
   (b) If v_i is a 1-1, 2-1, or 3-1 vertex then
      - $j := 0$;
      - goto (e);
   (c) If v_i is a 1-2 or 1-3 vertex then
      - form a pair containing vertex v_{i-1} and v_i
      - $j := 1$;
      - goto (e);
   (d) If v_i is a 2-2 vertex then
      i. $j := 1$;

ii. While v_{i-j} is a 1-1, 2-1 or 3-1 vertex and v_{i-j} is not a predecessor of v_i do

    $- j := j + 1;$

iii. End_While

iv. form a pair containing vertex v_{i-j} and v_i

(e) $i := i - j - 1;$

3. End_While

This algorithm assigns every 2-2, 1-2 and 1-3 vertex v_i where $3 \leq i \leq n' - 1$ to one pair. Vertex v_2 (which is a 1-2 or 1-3 vertex) might or might not be paired with another vertex and this depends on the graph and the st-numbering. Every 1-2 or 1-3 vertex v_i is always paired with vertex v_{i-1}, when v_i is the next vertex that Algorithm Form_pairs considers in line (a). If the next vertex v_i to be considered is a 2-2 vertex, then Algorithm Form_pairs looks for the highest $j < i$ so that vertex v_j is one of the following types, and pairs v_i with v_j. Vertex v_j may be: a 2-2 vertex, or a 1-2 vertex, or a 1-3 vertex, or a 1-1 vertex which is also a predecessor of v_i, or a 2-1 vertex which is also a predecessor of v_i, or a 3-1 vertex which is also a predecessor of v_i.

Since v_2 and v_3 are not 3-1 vertices, the algorithm will pair all 2-2, 1-2 and 1-3 vertices except possibly for v_2. Let us assume that vertex v_i is paired with vertex v_j ($j < i$) as a result of Algorithm Form_pairs. v_j might be a predecessor of v_i, or the two vertices might not have a predecessor-successor relationship. If the latter is the case, they are called *independent*. For different types of pairs, we draw the vertices of the pair in a different fashion. The pair $< v_i, v_j >$ can be one of the following types:

1. If v_i is a 2-2 vertex, we distinguish three cases for v_j:

   (a) v_j is a 2-2 vertex; we have a column pair. If v_j is v_i's predecessor, then a column which v_j's placement closes can be reused as shown in Fig. 1a. If v_i and v_j are independent, then we can always reuse one column (see Fig. 1b). Notice that in order for this column reuse to be possible, sometimes we might have to place v_i in a row that has lower y-coordinate than v_j's row. This placement is possible since the two vertices are independent, and can be depicted by Fig. 1b if we just swap the names of the two vertices shown.

   (b) v_j is a 1-1, 2-1 or 3-1 vertex and v_j is a predecessor of v_i. If v_j is 2-1 or 3-1 a column can be reused (i.e., we have a column pair), as shown in Fig. 1c. If v_j is 1-1, then the two vertices can share the same row (i.e., we have a row pair) as shown in Fig. 1d.

   (c) v_j is a 1-2 or 1-3 vertex. If v_j is a predecessor of v_i, we have a row pair, since v_i can be placed in the same row as v_j, at the intersection point between the edge coming from v_i's other predecessor and v_j's row, as shown in Fig 2a. Notice that v_i's other predecessor, say v_k, has to be such that $k < j$. If v_i and v_j are independent, we have a column pair since v_j can be placed in the row right above v_i's row and thus reuse one column (see Fig. 2b).

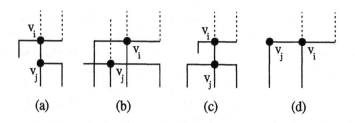

Fig. 1. v_i is a 2-2 vertex; a column is reused when: (a) v_j is 2-2 and v_i's predecessor, (b) v_j is independent from v_i, and (c) v_j is 2-1 or 3-1 and v_i's predecessor, (d) a row is shared when v_j is 1-1.

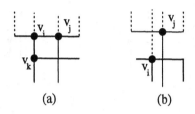

Fig. 2. (a) v_i and v_j share the same row, (b) v_j is placed in a row above v_i and a column is saved.

2. If v_i is a 1-2 or 1-3 vertex, it always pairs with vertex v_{i-1}. We distinguish four cases for v_{i-1}:

   (a) v_{i-1} is a 2-2, 2-1 or 3-1 vertex; we have a column pair. v_i is placed in a row above v_{i-1}'s row and a column is reused as described in Cases 1(a) and 1(b).

   (b) v_{i-1} is a 1-2 or 1-3 vertex and vertices v_i and v_{i-1} are independent. We have a row pair, and vertices v_i and v_{i-1} are placed in the same row as shown in Fig. 3a.

   (c) v_{i-1} is a 1-2 or 1-3 vertex and v_i's predecessor. If both of the following conditions hold (if not, see 2(d) below):

     – v_i is connected later to another vertex, say v_j, which is 1-1, 1-2 or 1-3. Or v_i is connected later to a 2-2 vertex v_j which is the second vertex of the pair of type 1(c) shown in Fig. 2a.

     – edge (v_{i-1}, v_i) has not absorbed any 1-1 vertices from the original graph G

   then we have a row pair and v_i and v_{i-1} are placed in the same row, as shown in Fig. 3b. In this case, we have to ensure that edge e (see Fig. 3b) will connect to v_j later in the drawing. This way, every edge is bent at most twice. Also notice that the total number of bends for both v_i and v_{i-1} is the same as if these two vertices were placed in two different rows.

(d) v_{i-1} is a 1-2 or 1-3 vertex and v_i's predecessor. If at least one of the two conditions described in pair type 2(c) does not hold, then v_i and v_{i-1} are placed in two different rows as shown in Fig. 3c. In this case, we make sure that the vertex (vertices) which is (are) supposed to be placed in the next row after v_i is (are) placed in the same row as v_i. Notice that this is possible since v_i together with the columns that it opens are placed entirely outside the boundaries of the current drawing. If new columns have to be opened in v_i's row, we open them in the middle of the current drawing, and to the immediate right or left of the vertex (vertices) that open(s) the new columns.

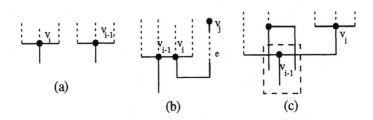

(a)

(b)

(c)

Fig. 3. (a) v_i and v_{i-1} are independent and share the same row, (b) v_i and v_{i-1} share the same row and edge e will connect to an appropriate vertex inserted later, (c) v_i shares the same row with the vertex that is next to be placed.

Let us assume that we have a pair of type 2(d) described in Fig. 3c, and the next pair to be placed is a column pair, say $< v_{i+1}, v_{i+2} >$. Let us also assume that there is an edge between v_i and v_{i+1}. Then, because v_i and v_{i+1} are placed in the same row, the savings for column pair $< v_{i+1}, v_{i+2} >$ really comes from reusing one of the columns that might have been used for the edge (v_i, v_{i+1}) had it not been drawn horizontally.

If it should happen that v_i is paired with a vertex v_j for which $j < i-1$, then this means that all vertices v_k where $j < k \leq i-1$ are 1-1, 2-1 or 3-1 vertices and they are not predecessors of v_i. Since this is the case, v_i can be placed in a row which is below the rows of vertices v_k, without affecting the pairing or the drawing algorithm.

Recall that in order to simplify our description, we absorbed some degree 2 vertices. After the drawing of G' is complete, we need to restore the degree 2 vertices of the original graph G which were absorbed at the beginning of the procedure. These vertices are placed primarily on bends, or on grid points (i.e., points of integer coordinates that do not have a crossing). In the extreme case where this is not possible, we introduce new rows as needed. Notice that all the other vertices of the drawing maintain their positions, that is the rows and columns in which they are placed.

We are now ready to present our algorithm formally.

Algorithm: 4_ORTHOGONAL
Input: A biconnected graph G of maximum degree 4.
Output: An orthogonal drawing of G.

1. Compute an st-numbering of G.
2. Produce a reduced graph G' and modify the st-numbering so that there are no gaps in the st-sequence.
3. Run Form_pairs on the reduced graph G'.
4. Place vertices v_1 and v_2 in the same row, if v_2 does not belong to a pair in which it shares a row with another vertex (see Fig. 4a). If v_1 and/or v_2 have degree less than 4, then the placement of v_1 and v_2 might require one or two rows. Figure 4b shows the case where v_1 had degree 3 and v_2 has degree 4. Notice that in this case there is only one bend along edge (v_1, v_2). If v_2 is assigned to a pair, we place v_1 as shown in Fig. 4c (if v_1 has degree 4). Vertex v_2 will be placed when its pair is considered.
5. REPEAT
   (a) Consider the next vertex v_i according to the st-numbering of G'.
   (b) If v_i has already been placed, then go to Step 6.
   (c) If vertex v_i is unassigned, then place v_i in a new row. Connect v_i with each vertex v_j $(j < i)$ such that (v_j, v_i) is a directed edge of G'. Add as many uncompleted edges as required, depending on v_i's outdegree.
   (d) If vertex v_i is assigned to a pair, then place v_i together with the other vertex in the same pair following the placement rules described above for the specific type of pair.
6. UNTIL the only remaining vertex is $v_{n'}$.
7. Insert $v_{n'}$ in a new row. If $v_{n'}$ is of degree 4, then there is an incoming edge that enters $v_{n'}$ from the top and bends twice. This edge is chosen to be the one that connects to $v_{n'-1}$.
8. Restore the degree 2 vertices of G that were absorbed in Step 2, as described above.
9. End.

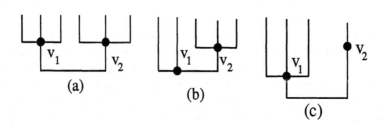

(a)

(b)

(c)

Fig. 4. v_1 and v_2 placed by Algorithm 4_ORTHOGONAL: (a) v_1 and v_2 can be placed in the same row, (b) the placement of v_1 and v_2 when v_1 has degree 3, and (c) the placement of v_1 when v_2 is assigned to a pair.

Notice that the difference between the given graph G and the reduced graph G' is that some degree 2 vertices of G have been (temporarily) removed, as discussed earlier in this section. In Step 3 of Algorithm 4_ORTHOGONAL we apply the pairing process to the vertices of G'. The pairing of the vertices of G' "transfers" to G since we can always assert that if two vertices participate in some pair in G', the same two vertices participate in the same pair in G.

Lemma 3. *Let us assume that there is a total of p_1 column pairs, p_2 1-1 vertices, p_3 2-1 vertices in G, and that $k_1 = p_1 + p_2 + \frac{p_3}{2}$. Then the width of the drawing of G will be at most $n + 1 - k_1$.*

Lemma 4. *Let us assume that there are k_2 row pairs of vertices in G. Then the height of the drawing will be $n + 1 - k_2$, when $k_2 \geq 1$, or n when $k_2 = 0$.*

Theorem 5. *Algorithm 4_ORTHOGONAL constructs an orthogonal drawing of an n-vertex degree 4 biconnected graph with area no more than $0.76n^2$. The total number of bends of the drawing is at most $2n + 4$, and no edge has more than two bends. The algorithm runs in $O(n)$ time.*

An important feature of the algorithm is that when we place vertices in the same row we save the potential crossings that would have been introduced if the vertices were placed in separate rows. Reusing columns also contributes to saving crossings significantly. Notice that in practice, the area which is typically required by the orthogonal drawing of a graph produced by 4_ORTHOGONAL is better than what the above theorem claims, for two reasons: the first is that, the total number of 1-1, 1-2, 1-3, 2-2 and 2-1 type vertices is usually much larger than $\frac{n}{2}$. The second reason is that when we form pairs, we typically expect some vertices of type 3-1 to participate as the second vertex in some column pairs of type 1(b) or 2(a). If this is the case, the total number of pairs increases.

In fact, experimentation of the performance of Algorithm 4_ORTHOGONAL was conducted on about 15 dense graphs of maximum degree 4. The size of these graphs varied from as small as 13 nodes to as big as 150 nodes. Each graph had a very small number (no more than six) of degree 3 vertices, whereas all the rest were vertices of degree 4. The set included both biconnected and non-biconnected graphs.

The first observation that we made is that the shape and area of the produced drawings depended heavily on the specific st-numbering that was employed. St-numberings that resembled DFS produced drawings in which the height was larger than the width, but the area was no more than $0.65n^2$. St-numberings that resembled BFS produced more squarish drawings, with shorter edges, but the column reuse was not as good as in the first case. As a result the area was a bit larger, but never more than $0.72n^2$. Finally, in all the cases, the number of bends was no larger than $2n$. We also noticed that for the larger graphs, the number of bends was significantly lower than $2n$.

Figure 5a shows a regular degree 4 graph with 13 vertices and Fig. 5b shows the orthogonal drawing which our algorithm produces for it. Notice that vertices 1 and 2, 3 and 4, and 6 and 7 are placed in the same row. Also, the pairs $< 10, 8 >$, $< 7, 6 >$, $< 5, 4 >$ and $< 3, 2 >$ save one column each. A total of 4 bends are saved in the rows where vertices 3 and 4, and 6 and 7 are placed. Our drawing has height 11 and width 10.

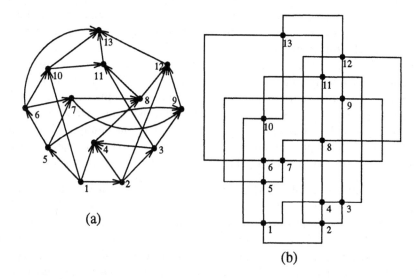

Fig. 5. (a) st-numbering of an example graph, (b) orthogonal drawing of the graph in Fig. 5a produced by Algorithm 4_ORTHOGONAL.

3 The One-Connected Case

In this section we extend our results to the case where the given graph is one-connected. The technique is based on breaking the graph into its biconnected components, which is also suggested in [3, 19]. In [3] it is shown that each component has to have no more than $2n - 1$ bends and $(n - 1) \times (n - 1)$ area in order for the final drawing (after merging the components together) to have at most $2n + 2$ bends and $n \times n$ area. Our technique is similar to the one in [3]. In order to apply this technique for one-connected graphs, we have to guarantee that, when Algorithm 4_ORTHOGONAL draws any biconnected graph of maximum degree 4, neither the height nor the width of the drawing can be larger than n. Moreover, the number of bends has to be at most $2n + 2$. These bounds guarantee that the height or width of any biconnected component is at most $n - 1$, and the number of bends is at most $2n - 1$.

Notice that it is possible for Algorithm 4_ORTHOGONAL to produce a drawing with $2n + 4$ bends or width $n + 1$. Here we describe a technique which forces a row pair in G's pairing which additionally saves one column and two bends. We scan the vertices starting from v_2 and following the st-numbering, until we find the first vertex v that is not a 1-3 vertex. We distinguish two cases for v:

1. v is a 2-2 vertex. Let u be v's highest numbered predecessor. Vertices v and u do not belong to the same pair, since in that case we would have a row pair of the kind we are trying to form. We break the pair that v is in, and pair v with u. This new pair is a pair of type 1(c) (the case shown in Fig. 2a). Notice that in order to form the new pair, we have to break the pair that u was previously assigned to. However, doing so will not increase the width or the total number of bends.

2. v is a 3-1 vertex. We break any pair that v might have been assigned to with another higher numbered vertex. We check v's highest numbered 1-3 predecessor, say u (notice that, since v has three incoming edges, u cannot be vertex v_2). Vertex u has to be assigned to a pair with another 1-3 vertex, and u is the second vertex in that pair, as a result of the running of Algorithm Form_pairs. This pair becomes a pair of type 2(d) (we disregard whatever kind this pair was before). In this case, v will be placed in u's row.

From the above it follows that, if we have a regular degree 4 biconnected graph G, we can always place one 2-2 or 3-1 vertex (vertex v in the above description) in the same row as its highest numbered predecessor (vertex u in the above description). We accomplish this by forming a new row pair of type either 1(c) (u is the first vertex and v is the second vertex of the pair), or 2(d) (u is the second vertex of the pair, v becomes an unassigned vertex and is placed in u's row). In either case, the resulting drawing has at most $2n+2$ bends; also, the width of the drawing is at most n. Notice that although we might have to break two existing pairs, the new row pair that we form saves one row, one column and two bends. Therefore we have:

Theorem 6. *Algorithm 4_ORTHOGONAL constructs an orthogonal drawing of an n-vertex maximum degree 4 biconnected graph with area no more than $0.76n^2$. The total number of bends of the drawing is at most $2n+2$, and no edge has more than 2 bends. The algorithm runs in linear time.*

We saw that the height of a drawing under Algorithm 4_ORTHOGONAL is $n+1-k_2$ (see Lemma 4), if vertices v_2 and v_3 form a row pair (in this case, $k_2 \geq 1$). For the rest of this section, we assume the same pairing process as described in the previous section, except that we will not count the row pair formed by v_2 and v_3 in k_2. In this way, we have a new k_2' for a biconnected graph, which is defined in the same way as in Lemma 4 and has the adjustment we just described. It follows that the height of the drawing of a biconnected graph is no more than $n-k_2'$ ($k_2' \geq 0$). From the discussion above, it holds that:

Lemma 7. *If we are given a biconnected graph G of maximum degree 4, then Algorithm 4_ORTHOGONAL can produce an orthogonal drawing of G whose size is as follows: the width of the drawing is at most $\min(n, n+1-k_1)$ (see also Lemma 3), and the height of the drawing is at most $n - k_2'$ ($k_2' \geq 0$).*

Algorithm 4_ORTHOGONAL can be extended to the case of one-connected graphs of maximum degree 4. Let us assume that we have such a graph G.

We split G into its biconnected components, produce G's block tree and apply Algorithm 4_ORTHOGONAL (see Theorem 6) on each biconnected component separately. Then we put the components together to form the final drawing. We use an inductive approach for producing the drawing of G, which is similar to [3]. The base case is always a biconnected graph. In the induction step we consider a subtree of G's block tree and we split the subtree into a biconnected component G_0 (i.e., the root of the subtree) and (not necessarily biconnected) subgraphs $G_1, G_2, \cdots G_{q+s}$. Each one of the G_i's is already drawn according to the induction hypothesis. The drawing of the subtree of G's block tree then reduces to drawing G_0 and merging the G_i's at their appropriate places.

We now give a description of our technique. There are q subgraphs of the G_i's that are connected to G_0 through a bridge, while the rest s of the G_i's are connected to G_0 via a cutvertex which is shared by both G_0 and a G_i. Clearly, G_0 has a total of $q+s$ cutvertices. For those subgraphs G_i which are connected to G_0 through a bridge, the edge representing the bridge is not part of the drawing of the subgraph. This edge will be added when the drawing of the corresponding subgraph is merged with the drawing of G_0. First, we insert subgraph G_1. The cutvertex shared between G_0 and G_1, or the one of the two cutvertices of the bridge connecting G_0 and G_1 is in the top row of the drawing of G_1. We continue from there with the drawing of G_0.

G_0 is drawn as a biconnected graph, making sure that none of its $q + s$ cutvertices is the "final" vertex (i.e., vertex drawn in the top row, locally, in the subgraph considered). G_0 has some other vertex (say w) as the final vertex of the drawing. Vertex w will be used to attach the produced drawing of G_0 and the G_i's (i.e., the drawing of the subtree of G's block tree) to the drawing of another biconnected component considered at a later induction level. For this reason, vertex w must be a cutvertex for graph G. The important thing to note here is that we regard G_0 and G_1 as "one" subgraph. This means that the vertices of G_0 are placed (forming row or column pairs) as if we were continuing the drawing of G_1. In other words, G_1's "final" vertex as well as the three first vertices of G_0 may form pairs which count towards reducing the number of rows and columns of the subgraph consisting of the union of G_0 and G_1 (see Fig. 6a).

When the time comes to place a cutvertex v that is also shared by some G_i, we do the following: We rotate the drawing of G_i (G_i was already drawn with v as the final vertex) and place it in such a way so that a total of at least three rows and/or columns of the current drawing of previous components are reused. Note that the rows and/or columns that are reused as a result of G_i's insertion are different from the row and the column that vertex v is placed in. If additional rows and/or columns need to open up to accommodate G_i's drawing, we do so now. In order for this row/column reuse to be possible, we make sure that G_1 is selected so that it has at least four rows or at least four columns. If no such subgraph exists, we select G_1 to be the next largest graph. In Fig. 6b we can see how four consecutive subgraphs can be inserted so that three columns are reused at each time. The drawing of each one of these subgraphs has four columns, while the shared cutvertices are in the corners of the drawings of subgraphs G_j and

G_{j+1}, and in one of the middle columns of the drawings of subgraphs G_{j+2} and G_{j+3} (see Fig. 6b). Notice that when subgraph G_{j+3} is inserted, edge e might acquire an extra bend which is saved later, since edge e' does not have a bend.

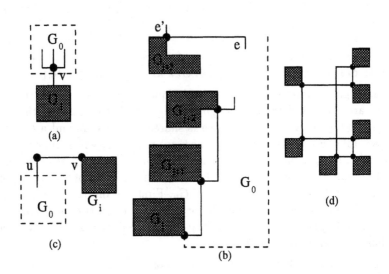

Fig. 6. G_0 continues the drawing of G_1, (b) examples of subgraph rotation, placement and column reuse, (c) reusing row(s) when G_i is connected to G_0 through a bridge, (d) drawing of a graph whose G_i's are small size graphs (each G_i here is a triangle).

We follow a similar procedure if subgraph G_i is connected to G_0 through a bridge. However, when such a subgraph G_i is inserted to the drawing, we tend to reuse mostly rows. For example, let b be the bridge that connects G_i with G_0, and let u be the cutvertex of G_0 that was incident to bridge b. We place b along u's row (right or left), and then we place the rest of the drawing of G_i (rotated appropriately). Note that G_i was already drawn with the vertex incident to bridge b as the final vertex. Both this vertex and u are placed in the same row (thus we have a row pair), so we only need to reuse two more rows. This example is illustrated in Fig. 6c, where G_i (whose drawing has three rows) is reusing three rows.

From the discussion above (especially Lemma 7 and Theorem 6) it follows that each biconnected component of G can fit within a grid of size $(n-1) \times (n-1)$ with $2n-1$ bends, when the component is drawn using our algorithm and with some vertex of maximum degree three as final vertex. Using a simple induction (similar to the one in [3]) we can show that the drawing of one-connected graph G requires area at most $n \times n$ and no more than $2n+2$ bends. We will not provide further arguments on the number of bends. However, since each component of G has area at most $0.76n^2$ (because of pairing within the component), our target is to show that the same upper bound holds for one-connected graph G. It turns out that in order for the induction to yield this upper bound, we have to reuse at least three rows and/or columns, as discussed above. This is formalized in the following invariant, which we maintain during the drawing process:

Invariant: Let G be a graph of maximum degree 4. We draw G using our approach and with some vertex of maximum degree 3 as the final vertex. Let K_1 and K_2 be all the savings we have along the x and the y axis respectively (i.e., columns and rows saved). It holds that: $K_1 + K_2 \geq \lceil \frac{n-4}{4} \rceil - 2$, $K_1 \geq 0$ and $K_2 \geq 0$. Additionally, the drawing of G requires at most $(n-1-K_1) \times (n-1-K_2)$ area.

Notice that when the G_i's are attached to G_0's appropriate places, they are typically rotated by $\frac{\pi}{2}$, π, or $\frac{3\pi}{2}$. The rotation of the G_i's does not affect the sum of $K_1 + K_2$, since a row saving will become a column saving and vice versa as a result of the rotation. We have described a technique that reuses at least three rows and/or columns, and that was sufficient in order to prove our bound for $K_1 + K_2$. However, a clever implementation of our algorithm will reuse as many rows and columns as possible, when subgraphs G_i merge with component G_0, yielding better bounds for $K_1 + K_2$ and thus drawings with even smaller area.

Theorem 8. *There is a linear time algorithm which constructs an orthogonal drawing of an n-vertex graph of maximum degree 4 with area no more than $0.76n^2$. The total number of bends is at most $2n + 2$, and no edge has more than 2 bends.*

Consider the example of Fig. 6d: each subgraph G_i is a simple triangle (we have a total of seven triangles), G_0 is a simple cycle, and the total number of vertices is $n = 21$. Drawing this graph using our technique, returns the drawing of Fig. 6d whose area is just $0.08n^2$; this result is mostly due to the very small size of these subgraphs, and not to the pairing process.

4 Drawing Maximum Degree 3 Graphs

In [14] we presented an algorithm for obtaining an orthogonal drawing of any biconnected graph (not necessarily planar) of maximum degree 3. This algorithm matched the upper bounds on the area and number of bends for planar graphs. In this paper, we enrich this algorithm with some more features so that it can draw maximum degree 3 graphs which are only one-connected. The new algorithm is called 3_ORTHOGONAL (explained in detail in [13]), and achieves bounds that are identical to the bounds in [14] in terms of area. Also, for biconnected graphs of maximum degree 3, the bounds of Algorithm 3_ORTHOGONAL in [13] are identical to the bounds in [14] in terms of the number of bends. More precisely, we have that:

Theorem 9. *[14] There exists a linear time algorithm for constructing an orthogonal drawing of a biconnected graph of maximum degree 3 with at most $\lfloor \frac{n}{2} \rfloor + 3$ bends. There is at most one edge that bends twice, while every other* *edge in the drawing has at most one bend. Moreover, the drawing requires at most $\frac{1}{4}n^2$ area.*

For graphs of maximum degree 3, Storer [17] has shown a family of graphs that require at least $\frac{n}{2} + 1$ bends under any orthogonal drawing. From this it follows that Algorithm 3_ORTHOGONAL for biconnected graphs [14, 13] is optimal in terms of the number of bends, within an additive constant of two.

Let us assume that we have a general n-vertex graph G of maximum degree 3, which is connected but not biconnected. In this case, we break G into its biconnected components and derive its block tree. Notice that for a graph with maximum degree 3, this can be done in linear time. Then, we obtain a global numbering for the vertices of G from 1 to n. This numbering implies an orientation of the edges of G in the same way as an st-numbering implies an orientation of the edges of a biconnected graph. In [13], we give a complete and detailed description of how Algorithm 3_ORTHOGONAL works. Due to space limitations, here we give only the main result.

Theorem 10. *There exists a linear time algorithm for constructing an orthogonal drawing of a general graph of maximum degree 3 with at most $\lfloor \frac{n}{2} + 2l + 1 \rfloor$ bends, where l is the number of the biconnected components of G that are leaves in G's block tree. Also, there is at most one edge that bends twice, while every other edge in the drawing has at most one bend. Moreover, the drawing requires at most $\frac{1}{4}n^2$ area.*

5   Open Problems

- Is it possible to obtain better area upper bounds than $0.76n^2$, for graphs of maximum degree 4?
- Develop algorithms to produce orthogonal drawings of graphs of arbitrary degree, and establish bounds on the area and number of bends. An issue that has to be studied here is the way that the vertices are represented.
- Develop interactive graph drawing algorithms. Some work for graphs of maximum degree four appears in [15].

References

1. Therese Biedl, "Embedding Nonplanar Graphs in the Rectangular Grid", *Rutcor Research Report 27-93*, 1993.
2. Therese Biedl, "New Lower Bounds for Orthogonal Graph Drawings", *Proc. of GD '95, Lecture Notes in Comp. Sci.*, 1027, Springer-Verlag, 1995, pp. 28-39.
3. T. Biedl and G. Kant, "A Better Heuristic for Orthogonal Graph Drawings", *Technical Report, Utrecht Univ., Dept. of Comp. Sci.*, UU-CS-1995-04. Prelim. version appeared in *Proc. 2nd Ann. European Symposium on Algorithms (ESA '94), Lecture Notes in Computer Science*, vol. 855, pp. 24-35, Springer-Verlag, 1994.

4. G. DiBattista, P. Eades, R. Tamassia and I. Tollis, "Algorithms for Drawing Graphs: An Annotated Bibliography", *Computational Geometry: Theory and Applications*, vol. 4, no 5, 1994, pp. 235-282. Also available via anonymous ftp from ftp.cs.brown.edu, gdbiblio.tex.Z and gdbiblio.ps.Z in /pub/papers/compgeo.

5. G. DiBattista, A. Garg, G. Liotta, R. Tamassia, E. Tassinari and F. Vargiu, "An Experimental Comparison of Three Graph Drawing Algorithms", *Proc. of ACM Symp. on Computational Geometry*, pp. 306-315, 1995. The version of the paper with the four algorithms can be obtained from http://www.cs.brown/people/rt.

6. G. DiBattista, G. Liotta and F. Vargiu, "Spirality of orthogonal representations and optimal drawings of series-parallel graphs and 3-planar graphs," *Proc. Workshop on Algorithms and Data Structures, Lecture Notes in Computer Science 709*, Springer-Verlag, 1993, pp. 151-162.

7. S. Even and G. Granot, "Rectilinear Planar Drawings with Few Bends in Each Edge", *Tech. Report 797, Comp. Science Dept.*, Technion, Israel Inst. of Tech., 1994.

8. S. Even and R.E. Tarjan, "Computing an st-numbering", *Theor. Comp. Sci. 2* (1976), pp. 339-344.

9. A. Garg and R. Tamassia, "On the Computational Complexity of Upward and Rectilinear Planarity Testing", *Proc. DIMACS Workshop GD '94, Lecture Notes in Comp. Sci. 894*, Springer-Verlag, 1994, pp. 286-297.

10. Goos Kant, "Drawing planar graphs using the lmc-ordering", *Proc. 33th Ann. IEEE Symp. on Found. of Comp. Science*, 1992, pp. 101-110.

11. F. T. Leighton, "New lower bound techniques for VLSI", *Proc. 22nd Ann. IEEE Symp. on Found. of Comp. Science*, 1981, pp. 1-12.

12. Charles E. Leiserson, "Area-Efficient Graph Layouts (for VLSI)", *Proc. 21st Ann. IEEE Symp. on Found. of Comp. Science*, 1980, pp. 270-281.

13. A. Papakostas and I. G. Tollis, "Algorithms for Area-Efficient Orthogonal Drawings", *Tech. Report UTDCS-06-95*, The University of Texas at Dallas, 1995. Also available on the WWW at http://wwwpub.utdallas.edu/~tollis.

14. A. Papakostas and I. G. Tollis, "Improved Algorithms and Bounds for Orthogonal Drawings", *Proc. DIMACS Workshop GD '94, Lecture Notes in Comp. Sci. 894*, Springer-Verlag, 1994, pp. 40-51.

15. A. Papakostas and I. G. Tollis, "Issues in Interactive Orthogonal Graph Drawing", *Proc. of GD '95, Lecture Notes in Comp. Sci. 1027*, Springer-Verlag, 1995, pp. 419-430.

16. Markus Schäffter, "Drawing Graphs on Rectangular Grids", *Discr. Appl. Math. 63* (1995), pp. 75-89.

17. J. Storer, "On minimal node-cost planar embeddings", *Networks 14* (1984), pp. 181-212.

18. R. Tamassia, "On embedding a graph in the grid with the minimum number of bends", *SIAM J. Comput. 16* (1987), pp. 421-444.

19. R. Tamassia and I. Tollis, "Planar Grid Embeddings in Linear Time", *IEEE Trans. on Circuits and Systems CAS-36* (1989), pp. 1230-1234.

20. R. Tamassia, I. Tollis and J. Vitter, "Lower Bounds for Planar Orthogonal Drawings of Graphs", *Information Processing Letters 39* (1991), pp. 35-40.

21. L. Valiant, "Universality Considerations in VLSI Circuits", *IEEE Trans. on Comp.*, vol. C-30, no 2, (1981), pp. 135-140.

Experimental and Theoretical Results in Interactive Orthogonal Graph Drawing*

Achilleas Papakostas, Janet M. Six and Ioannis G. Tollis

Department of Computer Science
The University of Texas at Dallas
Richardson, TX 75083-0688
email: {papakost,janet,tollis}@utdallas.edu

Abstract. *Interactive Graph Drawing* allows the user to dynamically interact with a drawing as the design progresses while preserving the user's mental map. This paper presents a theoretical analysis of Relative-Coordinates and an extensive experimental study comparing the performance of two interactive orthogonal graph drawing scenaria: *No-Change*, and *Relative-Coordinates*. Our theoretical analysis found that the Relative-Coordinates scenario builds a drawing that has no more than $3n - 1$ bends, while the area of the drawing is never larger than $2.25n^2$. Also, no edge has more than 3 bends at any time during the drawing process. To conduct the experiments, we used a large set of test data consisting of 11,491 graphs (ranging from 6 to 100 nodes) and compared the behavior of the above two scenaria with respect to various aesthetic properties (e.g., area, bends, crossings, edge length, etc) of the corresponding drawings. The Relative-Coordinates scenario was a winner over No-Change under any aesthetic measure considered in our experiments. Moreover, the practical behavior of the two scenaria was considerably better than the established theoretical bounds, in most cases.

1 Introduction and Preliminaries

Graph drawing addresses the problem of automatically generating geometric representations of abstract graphs or networks. For a survey of graph drawing algorithms and other related results see the annotated bibliography of Di Battista, Eades, Tamassia and Tollis [4]. An *orthogonal* drawing is a drawing in which vertices are represented by points on integer coordinates and edges are represented by polygonal chains consisting of horizontal and vertical line segments. Various algorithms have been introduced to produce orthogonal drawings of planar [2, 6, 10, 20, 22] or general [1, 2, 16, 19] graphs of maximum degree four, and maximum degree three [10, 16, 17]. All these algorithms run in linear time, except for the algorithm in [20]. For drawings of general graphs, the required area can be as little as $0.76n^2$ [16], the total number of bends is no more than $2n + 2$ [2, 16], and at most two bends can be on the same edge [2, 16].

* Research supported in part by NIST, Advanced Technology Program grant number 70NANB5H1162.

Upper and lower area bounds have been proved for orthogonal drawings of general graphs. Leighton [11] presented an infinite family of planar graphs which require area $\Omega(n \log n)$. Independently, Leiserson [12] and Valiant [25] showed that every planar graph of degree three or four has an orthogonal drawing with area $O(n \log^2 n)$. Valiant [25] showed that the orthogonal drawing of a general (non-planar) graph of degree three or four requires area no more than $9n^2$, and described families of graphs that require area $\Omega(n^2)$.

Many graph drawing algorithms have been implemented, and a number of results on their practical behavior have appeared in literature. Himsolt [8] presented his experimental findings when he compared the performance of twelve graph drawing algorithms (most of which were specialized for trees, graph grammars, Petri nets and planar graphs) in the GraphEd environment [9]. An extensive experimental work appeared in [5] where four general-purpose orthogonal graph drawing algorithms [2, 16, 21, 22] were implemented and compared with respect to their performance on various aesthetic properties including crossings, bends, area, running time etc. Their results have a great statistical value mainly because they are based on a very large data set including 11,582 graphs. The sizes of these graphs range from 10 to 100 vertices and are taken from "real-life" software engineering and database applications.

In most graph drawing algorithms a graph is given as an input and a drawing of this graph is produced. If an insertion (or deletion) is performed on the graph, then we have a "new" graph. Running the drawing algorithm again will result in a new drawing, which might be vastly different from the previous one. Obviously this is a waste of human resources to continually re-analyze the entire drawing and also of computational resources to re-compute the entire layout after each modification. Therefore it is important to efficiently produce a series of drawings which evolve with the structure while preserving the user's *mental map* [15]. The first systematic approach to dynamic graph drawing appeared in [3]: the target was to perform queries and updates on an implicit representation of the drawing, while maintaining its planarity. The insertion of a single edge however, could cause a planar graph to drastically change embedding, or even become non planar. An incremental approach to orthogonal graph drawing was presented in [14], where the focus was on routing edges efficiently without disturbing existing vertices or edges.

In [18] we discussed various features of interactive graph drawing systems, we introduced four scenaria for interactive graph drawing and presented a theoretical analysis of the performance of the No-Change scenario. All four scenaria were based on the assumption that the underlying drawing was orthogonal and the maximum degree of any vertex was four at the end of an update operation. The basic property of the No-Change scenario is that an update operation (i.e., a vertex insertion) does not alter the coordinates of any vertex or bend within the current drawing, since any vertex insertion or edge routing takes place around it. The analysis of the No-Change scenario of [18] is based on the assumption that the drawn graph is connected at all times.

Theorem 1. *[18] The "No-Change scenario" produces drawings with the following properties:*

1. *every insertion operation takes constant time,*
2. *every edge has at most three bends,*
3. *the total number of bends at any time t is at most $2.66n(t) + 2$, where $n(t)$ is the number of vertices of the drawing at time t,*
4. *the area of the drawing at any time t is no more than $(n(t) + n_4(t))^2 \leq 1.77n(t)^2$, where $n_4(t)$ is the number of vertices of local degree four which have been inserted up to time t, and*
5. *the upper bounds for both area and bends are tight.*

In this paper, we first analyze the performance of the Relative-Coordinates scenario by using linear programming to prove upper bounds on the area and the number of bends. More specifically, we show that an interactive graph drawing system under the Relative-Coordinates scenario builds a drawing that has no more than $3n(t) - 1$ bends, while requiring at most $2.25n(t)^2$ area. Moreover, no edge has more than 3 bends at any time during the drawing process. Then we compare the performance of the No-Change and Relative-Coordinates scenaria on a set of 11,491 "real-life" maximum degree four graphs, which were taken from the database of graphs used in [5]. Our experiments compare the quality of the drawings produced by the two scenaria, based on the following aesthetic measures: area, number of bends, number of crossings, aspect ratio, average edge length and maximum edge length. We also include results on the average number of new rows, columns and bends that are introduced in any drawing for different types of vertex insertions. Our experiments revealed:

- The practical behavior of the two scenaria is much better than their established theoretical bounds, in most cases.
- The Relative-Coordinates scenario exhibits better performance than No-Change under any aesthetic measure considered.

2 Analysis of the Relative-Coordinates Scenario

In this scenario, every time a new vertex is about to be inserted into the current drawing, the system makes a decision about the coordinates of the vertex and the routing of its incident edges. New rows and columns may be inserted anywhere in the current drawing in order for this routing to be feasible. The coordinates of the new vertex (say v) as well as the locations of the new rows and/or columns will depend on the following:

- v's degree (at the time of insertion).
- How many of v's adjacent vertices allow the insertion of a new incident edge towards the same direction (i.e., up, down, right, or left of the vertex).
- How many of v's adjacent vertices allow a new incident edge towards opposite directions.
- Whether or not the required routing of edges can be done utilizing segments of existing rows or columns that are free (not covered by an edge).
- Our optimization criteria.

When we use the Relative-Coordinates scenario in an interactive system, we can start from an existing drawing of a graph or from scratch, that is from an empty graph. In either case, we assume that the insertion of any vertex/edge under this scenario will not increase the number of connected components of the current graph. The only exception to this is when a single vertex is inserted into a currently empty graph. Any other vertex inserted during an update step will be connected to at least one other vertex of the current drawing. Let us assume that v is the next vertex to be inserted in the current graph during an update step. The number of vertices in the current graph that v is connected to, is called the *local degree* of v, and is denoted by *local_degree*(v).

From the discussion above it follows that we only consider the cases where an inserted vertex has local degree one, two, three or four, except for the first vertex inserted in an empty graph. If the user wishes to insert a new vertex that has local degree zero, then this vertex is placed in a temporary location and it will be inserted automatically in the future, when some newer vertices increase its (local) degree. Assume that vertex v is about to be inserted into the current graph. For each one of the vertices of the current drawing that is adjacent to v, the system checks the possible directions around these vertices that new edges may be inserted or routed. The target is to minimize the number of new rows or columns that have to open up in the current drawing, as well as the number of bends that appear along the routed edges.

There are many different cases because there are many possible combinations. In the example shown in Fig. 1a vertices u_1 and u_2 have a free edge (i.e., grid edge not covered by a graph edge) up and to the right respectively. In this case no new rows/columns are needed for the insertion of vertex v and no new bends are introduced. On the other hand however, in the example shown in Fig. 1b all four vertices u_1, u_2, u_3 and u_4 have pairwise opposite direction free edges. The insertion of new vertex v requires the insertion of three new rows and three new columns in the current drawing. Additionally, eight bends are introduced. Vertices u_1, u_2, u_3 and u_4 have general positions in Fig. 1b, and we can see that edge (v, u_4) has four bends. We can avoid the 4-bend edge, if we insert vertex v in the way shown in Fig. 1c. The total number of new rows, columns and bends is still the same, but the maximum number of bends per edge is now three. For a more even distribution of the bends of the edges adjacent to vertex v, we may choose to insert it in the way shown in Fig. 1d, where every edge has exactly two bends (three new rows and three new columns are still required). Notice, though, that the approach described in Fig. 1d for inserting vertex v, is not always possible (e.g., we cannot have this kind of insertion if vertices u_1, u_2, u_3 and u_4 are in the same row or column). At this point it is important to note that any one of the solutions presented in Fig. 1 (b, c and d) are acceptable: all add the same number of rows, columns and *total* number of bends. The characteristic of placing a maximum of three bends per edge is attainable, but ultimately is up to the implementor. The *total* number of bends added per insertion will always remain the same under this scenario.

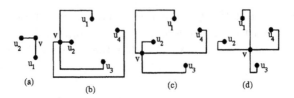

Fig. 1. Insertion of v: (a) no new row or column is required, (b),(c) and (d) three new rows and three new columns are required, with a maximum of 4 bends per edge in (b), 3 bends per edge in (c), and 2 bends per edge in (d).

Let v be the next vertex to be inserted. There are many cases, if one is interested in an exhaustive analysis. However, it is relatively easy to come up with all the cases for each insertion. Here, we distinguish the following main cases for vertex v:

1. v has local degree one. If u is the vertex of the current drawing that is adjacent to v, we draw an edge between u and v. Edge (u, v) uses a direction (up, right, bottom, or left) that is not taken by some other edge incident to u. This is depicted in Fig. 2a, and this insertion requires at most either a new row or a new column. No new bend is inserted.

2. v has local degree two. In the best case, the insertion requires no new rows, columns or bends as shown in Fig. 1a. In the worst case, though, two new rows and one new column, or one new row and two new columns (see Fig. 2b), and three new bends might be inserted.

3. v has local degree three. In the worst case, the insertion requires a total of four new rows and columns, and five new bends. In Fig. 2c we show an example of such an insertion that requires one new row, three new columns and five new bends.

4. v has local degree four. The worst case requires a total of six new rows and columns, and eight new bends. We have already discussed an example, which is depicted in Fig. 1c. In Fig. 2d we show another case, where two new rows, four new columns and eight new bends are introduced. Note that no more than four new rows or columns may be introduced when v has local degree four.

As discussed in the previous section, single edge insertions can be handled using techniques from global routing [13] or the technique of [14]. The easiest way to handle deletions is to delete vertices/edges from the data structures without changing the coordinates of the rest of the drawing. Occasionally, or on demand, the system can perform a linear-time compaction similar to the one described in [22], and refresh the screen.

In the rest of this section we assume that, when we use the interactive graph drawing scheme under the Relative-Coordinates scenario, we start from scratch. According to the discussion in the beginning of this section, the Relative-

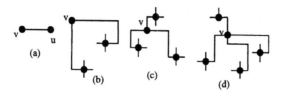

Fig. 2. Inserting v when its local degree is (a) one, (b) two, (c) three, and (d) four.

Coordinates scenario guarantees that the graph that is being built is always connected after any vertex insertion. Let $n_1(t)$, $n_2(t)$, $n_3(t)$ and $n_4(t)$ denote the number of vertices of local degree one, two, three and four, respectively, that have been inserted up to time t.

Theorem 2. *An interactive graph drawing system under the "Relative Coordinates scenario" produces drawings with the following properties:*

1. *after each vertex insertion, the coordinates of any vertex or bend of the current drawing may shift by a total amount of at most 6 units along the x and y axes,*
2. *there are at most 3 bends along any edge of the drawing,*
3. *the total number of bends is at most $3n(t) - 1$, and*
4. *the area of the drawing is at most $2.25n(t)^2$,*

where $n(t)$ is the number of vertices that have been inserted up to time t.

Sketch of Proof. The first property follows from the definition of the Relative-Coordinates scenario and from the fact that at most 6 new rows and new columns might open anywhere in the current drawing (see Figs. 1b, 1c, 1d, 2d) as a result of a vertex insertion. Figures 1 and 2 cover the worst cases in terms of rows, columns and bends required for a single vertex insertion, and for all possible local degrees of the inserted vertex. From these figures we observe the following: First there can be at most 3 bends along any edge of the drawing (see Fig. 1c). Second, the bends along an edge are introduced at the time of insertion of the vertex that is incident to that edge.

From Figs. 1 and 2 and from the discussion above, it follows that at most 3 new bends are introduced when a vertex of local degree 2 is inserted, at most 5 new bends when a vertex of local degree 3 is inserted, and at most 8 new bends when a vertex of local degree 4 is inserted. No new bend is introduced when a vertex of local degree 1 is inserted. In other words, if $B(t)$ is the total number of bends at time t, it holds that:

$$B(t) \leq 3n_2(t) + 5n_3(t) + 8n_4(t)$$

We want to compute the maximum value that $B(t)$ can take, in order to establish an upper bound on the number of bends of the drawing at time t. This is equivalent to solving the following linear program:

$$\text{maximize} \ : \ 3n_2 + 5n_3 + 8n_4$$

under the following constraints:

$$n_1 \geq 1$$
$$n_1 + n_2 + n_3 + n_4 = n - 1$$
$$n_1 + 2n_2 + 3n_3 + 4n_4 \leq 2n$$

The first constraint is an inequality on the number of local degree one insertions, the second is an equation on the number of vertices, and the third is an inequality on the number of edges of the graph, after n vertices have been inserted. Recall that the first vertex to be inserted has local degree 0, since it does not have any edges yet.

Solving this linear program with all three constraints leads to a non-integral solution. If we ignore the first constraint, the new linear program has an integral solution and the objective function is maximized (to $3n + 2$) when $n_1 = n_3 = 0$, $n_2 = n - 2$, and $n_4 = 1$. This solution implies that maximizing the number of bends depends solely on the number of vertices of local degree two and four. If we take into account the fact that the first two vertices inserted in an empty graph have local degrees 0 and 1 respectively, what we really have is that $n_2 = n - 3$ and $n_4 = 1$. This is the same solution as the one obtained from the first linear program after relaxing the solution into an integral one. We can also see that any other combination of values for n_2 and n_4 when $n_2 + n_4 = n - 2$ gives more than $2n - 1$ edges (recall that one edge is introduced by the second vertex, which has local degree 1). From the above analysis, it follows that the upper bound on the number of bends is $3n - 1$.

Regarding the area of the drawing at time t, we can infer from Figs. 1 and 2 that:

- when a vertex with local degree one is inserted, either a new row or a new column is required,
- when a vertex with local degree two is inserted, either two new rows and one new column are required, or one new row and two new columns are required,
- when a vertex with local degree three is inserted, we need a total of at most four new rows and new columns, and
- when a vertex with local degree four is inserted, we need a total of at most six new rows and new columns.

Let $h(t)$ and $w(t)$ denote the height and the width, respectively, of the drawing at time t. Then it holds that:

$$h(t) + w(t) \leq n_1(t) + 3n_2(t) + 4n_3(t) + 6n_4(t) \leq 2n(t) + n_2(t) + n_3(t) + 2n_4(t)$$

since $n_1(t) + 2n_2(t) + 3n_3(t) + 4n_4(t) \leq 2n(t)$. We want to maximize $h(t) + w(t)$. If we just multiply both sides of the last inequality (i.e., the one on the edges of the graph) by $\frac{3}{2}$, we obtain $h(t) + w(t) \leq 3n(t)$. However, this solution does not give us the values of the variables (i.e., $n_1(t)$, $n_2(t)$, etc), for which this upper bound is achieved. For this reason, we formulate this problem as a linear program, where the expression to be maximized is: $2n + n_2 + n_3 + 2n_4$, and the constraints are exactly the same as the ones in the above linear program.

Solving this new linear program, we have that $h(t) + w(t)$ is maximized when $n_1 = n_3 = 0$, $n_2 = n - 2$, and $n_4 = 1$, exactly as in the linear program we studied above, for the number of bends. According to the analysis we did for that linear program, these results really mean that $n_1 = 1$ (the second vertex to be inserted), $n_2 = n - 3$, and $n_4 = 1$. The maximum value of expression $2n + n_2 + n_3 + 2n_4$ that we wanted to maximize is now $3n$. This means that $h(t) + w(t) \leq 3n(t)$. It also holds that $h(t) \times w(t)$ is maximized when $h(t) = w(t) = \frac{h(t) + w(t)}{2} \leq 1.5n(t)$. In this case, the area of the drawing can be at most $2.25n(t)^2$. $\square$

Let us have a look at the expression giving the number of bends that we maximized in the linear program of the proof of Theorem 2. One might be tempted to believe that this expression is maximized when $n_4(t)$ is maximized (and this happens when $n_4(t) = \frac{n(t)}{3}$, if the graph is always connected). The result of the linear program was quite revealing. We discovered that this expression is maximized only under the following insertion sequence: insert the first two vertices with local degrees 0 and 1 respectively, followed by $n - 3$ vertices of local degree two, and conclude with the insertion of exactly one vertex of local degree four. In order to refresh the drawing after each update, the coordinates of every vertex/bend affected must be recalculated. Hence, it would take linear time.

3 Experimental Comparison of the Two Interactive Scenaria

3.1 Implementation and Experiments

Both the No-Change and Relative-Coordinates algorithms have been implemented in C++ (GNU C++ version 2.6.0) on a SPARC 5 running SUN OS Release 4.1.3. Our implementations are running on top of Tom Sawyer Software's Graph Layout Toolkit version 2.2 [23, 24]. We converted 11,491 of the graphs used in the experimental analysis of [5] for our set of experiments. This database of graphs is available by anonymous ftp from infokit.dis.uniroma1.it:public. A small number of edges and vertices were discarded in order to make all vertices maximum degree four. For each graph, *arbitrarily selected* vertices and their incident edges were inserted one at a time into an initially empty structure. The graph was connected at all times. The following standard measures of quality were determined for each drawing:

- Area: The area of the smallest box which can bound the drawing.
- Bends: The total number of bends.
- Crossings: The total number of crossings.
- Aspect Ratio: The width divided by the height of the drawing.
- Average Edge Length: The sum of all edge lengths divided by the number of edges in the drawing.
- Maximum Edge Length: The length of the single longest edge in the drawing.

In addition to these standard measures, we also count the number of rows, columns and bends added with each type of insertion (local degree one, two, three or four). These factors show behavior specific to interactive orthogonal drawing algorithms. All of these measures are plotted against the final number of vertices in the graph.

3.2 Performance Analysis Under Various Quality Measures

Our experimental results show that both No-Change and Relative-Coordinates scenaria behave better than their theoretical upper bounds. The most important aspect of our experimental work is the observation that the performance of the Relative-Coordinates approach for graph drawing is considerably better than that of No-Change with respect to the aesthetic measures we considered. Although both algorithms respected their theoretical bounds, Relative-Coordinates' average case behavior was consistently better than that of No-Change. In Fig. 3 we show the drawings of one graph from our experimental set with the same random order of vertex insertion, when drawn under the two different scenaria. As is evidenced in the pair of drawings, each scenario has its own distinctive style. The No-Change scenario placed the first two vertices at the top left corner of the drawing and grew in a south-easterly direction. Relative-Coordinates maintained the general shape of the drawing, but inserted a small number of rows and columns in order to facilitate the placement of the new vertex and the routing of its incident edges. Some of our experimental findings are summarized in Fig. 4. More specifically, we have:

Area: The area of graphs laid out by Relative-Coordinates is consistently smaller than that of No-Change. The theoretical upper bound of the No-Change scenario is $1.77n^2$ while the behavior is closer to $\frac{n^2}{2}$. Likewise, the theoretical upper bound of the Relative-Coordinates scenario is $2.25n^2$ and the experimental behavior is closer to $\frac{n^2}{4}$.

Bends: Relative-Coordinates produces drawings with fewer bends than No-Change. This happens because, under Relative-Coordinates, newly inserted vertices are expected to be placed closer to their adjacent vertices. In addition, the first invariant of No-Change forces us to place bends in a significant percentage of all degree one insertions. In other words, if a low or no bend edge exists, Relative-Coordinates will use it, but No-Change may not in order to comply with its own invariants. The theoretical upper bound for No-Change is $2.66n + 2$ while it behaves more like $\frac{n}{2}$. Likewise the theoretical upper bound for Relative-Coordinates is $3n - 1$ and it behaves closer to $\frac{n}{4}$.

Crossings: Again Relative-Coordinates performs significantly better than No-Change and this behavior is expected. Since the No-Change scenario allows no coordinate of any vertex or bend to change, the scenario must comply with its own invariants, so new edges are often forced to cross many old edges to reach their incident vertices. Contrariwise, Relative-Coordinates allows some change (in the form of row and column additions anywhere within the drawing) and places vertices closer to their adjacent vertices.

Aspect Ratio: As can be seen in the plot, No-Change and Relative-Coordinates behave in a very similar fashion. Both algorithms produce rather squarish drawings. It is important to note that certain modifications within the implementation will allow different behaviors: either algorithm could produce more rectangular drawings to comply with some requirement.

Average Edge Length: The Average Edge Length and Maximum Edge Length plots show a very important difference between the two scenaria. By the very nature of No-Change, newer vertices are forced to be far from their adjacent vertices if they were inserted at a much earlier point in the lifetime of the drawing. This factor, of course, causes the average edge length to be high. Relative-Coordinates adds a reasonable number of rows and columns as necessary to allow "good" placement of new vertices close to their adjacent vertices and edges with few bends.

Maximum Edge Length: The No-Change Algorithm produces long edges when a new vertex is connected to another vertex which was placed at, or near, the beginning of the lifetime of the graph. This is also an expected result.

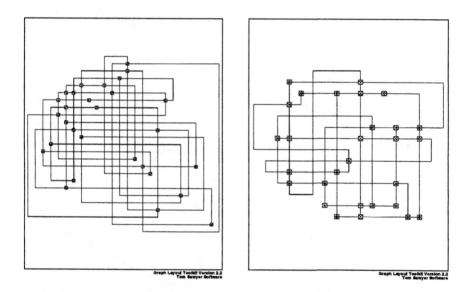

Fig. 3. Drawings of the same 29-vertex graph: No-Change on the left, Relative-Coordinates on the right.

At this point it is important to notice that Relative-Coordinates performs well even as a non-interactive algorithm. Although it is unfair to directly compare our results with those in [5] because we limit the test graphs to degree four, it is interesting that the average area of a Relative-Coordinates drawing is only slightly larger than the Giotto drawings. The same phenomenon is observed with crossings and average edge length. For these experiments we used an arbitrary

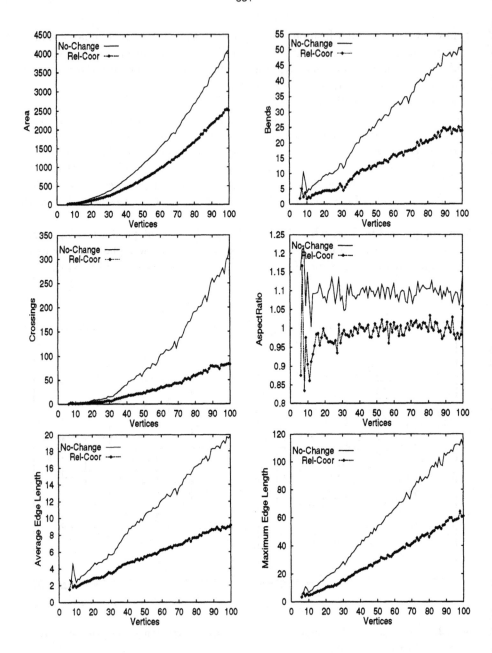

Fig. 4. Graphs of our experimental findings.

insertion sequence. If we find a "good" insertion sequence then we should obtain even better results. An additional refinement phase should produce data curves (representing area, bends, crossings, average and maximum edge lengths) that are very similar to those of Giotto.

3.3   Performance After a Single Update Step

Our second set of experimental data is pertinent to the interactive nature of the two scenaria. Our data show that Relative-Coordinates consistently outperforms No-Change in any measure considered. The plots of our results are shown in Fig. 5 and Fig. 6.

Rows and Columns Added per Degree One Insertion: No-Change adds at most one new row and one new column per degree one insertion while Relative-Coordinates adds either one new row or one new column. Therefore we expected the average number of rows and columns for degree one insertions to be slightly higher for No-Change. The collected data reflects this characteristic. The average number of rows and columns added per degree one insertion is $0.551 + 0.605 = 1.156$ for No-Change and $0.515 + 0.485 = 1.000$ for Relative-Coordinates.

Bends Added per Degree One Insertion: No-Change adds a bend to the edge if it is placed to the North or West of the old vertex while Relative-Coordinates never adds a bend during a degree one insertion. The average number of bends for No-Change is 0.118 and 0 for Relative-Coordinates.

Rows and Columns Added per Degree Two Insertion: No-Change adds at most one new row and one new column per degree two insertion. On average, No-Change adds $0.926 + 0.953 = 1.879$ rows and columns per degree two insertion. Notice that the sum is very close to the worst case. Relative-Coordinates adds a total of three new rows and columns in the worst case and the average is a much better $0.377 + 0.435 = 0.812$.

Bends Added per Degree Two Insertion: In the worst case, No-Change adds four bends while Relative-Coordinates adds three. The average number of bends inserted is 2.050 for No-Change and 0.812 for Relative-Coordinates. This number is so low because Relative-Coordinates makes good use of any free rows, columns and open degrees of freedom.

Rows and Columns Added per Degree Three Insertion: Theoretically at most two new rows and two new columns are added for a No-Change degree three insertion, and at most a total of four new rows and columns are needed for Relative-Coordinates. In our experiments we found that an average of $1.102 + 1.135 = 2.237$ rows and columns are added in the No-Change implementation and only $0.707 + 0.740 = 1.447$ rows and columns in Relative-Coordinates.

Bends Added per Degree Three Insertion: In the worst case, No-Change adds five bends and the average behavior is 3.205. Relative-Coordinates takes better advantage of available rows and columns and has an average of 2.447.

Rows and Columns Added per Degree Four Insertion: Degree four insertions are handled very similarly by both No-Change and Relative-Coordinates.

At most, a total of six new rows and columns are added by each algorithm. According to the data an average of $1.660+1.730 = 3.390$ rows and columns is added by the No-Change algorithm. Relative-Coordinates adds $1.343 + 1.373 = 2.716$. **Bends Added per Degree Four Insertion**: Both algorithms add eight bends in the worst case, and No-Change has been experimentally found to produce an average of 5.022 bends per degree four insertion while Relative-Coordinates introduces an average of 3.987 bends.

Remember in Sect. 2 we proved that the area and number of bends within a Relative-Coordinates drawing is contingent on the number of degree two insertions. Therefore it is quite interesting and important to note that the experimental behavior of degree two insertions is so much better than the worst case. This explains why the behavior (with respect to area and bends) of Relative-Coordinates is so much better.

4 Conclusions and Open Problems

The Relative-Coordinates scenario maintains the general shape of the current drawing after an update (vertex/edge insertion/deletion) takes place, and does not affect the number of bends of the current drawing even if the update operation is a vertex insertion. We used linear programming in order to establish an upper bound for the performance of this scenario. A comparison of the practical behavior of the No-Change and Relative-Coordinates scenaria was presented. The two scenaria were tried on a very large set of "real-life" graphs, and results were reported with respect to their performance under various aesthetic measures. The Relative-Coordinates scenario was consistently better, while both scenaria respected their theoretical bounds.

It is an interesting open problem to develop a theory that enables the efficient insertion, movement or deletion of more than one vertex simultaneously (that is a block of vertices) in the current drawing. Also, techniques for interactive graph drawing in other standards (straight line, polyline, etc.) are needed, and should be explored.

References

1. Therese Biedl, Embedding Nonplanar Graphs in the Rectangular Grid, *Rutcor Research Report 27-93*, 1993.
2. T. Biedl and G. Kant, A Better Heuristic for Orthogonal Graph Drawings, *Proc. 2nd Ann. European Symposium on Algorithms (ESA '94), Lecture Notes in Computer Science, vol. 855*, pp. 24-35, Springer-Verlag, 1994.
3. R. Cohen, G. DiBattista, R. Tamassia, and I. G. Tollis, Dynamic Graph Drawings:Trees, Series-Parallel Digraphs, and Planar st-Digraphs, *SIAM Journal on Computing*, vol. 24, no. 5, pp. 970-1001, 1995.
4. G. DiBattista, P. Eades, R. Tamassia and I. Tollis, Algorithms for Drawing Graphs: An Annotated Bibliography, *Computational Geometry: Theory*

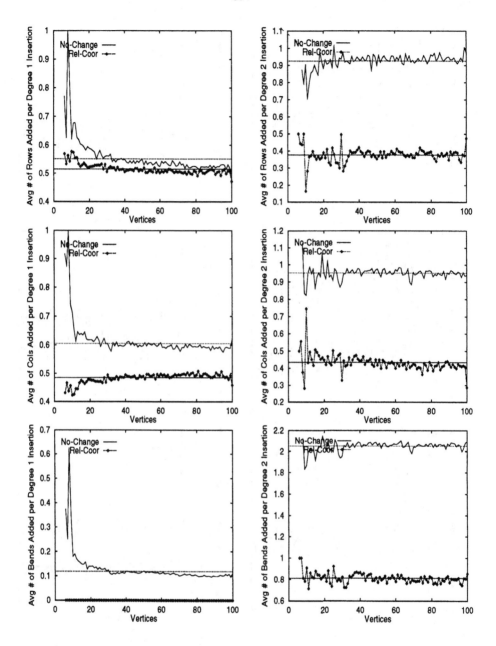

Fig. 5. Average number of rows, columns and bends added per degree one and two vertex insertion.

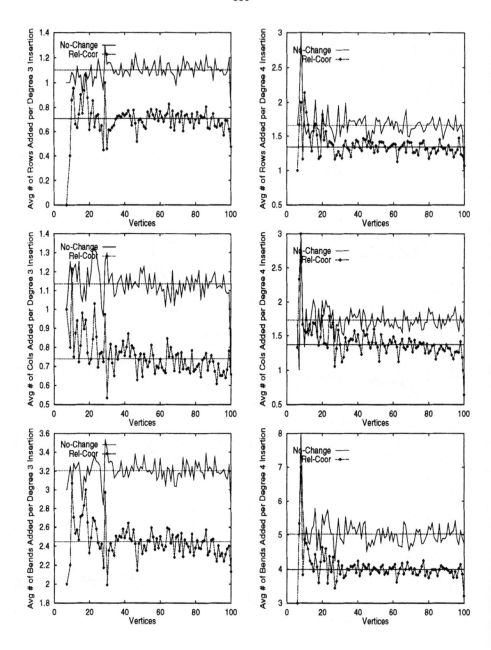

Fig. 6. Average number of rows, columns and bends added per degree three and four vertex insertion.

and Applications, vol. 4, no 5, 1994, pp. 235-282. Also available via anonymous ftp from ftp.cs.brown.edu, gdbiblio.tex.Z and gdbiblio.ps.Z in /pub/papers/compgeo.

5. G. DiBattista, A. Garg, G. Liotta, R. Tamassia, E. Tassinari and F. Vargiu, An Experimental Comparison of Three Graph Drawing Algorithms, *Proc. of ACM Symp. on Computational Geometry*, pp. 306-315, 1995. The version of the paper with the four algorithms can be obtained from http://www.cs.brown/people/rt.

6. S. Even and G. Granot, Rectilinear Planar Drawings with Few Bends in Each Edge, *Tech. Report 797, Comp. Science Dept.*, Technion, Israel Inst. of Tech., 1994.

7. S. Even and R.E. Tarjan, Computing an st-numbering, *Theor. Comp. Sci. 2* (1976), pp. 339-344.

8. Michael Himsolt, Comparing and evaluating layout algorithms within GraphEd, *J. Visual Languages and Computing*, vol. 6, no. 3. pp.255-73, 1995.

9. Michael Himsolt, GraphEd: a graphical platform for the implementation of graph algorithms, *Proc. DIMACS Workshop GD '94, Lecture Notes in Comp. Sci. 894*, Springer-Verlag, 1994, pp. 182-193.

10. Goos Kant, Drawing planar graphs using the lmc-ordering, *Proc. 33th Ann. IEEE Symp. on Found. of Comp. Science*, 1992, pp. 101-110.

11. F. T. Leighton, New lower bound techniques for VLSI, *Proc. 22nd Ann. IEEE Symp. on Found. of Comp. Science*, 1981, pp. 1-12.

12. Charles E. Leiserson, Area-Efficient Graph Layouts (for VLSI), *Proc. 21st Ann. IEEE Symp. on Found. of Comp. Science*, 1980, pp. 270-281.

13. Thomas Lengauer, *Combinatorial Algorithms for Integrated Circuit Layout*, John Wiley and Sons, 1990.

14. K. Miriyala, S. W. Hornick and R. Tamassia, An Incremental Approach to Aesthetic Graph Layout, *Proc. Int. Workshop on Computer-Aided Software Engineering (Case '93)*, 1993.

15. K. Misue, P. Eades, W. Lai and K. Sugiyama, Layout Adjustment and the Mental Map, *J. of Visual Languages and Computing*, June 1995, pp.183-210.

16. A. Papakostas and I. G. Tollis, *Algorithms for Area-Efficient Orthogonal Drawings, Technical Report UTDCS-06-95*, The University of Texas at Dallas, 1995.

17. A. Papakostas and I. G. Tollis, Improved Algorithms and Bounds for Orthogonal Drawings, *Proc. DIMACS Workshop GD '94, LNCS 894*, Springer-Verlag, 1994, pp. 40-51.

18. A. Papakostas and I. G. Tollis, Issues in Interactive Orthogonal Graph Drawing, *Proc. of GD '95, LNCS 1027*, Springer-Verlag, 1995, pp. 419-430.

19. Markus Schäffter, Drawing Graphs on Rectangular Grids, *Discr. Appl. Math. 63* (1995), pp. 75-89.

20. J. Storer, On minimal node-cost planar embeddings, *Networks* 14 (1984), pp. 181-212.

21. R. Tamassia, On embedding a graph in the grid with the minimum number of bends, *SIAM J. Comput.* 16 (1987), pp. 421-444.

22. R. Tamassia and I. Tollis, Planar Grid Embeddings in Linear Time, *IEEE Trans. on Circuits and Systems CAS-36* (1989), pp. 1230-1234.

23. Tom Sawyer Software Corp. GLT development group, *Graph Layout Toolkit User's Guide*, Berkeley, California, 1995.

24. Tom Sawyer Software Corp. GLT development group, *Graph Layout Toolkit Reference Manual* Berkeley, California, 1995.

25. L. Valiant, Universality Considerations in VLSI Circuits, *IEEE Trans. on Comp.*, vol. C-30, no 2, (1981), pp. 135-140.

An Interactive System for Drawing Graphs

Kathy Ryall[1], Joe Marks[2], and Stuart Shieber[1]

[1] Aiken Computation Lab
Harvard University
Cambridge, MA 02138, U.S.A.
E-mail: {kryall,shieber}@eecs.harvard.edu
[2] MERL
201 Broadway
Cambridge, MA 02139, U.S.A.
E-mail: marks@merl.com

Abstract. In spite of great advances in the automatic drawing of medium and large graphs, the tools available for drawing small graphs exquisitely (that is, with the aesthetics commonly found in professional publications and presentations) are still very primitive. Commercial tools such as Claris Draw or Microsoft's Powerpoint provide minimal support for aesthetic graph layout. At the other extreme, research prototypes based on constraint methods are overly general for graph drawing. Our system improves on general constraint-based approaches to drawing and layout by supporting only a small set of "macro" constraints that are specifically suited to graph drawing. These constraints are enforced by a generalized spring algorithm. The result is a usable and useful tool for drawing small graphs easily and nicely.

1 Introduction

Most small graphs (those with fewer than about 30 nodes) that appear in publications or presentations are still drawn with the aid of fairly primitive commercial drawing tools like Microsoft's PowerPoint or Claris Draw. Why do these tools not incorporate some of the advanced techniques that have been developed by either the graph-drawing or constraint-based-layout communities? One reason is that most graph-drawing algorithms cannot support the exquisite symmetries, spacings, and alignments that graphic designers utilize in professional-grade work. This kind of layout detail can be achieved in some constraint-based drawing systems, but the very general capabilities of such systems tend to make them cumbersome for the specific task of graph drawing.

Our system is based on constraints, but ones that are designed specifically for drawing graphs, not general graphics. These "macro" constraints, or *Visual Organization Features (VOFs)* [2], are listed in Figure 1; the application of each VOF is illustrated by "before" and "after" layouts. In our drawing tool, VOFs can be applied and removed interactively. Furthermore, the tool enforces syntactic constraints, such as preventing nodes overlapping other nodes and edges. The VOF and syntactic constraints are enforced by a generalized spring algorithm,

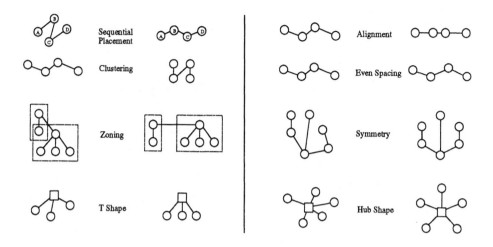

Fig. 1. Visual organization features (after [2]).

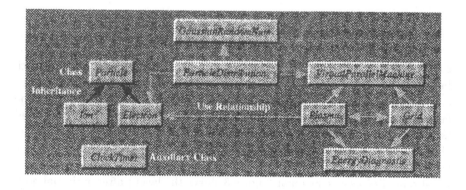

Fig. 2. Scanned diagram on which the example was based (from [3]).

an improved version of the one described by Dengler et al. [1]. We describe below how our tool can be used to replicate the drawing in Figure 2 [3]. The sample interactions shown illustrate the aptitude and convenience of our tool for drawing small graphs.

2 Example Interaction

Figures 3-9 show snapshots at various stages in the process of drawing a given graph. A screen dump of the entire interface to our system is shown in Figure 9; other figures show only the canvas area. The existence of active VOFs is indicated visually in our system by a user-responsive highlighting mechanism that

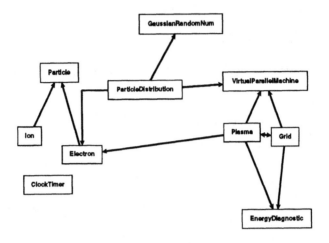

Fig. 3. The initial layout, with labeled nodes and edges.

cannot be replicated in static images. We therefore use shades of gray to indicate different VOFs in the figures.

The first step in the drawing process is typically to place the desired number of nodes in approximately the desired layout. To create a node, the user clicks the middle mouse button on the canvas area. Node characteristics (label, shape, font, background color, foreground color, border color, dimensions) can be modified via an edit dialog window. If a node's label becomes larger than the node itself, the system will automatically enlarge the node to accommodate its new label. Edges are added by dragging with the left mouse button from origin to destination node. Edges can be undirected, unidirectional, or bidirectional; they can also be direct or orthogonal. Figure 3 shows the layout after an initial node- and edge-placement phase.

To add a VOF to the layout, the user first selects a set of nodes using standard mouse techniques such as clicking or region-dragging. The user may then apply one or more VOFs to the set by pressing the appropriate push buttons, located on the right of the window in Figure 9. In Figure 4, the user has applied a horizontal *Alignment* VOF to the second, third, and fourth rows of nodes.

The system converts each VOF instance into a set of constraints, which it attempts to satisfy using a generalized spring algorithm. For example, an *Alignment* VOF mandates a set of zero-rest-length springs connecting each node to an axis-parallel line through the centroid of the points; a *Cluster* VOF places springs pairwise among the nodes with a short rest length. The physical simulation of the mass-spring model is continuously animated, indicating to the user the influence of the chosen VOFs. The user may move nodes and groups of nodes while the simulation proceeds in order to aid the system in finding better global solutions to the implicit constraint-satisfaction problem.

The use of a constraint-satisfaction scheme (mass-spring simulation) that

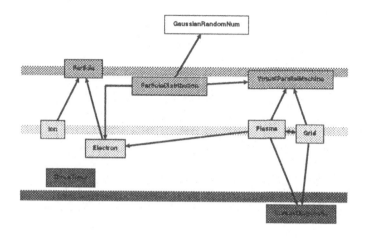

Fig. 4. User adds an *Alignment* VOF to each row.

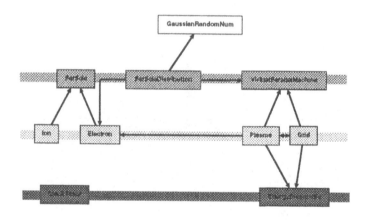

Fig. 5. A stable configuration, satisfying the *Alignment* VOFs.

is intuitive and predictable, rather than one better at finding global solutions, is deliberate. The tool is not intended to be good at globally satisfying the VOFs by itself. Rather, it is intended to provide an interface that allows a useful *collaboration* between user and computer in solving the layout problem. For this purpose predictability, simplicity, and the compelling nature of the animation are far more important than global optimality.

Figure 5 shows the stable configuration that ensues after the three *Alignment* VOFs have been satisfied.

Continuing with the example, the user adds three more VOFs, as illustrated in Figure 6. On the right, the user has added a *Hub Shape* VOF, indicated in

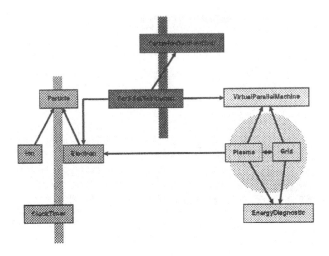

Fig. 6. The user adds three more VOFs: *Symmetry, Alignment,* and *Hub Shape.*

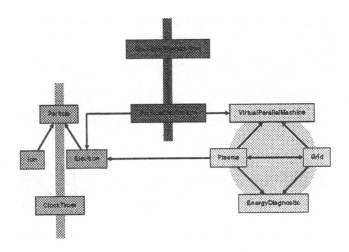

Fig. 7. The system satisfies both new and old constraints.

light gray. The center two nodes (dark gray) are to be vertically aligned. Finally, the four nodes on the left have had a *Symmetry* VOF applied to them.

Once again, the system converts each VOF instance into a set of constraints. It attempts to satisfy all constraints, both old and new, in determining node placement. Figure 7 shows the updated node positions. Each row is still aligned, and the new VOFs have been satisfied as well.

As a final step, the user adds text labels to the layout. Text labels are nodes

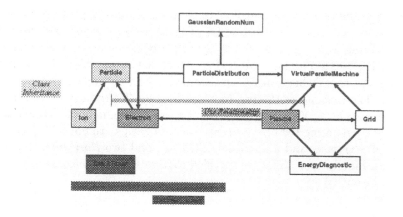

Fig. 8. The user adds three text labels, and three more VOFs: *Clustering, Even Spacing,* and *Alignment.*

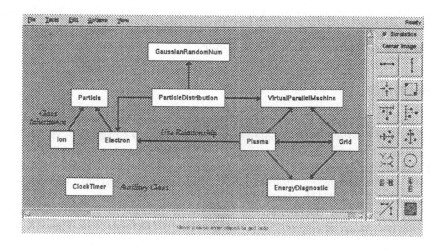

Fig. 9. The final layout (showing full system interface).

that have no background or border color, so that only the text string is visible. Any node can be designated a text label by selecting it and then pressing the *Text Label* push button. Text labels can also participate in VOFs, just like regular nodes. In Figure 8, the user has added three text labels, and applied a VOF to each one. The light-gray nodes on the left are subject to a *Cluster* VOF. The dark-gray nodes on the bottom are to be horizontally aligned. The middle three nodes, shaded medium gray, are to be evenly spaced. Figure 9 shows the final layout embedded in the system interface.

As an example of the flexibility of the system, Figure 10 presents successive snapshots of quiescent states of the interface as a user develops an alternative layout starting from the same initial layout (a) of the previous example graph. The user imposes a *T-shape* VOF on four nodes at the top. After the simulation stabilizes (b) the T-shape has been enforced. The user adds another *T-shape* VOF (c). The user moves the two root nodes of the T's roughly as indicated in (d) in order to invert the direction of the T's; the system maintains the T-shape constraints generating the layout in (e). The user swaps two nodes (f), and adds a *Vertical Alignment* and a *Symmetry* VOF (g), and two *Horizontal Alignment* VOFs (h). Finally, four nodes are specified to have *Equal Spacing* (i) to generate the final layout (j).

3   Conclusion

The VOFs supported by our system provide a natural and powerful vocabulary whereby users can express easily the desired characteristics of a graph layout, and the intuitive constraint-satisfaction method allows for a collaborative interaction between user and computer in solving graph-layout problems. An informal comparison with commercial drawing programs shows our system to be markedly superior for drawing small, aesthetic graphs.

References

1. E. Dengler, M. Friedell, and J. Marks. Constraint-driven diagram layout. In *Proceedings of the 1993 IEEE Symposium on Visual Languages*, pages 330–335, Bergen, Norway, August 1993.
2. C. Kosak, J. Marks, and S. Shieber. Automating the layout of network diagrams with specified visual organization. *IEEE Transactions on Systems, Man, and Cybernetics*, 24(3):440–454, March 1994.
3. C. D. Norton, B. K. Szymanski, and V. K. Decyk. Object-oriented parallel computation for plasma simulation. *CACM*, 38(10):88–100, October 1995. Figure 3.

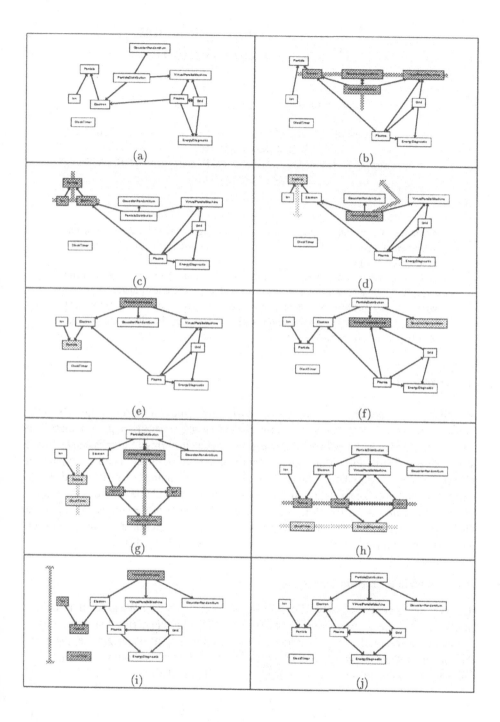

Fig. 10. Development of an alternative layout.

Automatic Graph Clustering
(System Demonstration)

Reinhard Sablowski[1], Arne Frick[2] *

[1] Universität Karlsruhe, Institut für Programmstrukturen und Datenorganisation, EMail:
RSablowski@aol.com
[2] Tom Sawyer Software, 804 Hearst Avenue, Berkeley CA 94710, EMail:
africk@tomsawyer.com

Abstract. We present a new, easy to understand algorithm and programming environment allowing for the interactive or automatic clustering of graphs according to several heuristics.

Our approach is based on graph structure only and can be implemented to run efficiently with a personal computer. It is capable of efficiently clustering graphs with > 3000 vertices. We shall demonstrate the interactive user environment for automatic clustering. As an application, we consider the clustering of large WWW connectivity graphs.

1 Introduction

Clustering is the process of grouping information to achieve a more recognizable presentation of source data. The computation of a clustering generally requires a metric on the data to determine the closeness of data points. The clustering of graphs can be based on either graph structure, or on some semantic properties of the application domain. In this paper, we do not make any assumptions about the "meaning" of vertices and edges, but focus solely on the structure. In this context, the only available metric is the *graph-theoretic distance* of vertices, which is defined as the length of the shortest path joining them, if one exists. If a drawing of a graph is known, the Euclidean distance between the vertex positions can be used instead of their graph-theoretic distance.

1.1 Criteria for a good Clustering

Given these assumptions, what are criteria for good clustering? We argue that the importance of a *cluster vertex*, i.e. a vertex in the resulting clustering that represents a group of vertices in the original graph, should be a function of its *weight*, i.e. the number of vertices and edges represented by it. The cluster weights should come close to a user-defined value to achieve a clustering of desired granularity.

$$GA = \sum_{i=1}^{\|v\|}(g_i - G_{\text{predef}})^2 \tag{1}$$

---

* This research was performed while the author was working at Universität Karlsruhe, Institut für Programmstrukturen und Datenorganisation.

Also, the cluster sizes should be approximately equal. In other terms, the standard deviation of cluster sizes should be minimal.

$$\sigma = \sqrt{\frac{1}{\|v\|} \sum_{i=1}^{\|v\|} g_i - \overline{g}} \tag{2}$$

Connectivity within a cluster (*intra-connectivity*) should generally be stronger than the connectivity between clusters (*inter-connectivity*). Therefore, the total number of inter-cluster edges should be minimized:

$$EH = \|\{e \in E | e \text{ is inter-cluster edge}\}\| \tag{3}$$

Realizing that these criteria may conflict each other, a global optimization strategy should be used, combining these criteria by assigning weights to each of them. This results in a global energy function

$$E = w_1 GA + w_2 \sigma + w_3 EH \tag{4}$$

that is supposed to represent the overall quality of a clustering.

2 Related work

The problem of clustering graphs is closely related to that of partitioning graphs, for which there is a variety of literature [2]. Little appears to be known about the problem of how to efficiently cluster large graphs in an interactive environment, although the problem is closely related to the task-scheduling problem for parallel processors. The requirement there is to have partitions of approximately equal size and sparse interconnectivity, which is similar to our notion of a well-clustered graph.

The well-known Basic-ISODATA algorithm for arbitrary clustering problems in Euclidean spaces is quite fast in practice, but it requires *a priori* knowledge of the number of clusters to partition into. Other general strategies for cluster analysis in Euclidean spaces are discussed in [1].

The idea of visualizing WWW graphs seems to have appeared first in [6]. However, this approach is based on semantic information.

3 Cluster Drawings

In order to reveal information about the graph using clusterings, we need to find ways to communicate the clustering. An intuitive technique is to visualize these clusters using graph drawing techniques.

In addition to the well-known global aesthetics criteria employed by spring-embedder algorithms, drawings of clusterings should display vertex and edge weights, and hide intra-cluster edges.

4 Idea

4.1 Basics

Our approach for clustering large graphs is based on the successive identification of structural elements (*patterns*) in the graph. About the simplest pattern occurring in a graph are *leaves*, i.e. nodes with in-degree 1 and out-degree 0. Other suitable patterns include *paths* (nodes with in-degree=out-degree=1), triangles and similar simple structures.

The existence of such groups of vertices indicates a close relationship between their constituent nodes and should therefore be clustered. It is important that patterns can be identified efficiently.

A clustering step consists of clustering of all vertices satisfying one or more patterns. This process can be iterated under the user's control to yield a hierarchical clustering of desired weight and density.

This approach is similar to a stepwise parsing of the graph according to a graph grammar [3, 7], where patterns are *non-terminals* and vertices are *terminals*.

4.2 Data Structures

Although the basic idea is straightforward, it remains a challenge to implement efficiently. A data structure suited for the hierarchical clustering of graphs should support the following basic operations efficiently: Insertion and deletion of nodes and edges as well as recursive clustering and unclustering nodes.

The need for efficient data structures is motivated by the following scenario. Assume first that a dense graph has to be grouped into two clusters. After forming the first cluster, all edges from a node outside the cluster that are connected to two or more nodes within the cluster need to be drawn as a single new, heavier edge. After the clustering step is completed, there will be only a single, very heavy edge connecting both clusters. In this situation, unclustering the first group has to reveal all edges to vertices in the second cluster, and this should be efficient.

Our implementation of a suitable data structure is based on an object-oriented design that defines ClusterVertex as a subtype of Vertex and therefore allows for the polymorphic use of ClusterVertex, wherever the abstract data type Graph requires a Vertex. Similarly, ClusterEdge is a subtype of ClusterEdge. The type ClusterEdge is based on two ClusterVertex stacks. Whenever a vertex v is inserted into a cluster c, c is pushed upon the corresponding edge stacks of all incident edges, thus providing a fast and information-preserving method for retrieval of cluster information.

5 Result characteristics

Our pattern-based approach does not create a predefined number of clusters. Instead, the process is under the user's control, who chooses graph patterns from an extensible set of standard patterns such as leaves, paths, triangles etc.

By iteratively reapplying this pattern-based compression, the resulting graphs may contain original nodes as well as small and large clusters. The number of resulting nodes depends heavily on the input graph. We have found that in general, the final node count represents a compression by a factor of 20–80 for *WWW-Graphs* (see Sect. 6).

One possible technique to visualize clusterings is to apply a spring-embedder algorithm [4]. This is actually a quite natural way of drawing clusterings, as one of the main aesthetics criteria employed by spring-embedder algorithms is that vertices connected by an edge should be close together. This paradigm coincides with our criteria for good clusterings, that would require the nodes within a cluster to be drawn close together, while clusters themselves should be separated in a drawing. The drawings resulting from applying spring-embedder a algorithm generally reveal a lot of the original graph's structure (cf. Fig. 3).

Conversely, spring-embedder algorithms may be helpful to *compute* clusterings in the first place, too. As spring-embedder drawings tend to have related nodes close to each other, geometric proximity and statistical techniques [5] may be applied to compute clusterings from spring-embedder drawings as well.

6 An Application: WWW graphs

As an application domain in which very large real-life graphs occur naturally, we have chosen to consider *WWW graphs*, whose vertices are defined by the URL's of WWW documents, and whose edges are defined by the hyper-links between them. A so-called web robot was used to automatically extract graphs of this nature, starting from a single URL.

One particular graph gathered this way has about 3000 nodes and 3500 edges. Fig. 1 shows this graph in its raw, unclustered state. This example can be clustered well, revealing interesting structural properties (cf. Fig. 2), with little interaction in under two minutes of runtime on a regular IBM-compatible personal computer (based on Intel's Pentium P60 processor). The final result is shown in Fig. 3.

References

1. M. R. Anderberg. *Cluster Analysis for applications*. Academic Press, 1973.
2. Bosak. *Graph Partitioning*. Kluwer, 1990.
3. F. J. Brandenburg. Designing graph drawings by layout graph grammars. In Roberto Tamassia and Ioannis Tollis, editors, *Proceedings of Graph Drawing '94*, volume 894 of *Lecture Notes in Computer Science*, pages 416–427. DIMACS Workshop on Graph Drawing, Springer Verlag, 1995.
4. P. Eades. A heuristic for graph drawing. *Congressus Numerantium*, 42:149–160, 1984.
5. J. Hartigan. *Clustering Algorithms*. J. Wiley and Sons, 1975.
6. S. Mukherjea, J. D. Foley, and S. Hudson. Visualizing complex hypermedia networks through multiple hierarchical views. Technical Report 95-08, Georgia Institute of Technology, Graphics, Visualization and Usability Center, College of Computing, Atlanta, GA 30332-0280, 1995. Also appeared in the Proceedings of the ACM SIGCHI CHI'95, May 1995, Denver, Colorado.
7. G. Zinssmeister and C. McCreary. Drawing graphs with attribute graph grammars. In *Graph Grammar Workshop*, pages 355–360, Williamsburg, 1994.

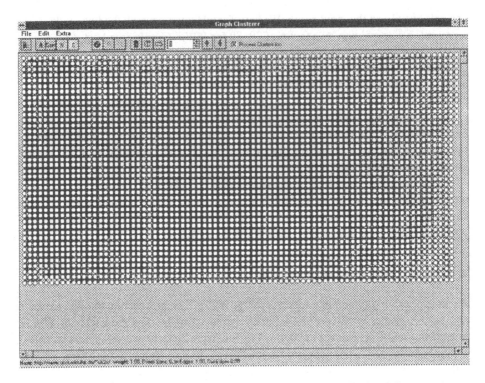

Fig. 1. Original, unclustered graph. No structure is recognizable at all.

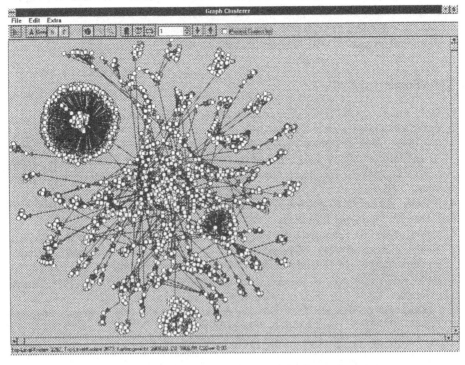

Fig. 2. The process of clustering may reveal rich internal structure.

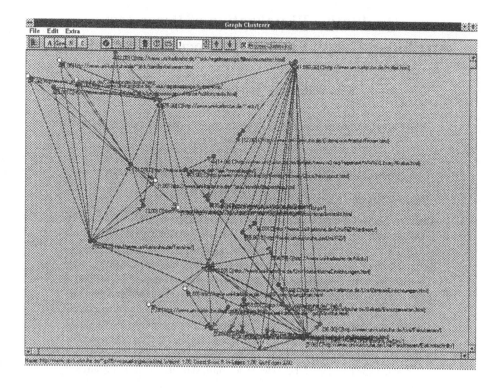

Fig. 3. Final clustering.

Qualitative Visualization of Processes: Attributed Graph Layout and Focusing Techniques

Kent Wittenburg[1] and Louis Weitzman[2]

[1]Bell Communications Research, 445 South St., Morristown, NJ 07960 USA
kentw@bellcore.com

[2]Meta Media Design, 49 Melcher St., Boston, MA 02210 USA
weitzman@media.mit.edu

Abstract. Many techniques and algorithms have been proposed for layout and interactions with basic node and link graphs. Here we consider layout and focusing techniques for attributed graphs and show an application of these techniques in a Bellcore system for support of business process design and reengineering called ShowBiz. ShowBiz supports two primary visualization techniques for qualitative analysis of flowgraphs. The first involves the use of parsing to perform graph reduction for encapsulation and focusing. The second is an attributed graph layout technique we have called RB-layout in recognition of its genesis in Rummler-Brache methodology for business process reengineering [7]. The user selects attributes that partition the set of nodes in the graph and the system generates layouts in which the position of the nodes is determined by the value of the attribute in question.

1 Introduction

Despite the large literature on layout and focusing techniques for basic node and link graphs [4][6], there have been relatively few general techniques proposed specifically for attributed graphs. Standard practice treats attribute information as "adornment," i.e., drawing nodes and/or links with differing icons, color, or size. Here we present novel layout and focusing techniques specific to attributed graphs that have been motivated by an application in business process design and reengineering. We begin with an overview of the ShowBiz tool, followed by illustrations of (1) our use of grammars for support of flowgraph abstraction and focusing and (2) a layout technique that enables visualization and editing of attribute spaces. We then discuss these techniques in the context of other methods proposed for graph layout and focusing.

2 ShowBiz

The process flow modeling tool called ShowBiz has been created by Bellcore Applied Research for internal customers in the Bellcore Professional Services organization. The primary users of this tool consult with clients to improve the efficiency of existing business processes or else design new business workflows occasioned by the introduction of new technology or new forms of business. To date, work has focused on activities in the work centers of large telephony concerns such as the Regional Bell Operating Companies and also on the publishing and fullfillment processes surrounding premier World Wide Web sites. An important function of such a modeling tool is to facilitate rapid qualitative understanding of complex processes. The typical life-cycle of a project for these consultants involves iterating with their customers on representations of current methods of operations. Once the basic representations are agreed upon, subsequent stages may require data gathering to support quantitative analysis and simulation. Then improved future methods of operations are designed and ultimately implemented, but not necessarily with the involvement of the same part of the Bellcore organization.

ShowBiz is designed to allow users quickly and easily to build customizable process models. It supports a transition from the early stages of a design problem to more structured uses of the information in which data can be exported to external tools for simulation, analysis, and/or code generation. It also includes one-step World Wide Web publishing capabilities: whatever view the user can construct can be exported for information sharing on the WWW in the form of an imagemap along with automatically generated and linked html files that contain attribute information for the underlying task objects.

3 Dynamic Aggregation of Subflows

Visualization plays an important role in supporting rapid qualitative understanding of workflows. It is important to be able to easily hide and reveal levels of detail and construct alternative views of complex operations. In order to create more highly structured and concise models, reusable workflow components should be identified. Just as well-structured programming languages facilitate understanding and design, well-structured information models do the same. Aggregation operations in ShowBiz are designed to facilitate the creation of well-structured models and views in support of qualitative understanding of workflows.

Existing commercial flowcharting and process modeling tools standardly support the feature of hierarchically structured flowgraphs, where a single node in a graph can be expanded into another window, in which more detail is shown. However, these hierarchical structures must be assembled by hand and, once created, they are not easily changed. A distinguishing feature of ShowBiz is that users can form alternative hierarchies quickly and easily for the purposes of encapsulation or viewing.

Figure 1 shows a screen dump of a "home view" for a ShowBiz flowgraph, in which users can enter process flow information. The nodes with dog-eared corners represent reduced subflows that can either be expanded into the current diagram or viewed in another window. As users iterate on their models or construct views to focus on aspects of the representation, they can freely form new graph aggregations by invoking a parsing operation. The parser expects there to be one or more nodes pre-selected. It then expands out from the selected items seeking a derivation that includes the selected items and that constitutes a procedural block as defined by the flowgraph grammar. See Wittenburg and Weitzman [12] for a detailed discussion of the flowgraph grammar used in the application. The parser terminates with the first such derivation found and selects the subgraph contained within. The user can then operate upon the selected subflow for viewing or data organization purposes, replacing the selected subgraph with one of the dog-eared nodes.

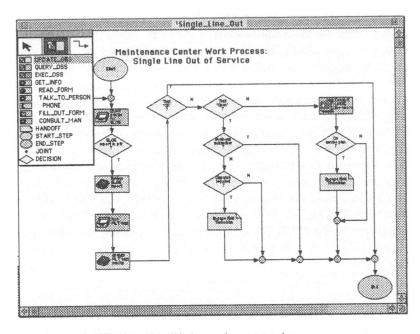

Fig. 1. A ShowBiz home view screen shot.

For example, let us presume that the user has selected the node labeled "Business Subscriber?" appearing near the center of Figure 1. A menu item labeled "Extend Selection to Subflow Group" would then be enabled, and, if chosen, would invoke a parsing operation. The result of such a parse in this case would be the group selection of the subflow nodes shown with heavy outlines in Figure 2. (The selection of arcs connecting a selected group of nodes is implicit.) It is guaranteed through the grammar that such a subflow meets all the requirements for encapsulation. The user may continue to invoke this grouping operation until a subgraph of the appropriate size is selected.

Once the larger group has been selected, other operations are enabled as shown in the menu in Figure 2. Users may reduce the subflow to a node in the same diagram or encapsulate the subflow for subsequent reuse by moving it to a separate file. This operation is similar to cut and paste, except that in addition it splices in an encapsulated subflow node in place of the subflow graph being removed and takes care of some bookkeeping operations such as adding *Start* and *End* nodes to the newly created subflow diagram.

Reduction of visual complexity through such aggregation operations is only part of the story towards creating effective visual presentations. The space freed up by reducing a subgraph should now be utilized by other graph elements still present. And of course the complementary insertion operations need to somehow make room for subgraphs that may require far more graphical real estate than the original encapsulated subflow node. While we have not achieved an automated solution to this space mangement problem when users are in the manual layout mode shown in Figures 1 and 2, reductions or expansions do trigger new layouts when users are in the automated layout

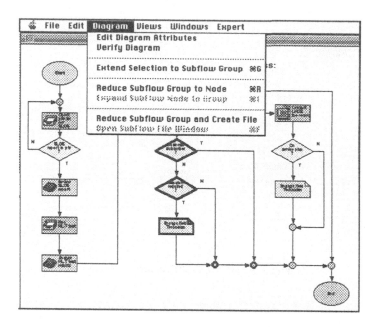

Fig. 2. Extending a selection to a subflow group

mode described in the next section. As has been noted by others who have experimented with sequences of user-directed operations that occasion new graph layouts (e.g., [5]), a remaining issue here is one of visual continuity. It is easy for users to become momentarily disoriented as an entire graph is erased and replaced. We expect to address this issue in future work.

4 Attribute-based Layout

In the views menu of ShowBiz, users are offered the option of specifying a "Rummler-Brache" layout, which places the system in automatic layout mode. In Figure 3 we show the sequence of dialogs for specifying such a layout. The user first has to choose an attribute that will partition the flowgraph's nodes. This choice list is determined by the attribute type definitions in the model. Next a dialog box pops up that lists the value choices for that attribute type, in this case, the type is *database*. Choices are preselected that currently are instantiated somewhere in the graph, but the user may override this default if, for example, editing actions are intended or some alternative focus is desired.

Figure 4 shows the layout that results from the specification in Figure 3. The left-hand pane lists the chosen values of the attribute, which determine the number and order of rows in the layout. The inital row is for "not applicable," used when either the node's class definition does not contain that attribute or else the attribute's value is unbound. These views project the process flow onto an attribute space that immediately reveals attribute values in the context of the overall control flow of the process. Such views can reveal important aspects of a process flow, such as all the points at which handoffs oc-

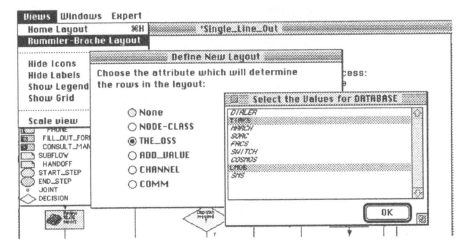

Fig. 3. Specifying an attribute-based layout.

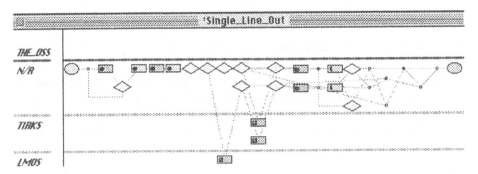

Fig. 4. A layout specfied by the choices in Figure 3.

cur across organizational boundaries or control is passed between systems. The Rummler-Brache methodology proposes creating such a process map to reveal organizational handoffs in a business process flow. Good design for business processes encourages minimizing such handoffs and maximizing accountability of a particular organization in the complete process.

In Figure 5 the user has chosen a different focus through a layout based on *node-class*. (Technically, node-class is not an attribute, but we can treat it as such in the layout.) Note the indentation in the value pane -- these values form an object-oriented hierarchy. The user could choose to consolidate values higher in the hierarchy, in which case determining row position for nodes in the main pane would make use of inheritance. The user has also simplified the diagram through the aggregation methods discussed in Section 3 -- thus we see fewer nodes than in Figure 4. The user is also in the middle of editing. Dragging a node from one lane to another is interpreted by the system as an editing action. In this case the user is changing the node class of the node under the cursor.

The algorithm employed for attribute-based layout is roughly as follows. A variant

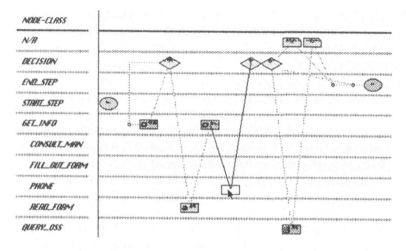

Fig. 5. Editing hierarchical values by drag and drop.

of the initial phase of the Sugiyama algorithm [10] determines the topological layers in the graph, which in our case sets the x-values of the node positions. We assign the x layers starting from the source of the graph rather than from the sinks in order for the node to appear at the least x value consistent with topology. An initial y-position for each of the attribute value rows is then computed and assigned to the nodes based on the node's value of the attribute. The initial y-value assigment is based on a simple linear ordering that derives from the type definition of the attribute. If the attribute's type is hierarchical, the order is arbitrary within a given level of the hierarchy tree. The nodes are then sorted first by y-value and then secondarily by x-value. We iterate through the sorted node list looking for ties of x/y values. When a tie is found, the y-value of each tying node is incremented, which has the effect of widening the swimming lane in the overall layout and incrementing the y of each succeeding attribue value in the ordering.

5 Discussion

Our approach to this interface is consistent with some of our earlier work in interfaces for design support [11]. We have attempted to allow the system to support the user in designing effective views of underlying data while allowing the user to interactively fine-tune the layout and override system decisions. Thus while a user drags a node or a node group, the system incrementally computes external arc positions within a certain distance from the changed nodes. The system constantly gives feedback to the user of what the result will be once the drag terminates. However, the system is not always successful (particularly since our algorithm adds at most two bends to each arc), and thus the interface adds a feature whereby the user can manually alter the position to which an arc connects to a node and then lock the arc to this position. A limitation of the current implementation is that the user cannot similarly override the system's decision of where to position arc bends. There is room for improvement as well in the arc layout methods that the system does use.

The use of parsing for aggregation is comparable to graph aggregation techniques found in many tools presented at previous Graph Drawing workshops such as VCG [8], daVinci [3], or Kimmelman et al. [5]. The difference is that we employ a declarative grammar-based description. Reducible clusters can be defined topologically but also by utilizing relations and predicates determined by attribute information. One of the user-interface issues that is not addressed by some of the other systems is how to define natural clusters that users can easily understand and specify. Our impression is that, for well-structured flowgraphs, users can easily work with the hierarchical decomposition that is a natural outcome of a context-free grammatical description of the visual language. Although graph grammars have been proposed elsewhere in the context of graph layout (e.g., Brandenburg [2]), we are not aware of any other work that employs higher-dimensional grammars in the service of aggregation for information modeling and focusing.

Alternative visual focusing methods, surveyed by Noik [6], include generalized zooming interfaces such as Pad++ [1] and fisheye transformations [9] as well as a variety of abstraction operations including node and arc hiding and/or ghosting [5]. There is a need for both geometric-based operations in support of visual focusing as well as for logically-based aggregation operations. While one may want to verify that a subflow has all the necessary structure to be encapsulated, one does not always need to do local rule-based aggregation for visual focusing. For example, one may want to de-emphasize all nodes of a certain type whether or not they are connected or form a logically coherent group. We see the logical and the geometric methods as largely complementary.

Note that the focusing techniques just discussed are confined to the adding, deletion, magnification, or demagnification of nodes and arcs. The attribute-based layout technique we presented, however, is another type of focusing mechanism. It allows a user to focus on attribute spaces that cut across all visible graph elements. The value of Rummler-Brache process maps is well-established as a pencil-and-paper method for visualizing organizational participation in business processes. We have automated the technique and applied it to attributed graphs more generally. The algorithm requires only that graphs be directed and that eligible attributes partition the nodes of the graph. It is useful not only for visualization but also for drag-and-drop editing.

Ongoing work includes application of these techniques to other domains such as World Wide Web management and information access. An open problem is the use of grammatical descriptions for incremental editing -- the aggregation methods mentioned here succeed only if the (sub)graphs conform to the grammatical descriptions, but satisfactory imcremental parsing methods are still needed when dealing with two-dimensional visual languages such as directed attributed graphs [13].

6 Acknowledgments

Contributers to the ShowBiz implementation include Cliff Behrens, Ron Dolin, Mark Rosenstein, Loren Shih, and Larry Stead. We are also indebted to Rodney Fuller, Catherine Hanson, Bob Root, and Laurie Spiegel. The research has been supported in part by ARPA grant N66001-94-C-6039.

References

1. Bederson, B., and J. Hollan. (1994) Pad++: A Zooming Graphical Interface for Exploring Alternate Interface Physics. Proceedings of ACM UIST `94 (Marina Del Rey, CA, USA), ACM Press, pp. 17-26.

2. Brandenburg, F. (1995) Designing Graph Drawings by Layout Graph Grammars. In R. Tamassia and I. G. Tollis, eds., Graph Drawing: DIMACS International Workshop, Princeton, NJ, USA, October 1994, Proceedings, Lecture Notes in Computer Science 894, Springer-Verlag, pp. 416-427.

3. Froelich, M., and M. Werner (1995) Demonstration of the Interactive Graph Visualization System da Vinci. In R. Tamassia and I. G. Tollis, eds., Graph Drawing: DIMACS International Workshop, Princeton, NJ, USA, October 1994, Proceedings, Lecture Notes in Computer Science 894, Springer-Verlag, pp. 266-269.

4. Di Battista, G., P. Eades, R. Tamassia, and I. Tollis (1994) Algorithms for Drawing Graphs: an Annotated Bibliography. Computational Geometry: Theory and Applications 4:235-282. [Updated version at http://www.cs.brown.edu/people/rt/gd-biblio.html.]

5. Kimelman, D., B. Leban, T. Roth, and D. Zernik. (1995) Reduction of Visual Complexity in Dynamic Graphs. In R. Tamassia and I. G. Tollis, eds., Graph Drawing: DIMACS International Workshop, Princeton, NJ, USA, October 1994, Proceedings, Lecture Notes in Computer Science 894, Springer-Verlag, pp. 218-225.

6. Noik, E.G. (1994) A Space of Presentation Emphasis Techniques for Visualizing Graphs. In Graphics Interface '94, pp. 225-234.

7. Rummler, G. A., and A. P. Brache (1990) Improving Performance: How to Manage the White Space on the Organization Chart, Jossey-Bass Publishers.

8. Sandar, G. (1995) Graph Layout Through the VCG Tool. In R. Tamassia and I. G. Tollis, eds., Graph Drawing: DIMACS International Workshop, Princeton, NJ, USA, October 1994, Proceedings, Lecture Notes in Computer Science 894, Springer-Verlag, pp. 194-205.

9. Sarkar, M., and M. Brown. (1992) Graphical Fisheye Views of Graphs. In Proceedings of CHI '92, Monterey, CA, May, 1992, pp. 83-91.

10. Sugiyama, K., S. Tagawa, and M. Toda (1981) Methods for Visual Understanding of Hierarchical System Structures. IEEE Transactions on System, Man, and Cybernetics 11:1047-1062.

11. Weitzman, L., and Wittenburg, K. (1993) Relational Grammars for Interactive Design. In Proceedings of IEEE Symposium on Visual Languages, August 24-27, Bergen, Norway, pp. 4-11.

12. Wittenburg, K., and L. Weitzman (1996) Relational Grammars: Theory and Practice in a Visual Language Interface for Process Modeling. In Proceedings of Workshop on Theory of Visual Languages, May 30, 1996, Gubbio, Italy. [http://www.cs.monash.edu.au/~berndm/TVL96/tvl96-home.html.]

13. Wittenburg, K. (1995) Visual Language Parsing: If I Had a Hammer... In Proceedings of CMC/95: International Conference on Cooperative Multimodal Communication, Theory and Applications, Eindhoven, The Netherlands, May 24-26, 1995, pp. 17-33.

Author Index

Lecture Notes in Computer Science

For information about Vols. 1–1113

please contact your bookseller or Springer-Verlag